Natural Language Generation in Interactive Systems

An informative and comprehensive overview of the state-of-the-art in natural language generation (NLG) for interactive systems, this guide introduces graduate students and new researchers to the field of natural language processing and artificial intelligence, while inspiring them with ideas for future research. Detailing the techniques and challenges of NLG for interactive applications, it focuses on research into systems that model collaborativity and uncertainty, are capable of being scaled incrementally, and can engage with the user effectively. A range of real-world case studies is also included.

The book and the accompanying website feature a comprehensive bibliography, and refer the reader to corpora, data, software, and other resources for pursuing research on natural language generation and interactive systems, including dialogue systems, multimodal interfaces, and assistive technologies. It is an ideal resource for students and researchers in computational linguistics, natural language processing, and related fields.

Amanda Stent is a researcher at AT&T Labs – Research. Her areas of interest are at the intersection of dialogue systems and natural language generation and include coreference, sentence planning, alignment in dialogue, and applications of NLG to topics ranging from computer security to assistive technology. Dr. Stent has authored over 70 research publications and holds several patents.

Srinivas Bangalore has been at AT&T Labs – Research since 1997 and has worked on many areas of natural language processing, including Spoken Language Translation, Multimodal Understanding, Language Generation, and Question-Answering. He has co-edited a book on Supertagging, authored over 100 research publications, and holds 45 patents in these areas. He has been awarded the AT&T Outstanding Mentor Award and the AT&T Science & Technology Medal.

Natural Language Generation in Interactive Systems

Edited by

AMANDA STENT AND SRINIVAS BANGALORE
AT&T Research, Florham Park, New Jersey, USA

CAMBRIDGE
UNIVERSITY PRESS

CAMBRIDGE
UNIVERSITY PRESS

University Printing House, Cambridge CB2 8BS, United Kingdom

Published in the United States of America by Cambridge University Press, New York

Cambridge University Press is part of the University of Cambridge.

It furthers the University's mission by disseminating knowledge in the pursuit of
education, learning and research at the highest international levels of excellence.

www.cambridge.org
Information on this title: www.cambridge.org/9781107010024

© Cambridge University Press 2014

First published 2014

Printed in the United Kingdom by TJ International Ltd., Padstow Cornwall

A catalogue record for this publication is available from the British Library

Library of Congress Cataloguing in Publication data
Natural language generation in interactive systems / [edited by] Amanda Stent, AT&T Research,
Florham Park, New Jersey, Srinivas Bangalore, AT&T Research, Florham Park, New Jersey.
    pages   cm
Includes bibliographical references and index.
ISBN 978-1-107-01002-4 (hardback)
1. Natural language processing (Computer science)   2. Interactive computer systems.
I. Stent, Amanda, 1974–   II. Bangalore, Srinivas, 1969–
QA76.9.N38N382 2014
006.3′5–dc23   2013044149

ISBN 978-1-107-01002-4 Hardback

Contents

Contributors

Elisabeth André

Elisabeth André is a Professor of Computer Science at Augsburg University and Chair of the Laboratory for Multimedia Concepts and their Applications. Prior to that, she worked as a researcher at DFKI GmbH. Her current research interests include affective computing, multimodal user interfaces, and synthetic agents. She holds a Ph.D. from Saarland University.

Anja Belz

Anja Belz is a Reader in Computer Science at the University of Brighton and a Visiting Senior Research Fellow at the University of Sussex. She has also held a position at SRI International. Her research interests include probabilistic natural language generation and the evaluation of generation systems. She holds a Ph.D. from the University of Sussex.

Rolf Black

Rolf Black is a researcher at Dundee University. He has a clinical background in rehabilitation engineering and extensive working experience with children with complex disabilities, their families, and their school environments. His research area is personal narrative and literacy acquisition for children with complex communication needs.

Nate Blaylock

Nate Blaylock is a NLP Research Manager at Nuance Communications. He has worked at Saarland University in Saarbrucken, Germany, where he helped build the bilingual German/English SAMMIE in-car dialogue system; at Cycorp; and at IHMC, where he worked on dialogue systems, plan and intent recognition, and natural language understanding. He holds a Ph.D. from the University of Rochester.

Heriberto Cuayáhuitl

Heriberto Cuayáhuitl is a senior research fellow in the school of mathematical and computer sciences at Heriot-Watt University. In the past, he has held positions at the German Research Center for Artificial Intelligence (DFKI GmbH), at the University of Bremen, at the University of Edinburgh, at the Autonomous University of Tlaxcala, and at SpeechWorks (now Nuance Communications). He is interested in applying machine learning methods to develop adaptive dialogue systems and intelligent robots. He has twice co-organized workshops on machine learning for interactive systems.

Nina Dethlefs

Nina Dethlefs is a research associate at the Interaction Lab at Heriot-Watt University. Her research interests center on robust, adaptive, and flexible NLG methods for situated interactive systems. She has a special focus on the application of reinforcement learning and graphical models. She has twice co-organized workshops on machine learning for interactive systems. She holds a Ph.D. from the University of Bremen.

David DeVault

David DeVault is a Research Assistant Professor in the Department of Computer Science at the University of Southern California, and a Research Scientist at the Institute for Creative Technologies. His research focus is on enabling dialogue systems to respond with naturalness, robustness, and flexibility to the uncertainties of interaction. He holds a Ph.D. from Rutgers, the State University of New Jersey.

Birgit Endrass

Birgit Endrass is a Lecturer at the Augsburg University, where she also did her Ph.D. Her research focuses on behavior modeling for virtual agents, cultural computing, and interactive storytelling.

Martijn Goudbeek

Martijn Goudbeek is a postdoctoral researcher at Tilburg University and previously worked at the Swiss Center for Affective Sciences. His research interests include the production and perception of emotion, non-verbal communication, speech perception and speech acquisition, and the production of referring expressions. He holds a Ph.D. from the Radboud University of Nijmegen.

Eleni Gregoromichelaki

Eleni Gregoromichelaki is a Visiting Research Fellow at King's College London, where she also did her Ph.D. Her principal research interests lie in the syntax/semantics/pragmatics interface, in particular anaphora and ellipsis. She has also done work on conditionals, relative clauses, quantification, and clitics and she has explored extensively the philosophical and psychological implications of various views on language.

Helen Hastie

Helen Hastie is a Lecturer and member of the Interaction Lab at Heriot-Watt University, Edinburgh. She has also held positions at Edinburgh University, at Lockheed Martin, and at AT&T Labs–Research. Her research interests include incremental spoken dialogue systems and NLG, and health informatics. She holds a Ph.D. from the University of Edinburgh.

Julian Hough

Julian Hough is a doctoral candidate in the Interaction, Media and Communication group in the School of Electronic Engineering and Computer Science at Queen Mary, University of London. His research centers around the computational modeling of incremental processing in dialogue, particularly focusing on the generation of self-repair phenomena.

Kiwaki Ito

Kiwaki Ito is a senior researcher in the Department of Linguistics at Ohio State University. Her research focuses on the effect of prosody on the comprehension of speech in spontaneous dialogue, and how children learn to use prosodic prominence in discourse. She holds a Ph.D. from the University of Illinois at Urbana-Champaign and has previously worked at the RIKEN Brain Science Institute in Japan.

Srinivasan Janarthanam

Srinivasan Janarthanam is a research associate at the Interaction Lab, Heriot-Watt University, Edinburgh. His research interests include dialogue management, natural language generation, and user modeling in the context of spoken dialogue systems. He holds a Ph.D. from the University of Edinburgh.

Emiel Krahmer

Emiel Krahmer is a Professor at Tilburg University, where he also did his Ph.D. He has also held positions at Eindhoven University of Technology and at the University of Brighton. He works on referring expression generation, emotion, gesture, and audiovisual prosody.

Oliver Lemon

Oliver Lemon is a Professor in the School of Mathematical and Computer Sciences at Heriot-Watt University, Edinburgh, where he is head of the Interaction Lab. His research interests span machine learning, speech and language technology, human–robot interaction, and technology-enhanced learning. He previously worked at Stanford University (2000–2002) and at the School of Informatics, Edinburgh University (2003–2009). He is co-author of the book *Reinforcement Learning for Adaptive Dialogue Systems* (2011) and co-editor of *Data-Driven Methods for Adaptive Spoken Dialogue Systems* (2012). He holds a Ph.D. from Edinburgh University.

François Mairesse

François Mairesse is a Senior Research Engineer at Nuance Communications. He does research on natural language generation and natural language understanding for dialogue systems. He holds a Ph.D. from the University of Sheffield.

Matthew Purver

Matthew Purver is a Lecturer in the School of Electronic Engineering and Computer Science at Queen Mary, University of London. His research examines the modeling of conversational context to understand and generate dialogue; while at Queen Mary, Stanford University, and King's College London he has applied this to spoken dialogue systems, business meeting summarization, incremental language processing, and doctor–patient consultation mining. He is co-founder of Chatterbox, a commercial venture applying dialogue research to analyze conversations on social media.

Rajakrishnan Rajkumar

Rajakrishnan Rajkumar is an Assistant Professor at the Indian Institute of Information Technology and Management-Kerala. His research interests center on broad-coverage surface realization and evaluation of surface realizers. He holds a Ph.D. from Ohio State University.

Ehud Reiter

Ehud Reiter is a Professor of Computing Science at the University of Aberdeen, and founder of the Aberdeen Natural Language Generation research group. He has also worked at CoGenTex (a small US-based NLG company) and is Chief Scientist of Data2Text, a company that was spun out of Aberdeen University to commercialize its NLG work. He holds a Ph.D. from Harvard.

Verena Rieser

Verena Rieser is a Lecturer in Computer Science at Heriot-Watt University, Edinburgh. She previously worked at Edinburgh University in the Schools of Informatics (2008–2010) and Geosciences (2010–2011), performing research in data-driven statistical methods for multimodal interfaces. She has applied these techniques in two main areas: language-based interfaces (e.g., natural language generation), and spatial decision support tools (e.g., sustainable land use change). She is co-author of the book *Reinforcement Learning for Adaptive Dialogue Systems* (2011). She holds a Ph.D. from Saarland University.

Shari Speer

Shari Speer is a Professor and Chair in the Department of Linguistics at Ohio State University. Her research interests center on the relationship between language production and comprehension, particularly in spoken language, and the influence of prosodic structure on these processes. She holds a Ph.D. in Psychology from the University of Texas at Austin and has held positions at the University of Kansas, Northeastern University, Yale University, and the University of Massachusetts at Amherst.

Matthew Stone

Matthew Stone is an Associate Professor in the Computer Science Department and Center for Cognitive Science at Rutgers, the State University of New Jersey. He has also held visiting positions at the University of Edinburgh and the Universität Potsdam. He works toward computer systems that mean what they say to you and understand what you say to them. He holds a Ph.D. from the University of Pennsylvania.

Mariët Theune

Mariët Theune is an Assistant Professor at the University of Twente. Her research interests include interactive digital storytelling, natural language generation, embodied conversational agents, and multimodal information presentation. She holds a Ph.D. from Eindhoven University of Technology.

Nava Tintarev

Nava Tintarev is a Research Fellow in Natural Language Generation at the University of Aberdeen, where she also did her Ph.D. She has previously held a position at Telefonica Research. Her research interests include natural language generation and assistive technology, in particular generation of personal narrative from sensor input and multimedia.

Kees van Deemter

Kees van Deemter is a Professor at the University of Aberdeen. He works at the intersection of computational semantics and natural language generation; his research centers around the question of how information can be expressed in a way that is most suitable for human recipients. He has recently authored the book *Not Exactly: In Praise of Vagueness* (2010). He holds a Ph.D. from the University of Leiden.

Annalu Waller

Annalu Waller holds a personal chair in Human Communication Technologies in the School of Computing at the University of Dundee, where she also did her Ph.D. She has worked in the field of Augmentative and Alternative Communication (AAC) since 1985, designing communication systems for and with non-speaking individuals. She established the first AAC assessment and training center in South Africa in 1987. Her primary research areas are personal narrative and assistive technology, human–computer interaction, and natural language processing. In particular, she focuses on empowering end users, including disabled adults and children, by involving them in the design and use of technology.

Michael White

Michael White is an Associate Professor in the Department of Linguistics at Ohio State University. His research interests are primarily in natural language generation, paraphrasing, and spoken dialogue systems. Much of his research uses OpenCCG. He has previously held positions at the University of Edinburgh and CoGenTex, Inc., an NLG startup. He holds a Ph.D. from the University of Pennsylvania.

1 Introduction

Amanda Stent and Srinivas Bangalore

The story of Pinocchio is the story of a man who creates a puppet with the ability to speak using human language, and to converse naturally – even to lie. Since the early days of the industrial revolution, people have imagined creating agents that can emulate human skills, including language production and conversation (Mayer, 1878; Wood, 2003). A modern example of this is Turing's test for artificial intelligence – a machine that is indistinguishable from a human being when evaluated through disembodied conversation (Turing, 1950).

This book concerns the intersection of two areas of computational research involved in the production of conversational agents – **natural language generation** and **interactive systems**.

1.1 Natural language generation

Natural language generation (NLG) systems are systems that produce human language artifacts (including speech, text, and language-rich multimedia presentations). NLG systems differ in the inputs they take, the types of output they produce and the degree to which they support interactivity. However, there are some common challenges across all types of NLG system. All end-to-end NLG systems perform the following three tasks: **content selection**, or determination of "what to say"; **surface realization**, or determination of "how to say it" (including assignment of content to media, selection of words, and arrangement of the content); and **production**, or the presentation/performance of the generated material (which in the case of an embodied agent may include production of body postures, gestures, and sign language). At the core of each NLG task is *choice*: for any decision, there may be numerous reasonable options, and the selection of a "right" option may depend on many factors including the intended recipient (the hearer or reader), the physical and linguistic context, and the persona of the agent itself. Consequently, NLG involves elements of creativity and subjectivity that are not typically thought of as key to computational tasks. This also complicates the evaluation of NLG systems, as there may be numerous reasonable outputs for a given input.

With respect to input, there are two broad categories of NLG system: **text-to-text** (which take text as input), and **data-to-text** (which start from non-linguistic inputs). There are interactive systems with NLG components of each type: for example, the MATCH dialogue system uses data-to-text generation to produce restaurant recommendations (Walker *et al.*, 2002), while the Have2eat system uses text-to-text generation for the same purpose (Di Fabbrizio *et al.*, 2010).

There are three broad categories of NLG system with respect to output: those that produce **language** only (text or speech), such as the COREF and PERSONAGE systems (Chapters 3 and 9); those that produce **multimedia** presentations, such as the ILEX (Cox *et al.*, 1999) and interactiveX (Molina *et al.*, 2011) systems; and those that generate the behaviors of **embodied conversational agents** (Cassell, 2000; André *et al.*, 2001).

The standard pipeline for NLG systems since the mid-1990s, which reflects a Leveltian model of language production (Levelt, 1989; Reiter and Dale, 2000; Rambow *et al.*, 2001), comprises four stages of processing corresponding to the three NLG tasks outlined earlier: content and discourse planning, sentence planning and surface realization, and production. While a strict pipeline architecture like this one may permit reuse of NLG components and simplify the implementation of each individual component, it also restricts the ability of NLG systems to be responsive and adaptive in very interactive contexts. This concern was expressed as early as 1995 (Kilger and Finkler, 1995), and recent years have seen a range of attempts to integrate the decision-making process across NLG tasks (for example, see Chapters 7 and 8).

1.2   Interactive systems

An interactive system is any computer system that engages in ongoing interaction with one or more human users. Although interaction can take many forms – an interactive photo album may only interact through images and gestures, a robot only through physical actions – in this book, we are concerned with systems that use human languages such as English, Korean, or sign language as a primary medium for interaction. Within this category, there is a wide range of types of system, both in terms of the amount and types of interaction supported and in terms of the flexibility and richness of the natural language produced by the system. At one extreme, there are systems that, while interactive, use very little natural language; an example is the baseline system for the recent GIVE challenges, which only said *warmer* and *colder* (Koller *et al.*, 2010). Then, there are systems that use rich models of natural language, but are not very interactive, such as the BabyTalk system (Reiter *et al.*, 2008). Finally, there are systems that are very interactive and use rich models of natural language, such as the ILEX museum guide (Cox *et al.*, 1999). The contributions in this book contain numerous examples of language-intensive interactive systems, including instruction-giving agents for virtual environments (Chapter 8), embodied conversational agents (Chapter 10), and assistive and augmentative communication systems that facilitate communication between humans (Chapter 11).

1.3   Natural language generation for interactive systems

Consider the writing of a movie review for a magazine. The author has some content, which must be structured into a discourse. The author also probably has in mind the audience for the story, and targets the content of the review and the language used to that type of reader. However, the reader is (a) amorphous (a group of people rather than an individual); and (b) separated in time and space from the author (consequently, unable to reply or give feedback as the story is being constructed).

Now consider that a friend of the author asks about the movie over dinner. The author may communicate the same basic content, but everything else about the discourse is different. First of all, the discourse is **collaboratively constructed**; both people contribute to the language, topics, and overall shape of the conversation. This collaboration means that, at any point in the conversation, each person has their own mental model of the discourse. This introduces **uncertainty**: neither person knows if their mental model of the discourse matches that of the other person. Fortunately, the uncertainty can be ameliorated through **grounding** and **alignment**. Grounding behaviors (such as asking clarification questions, providing acknowledgments, and backchanneling) allow the conversational participants to check each other's models of the discourse. Alignment behaviors (such as converging on similar lexical or syntactic choices) allow the conversational participants to tailor their contributions to each other to minimize misunderstandings. Finally, because the author knows the friend, the content and form of the author's contributions to the conversation can be adjusted to make them particularly relevant and interesting for the friend, i.e., to encourage **engagement**. So we see that interaction affects every aspect of the NLG process. The chapters in this book address each of these aspects of interactive NLG.

1.3.1   Collaboration

In an interaction, no participant's contributions stand alone – the process of constructing the discourse is a **collaboration**. (This is true even if the participants are disagreeing or attempting to deceive each other – collaboration on an underlying task is not necessary for collaboration in the interaction process.) Each contribution is shaped by the need to fit into what has come before, and to maximize the chances of success of what will follow. These constraints influence content selection, surface realization and production. In addition, the collaboration causes additional language behaviours not present in noninteractive situations – these include interaction maintenance behaviours such as turn-taking.

The three chapters in Part I of the book highlight particular aspects of collaboration relevant to NLG for interactive systems.

When we imagine a collaboration, we tend to visualize something like bricklaying – where all participants are contributing to a common infrastructure (such as a brick wall) that is available to each participant. However, this is not typically the case for language-rich interactions. In a conversation, each participant keeps their own representation of their own internal state and of what they imagine to be the "shared state," and these representations may not be perfectly aligned (Clark, 1996). This introduces **uncertainty**. The management and reduction of uncertainty in conversation is carried out through processes including **grounding** and **alignment**. Grounding is the process through which conversational participants validate their mental models for the "shared state" of the interaction (Clark, 1996). Grounding behaviors include acknowledgments (e.g., *okay*), check questions (e.g., *you mean Sally from the store?*), and direct indications of understanding or of lack of understanding (e.g., *I didn't follow that*). Grounding behaviors are driven by an agent's understanding, but produced by NLG systems; the need to generate grounding behaviors affects the content selection, discourse planning, and surface realization processes. Alignment is a set of behaviors that conversational participants use to encourage and reinforce the development of similar mental models for the "shared state" of the interaction. For example, conversational participants may adopt the same postures, follow each others' eye gaze, repeat each others' gestures, adapt to each others' prosody and choice of words, and adopt each others' means of information packaging (such as a preference for longer or shorter contributions). The need to generate alignment behaviors primarily affects the sentence planning, surface realization, and production processes, but also means that conversational agents should maintain rich models of the linguistic content of the interaction.

The importance of modeling communicative intentions for advanced dialogue modeling is highlighted by Blaylock in Chapter 2. The chapter starts with an extensive survey of types of dialogue model – finite-state, slot-filling, and plan based – comparing and contrasting their ability to model communicative intentions. The author then introduces a **collaborative problem solving** model of dialogue which explicitly models communicative intentions. While most of the chapter is concerned with establishing the need for a representation of communicative intention, defining the range of communicative intentions and extracting them from user utterances, the chapter ends with some significant challenges at the intersection of dialogue management and content selection in dialogue systems.

In Chapter 3, Devault and Stone describe the importance of **grounding** for a successful conversation. They present a prototype system, COREF, that treats grounding as problem solving. COREF addresses both the issue of uncertainty in the system's understanding of the user's contributions, and the possibility of ambiguity in system-generated contributions. By maintaining a *horizon graph*, the system can evaluate the space of its possible contributions at each turn for implicatures and ambiguities, and select either a contribution that is acceptable under all interpretations, or a minimally underspecified contribution (in the case of ambiguity).

In Chapter 4, Purver *et al.* present an interesting phenomenon in human–human communication, **compound contributions**. Compound contributions are utterances spoken by a conversational participant that continue or complete an utterance spoken by another participant and result in a single syntactic and propositional unit distributed over multiple turns. The authors present interesting real-world examples drawn from the British National Corpus. They discuss the implications for modeling such contributions from a natural language generation perspective in human–machine dialogue. They sketch models of incremental joint parsing and generation using the grammatical formalism of **dynamic syntax** that can account for compound contributions.

1.3.2 Reference

A problem that is key to any interactive system is the understanding and production of **references** to entities' states, and activities in the world or in the conversation. Reference influences or is influenced by every dialogue task, from recognition of user input through production of user output. Referring expression generation, in particular, is the core of the NLG problem, for everything that a dialogue participant does is aimed at describing the world as it is or as the participant wishes it to be.

In the first chapter in Part II, Chapter 5, van Deemter introduces the reference problem and lays out a detailed argument for rich, comprehensive referring expression generation, with a series of increasingly comprehensive algorithms for generating referring expressions of increasing complexity. He starts by looking at the generation of simple referring expressions consisting of conjunctions of atomic properties by means of a **description by satellite sets** algorithm. He extends this algorithm to handle relations, other logical operators (including disjunction and negation), and different types of quantification. Through the discussion, the reader is introduced to modern knowledge representation techniques and to the question of the **expressive power** of referring expression generation algorithms.

In Chapter 6, Krahmer, Goudbeek, and Theune examine the interaction of choice and **alignment** during referring expression generation. In human–human conversation, a speaker uses their model of their listener(s) to generate referring expressions that uniquely identify the target object(s), as well as potentially serving other communicative functions. The challenge for the speaker is to construct brief, informative, and unambiguous referring expressions in a small amount of time. In this chapter the authors present a graph-based algorithm for referring expression generation. They demonstrate empirically that this approach can balance a speaker's preferences for referring expression construction and alignment constraints based on the history of the conversation.

1.3.3 Handling uncertainty

Grounding and alignment behaviors help minimize uncertainty related to each conversational participant's model of the interaction so far, but do not drive the interaction further. There is another type of uncertainty relevant to NLG, and that is the construction of a "right" contribution to drive the next step of the interaction. This topic is addressed in the two chapters in Part III of the book. These chapters present emerging approaches to NLG that model uncertainty in the dialogue state and that jointly optimise across different stages of the NLG pipeline.

As we said earlier, natural language generation has traditionally been modeled as a planning pipeline involving content selection, surface realization and presentation. Early NLG systems were rule-based, and consequently quickly grew complex to maintain and adapt to new domains and constraints. In Chapter 7, Lemon *et al.* present a reinforcement learning approach to NLG. The authors argue that the attraction of this new approach is that it relies on very little training data, is provably optimal in its decisions, and can quickly be adapted to new domains and additional constraints. They demonstrate how to use their approach to bootstrap NLG systems from small amounts of training data.

It has been widely acknowledged that the stages of the NLG pipeline are inter-dependent, even though they are typically modeled in isolation (primarily due to the combinatorial growth of the choices at each stage of the generation process). In Chapter 8, Dethlefs and Cuayáhuitl present an approach to jointly optimizing choices across multiple stages of NLG. They propose a reinforcement learning-based method that exploits a hierarchical decomposition of the search space in order to achieve com-putationally feasible solutions. They demonstrate the utility of their approach using the GIVE Challenge testbed.

1.3.4   Engagement

Any agent that produces language, whether for interactive or non-interactive contexts, will be more successful if it is **engaging**. An agent will engage a user if it is infor-mative, relevant, clear, consistent, and concise (Grice, 1975). However, it is easier to be engaging for a particular (known) audience than for a general audience, and in a conver-sational interaction the participants are constantly giving each other information about themselves (what will best inform and engage them) and feedback about the success of the interaction so far. In particular, conversational participants give evidence of their culture and personality. In Part IV of the book we include three chapters that address the theme of **engagement** in NLG systems from diverse perspectives.

In the early 2000s, NLG researchers began to model parts of the language generation problem using **overgenerate and rank** techniques. Conceptually, these techniques use a (weak) model of language generation to produce a set of possible alternative real-izations, and then rank or score the realizations using a language model. In Chapter 9, Mairesse employs this technique to address the issue of **personality-driven** surface realization. He shows that the effectiveness of interactive systems can be substantially influenced by the system's personality, as manifested in the linguistic style of system contributions. He presents a system that employs the overgenerate-and-rank technique as well as parametric predictive models for personality-driven surface realization. He also highlights the challenges in data-driven stylistic language generation techniques.

As communication aids become increasingly multimodal, combining speech with image and video input, so also natural language generation is evolving to produce multi-modal contributions. One such multimodal generation mechanism is to use an **avatar** or a **virtual agent**. In the realm of multimodal communication, the intent of the message is carried not only by words, but also by the prosody of speech and by nonverbal means including body posture, eye gaze, and gestures. These paralinguistic means of commu-nication take on a different dimension when contextualized by the cultural backgrounds of the conversational participants. In Chapter 10, the authors highlight the influence of cultural background in human communication and address the challenging problem of formulating and modeling culture for human-machine dialogues.

Another area in which engagement affects the performance of dialogue systems is when they are used to provide assistive and augmentative communication services. In Chapter 11, Tintarev *et al.* present an assistive and augmentative communication aid for helping children with language disabilities communicate their thoughts through language and multimodal interfaces. Here, the system must represent one user to other users during conversation, so must be adaptive and engaging. The chapter presents a system called "How was School Today?" that allows children to engage in a dialogue with their caregiver to discuss the events of their school day. The content planning component of the system summarizes information about where and when the child was during the school day. The sentence planning and surface realization components are fine-tuned to the application setting. The authors discuss possible future research aimed at exploiting richer contextual information to provide a more effective communication aid.

1.3.5   Evaluation and shared tasks

The last two chapters in Part V of this collection address the crucial issues of evaluation and shared tasks in NLG for interactive systems. The first chapter is a detailed discussion of the possibilities of eye tracking as an objective evaluation metric for evaluating the speech produced at the end of an NLG system, while the second is a sweeping survey of shared tasks and evaluation in NLG and interactive systems.

With the increasingly multimodal nature of human–machine conversation, where humans interact with devices endowed with cameras, the efficacy of conversation can be hugely improved by taking into account different factors beyond just the speech input/output to/from the system. In Chapter 12, White *et al.* present a detailed analysis of the impact of prosody on speech synthesis. They propose to include the eye tracking movements of humans listening to synthesized speech as an additional metric to evaluate the prosody of speech synthesis. Their experimental results indicate that the fundamental frequency (F_0) feature of the synthesised speech plays an important role in the comprehension of the speech, particularly in a contrastive discourse context.

Since the early days, NLG researchers have embedded their components in real-world applications. However, since these applications are tightly tied to particular tasks, they do not permit comparison of different NLG techniques. In Chapter 13, Belz and Hastie provide a comprehensive study of recent efforts to provide evaluation frameworks and shared evaluation campaigns for a variety of NLG tasks, including surface realization, referring expression generation, and content planning. These campaigns allow the NLG community to compare the effectiveness of different techniques on shared data sets, thus decoupling NLG components from the context of an overall application. It also helps to lower the barrier for entry to the NLG field. However, as Belz and Hastie point out, NLG evaluation in interactive systems has not been a focus in these efforts. Given that conversation is collaborative, it is challenging to evaluate the success of an individual system contribution, and especially to separate task success (over a whole conversation) from linguistic success (of any individual NLG decision). This research direction in NLG evaluation is ripe for further exploration.

1.4   Summary

The contributors to this collection intend to spur further research on NLG for interactive systems. To this end, each chapter contains concrete ideas for research topics. The book is also accompanied by a website, www.nlgininteraction.com, containing pointers to software and data resources as well as a bibliography.

References

André, E., Rist, T., and Baldes, S. (2001). From simulated dialogues to interactive performances. In *Multi-Agent-Systems and Applications II*, pages 107–118. Springer, Berlin, Germany.

Cassell, J. (2000). Embodied conversational interface agents. *Communications of the ACM*, **43**(4):70–78.

Clark, H. (1996). *Using Language*. Cambridge University Press, Cambridge, UK.

Cox, R., O'Donnell, M., and Oberlander, J. (1999). Dynamic versus static hypermedia in museum education: An evaluation of ILEX, the intelligent labelling explorer. In *Proceedings of the Conference on Artificial Intelligence in Education*, pages 181–188, Le Mans, France. International Artificial Intelligence in Education Society.

Di Fabbrizio, G., Gupta, N., Besana, S., and Mani, P. (2010). Have2eat: A restaurant finder with review summarization for mobile phones. In *Proceedings of the International Conference on Computational Linguistics (COLING)*, pages 17–20, Beijing, China. International Committee on Computational Linguistics.

Grice, H. P. (1975). Logic and conversations. In Cole, P. and Morgan, J. L., editors, *Syntax and Semantics III: Speech Acts*, pages 41–58. Academic Press, New York, NY.

Kilger, A. and Finkler, W. (1995). Incremental generation for real-time applications. Technical Report DFKI Report RR-95-11, German Research Center for Artificial Intelligence – DFKI GmbH, Saarbrücken, Germany.

Koller, A., Striegnitz, K., Byron, D., Cassell, J., Dale, R., and Moore, J. D. (2010). The first challenge on generating instructions in virtual environments. In Krahmer, E. and Theune, M., editors, *Empirical Methods in Natural Language Generation*, pages 328–352. Springer LNCS, Berlin, Germany.

Levelt, W. J. M. (1989). *Speaking: From Intention to Articulation*. MIT Press, Cambridge, MA.

Mayer, A. M. (1878). On Edison's talking-machine. *Popular Science Monthly*, **12**:719–723.

Molina, M., Parodi, E., and Stent, A. (2011). Using the journalistic metaphor to design user interfaces that explain sensor data. In *Proceedings of INTERACT*, pages 636–643, Lisbon, Portugal. Springer.

Rambow, O., Bangalore, S., and Walker, M. (2001). Natural language generation in dialog systems. In *Proceedings of the Human Language Technology Conference (HLT)*, pages 270–273, San Diego, CA. Association for Computational Linguistics.

Reiter, E. and Dale, R. (2000). *Building Natural Language Generation Systems*. Cambridge University Press, Cambridge, UK.

Reiter, E., Gatt, A., Portet, F., and van der Meulen, M. (2008). The importance of narrative and other lessons from an evaluation of an NLG system that summarises clinical data. In *Proceedings of the International Conference on Natural Language Generation (INLG)*, pages 147–156, Salt Fork, OH. Association for Computational Linguistics.

Turing, A. M. (1950). Computing machinery and intelligence. *Mind*, **59**:433–460.

Walker, M., Whittaker, S., Stent, A., Maloor, P., Moore, J. D., Johnston, M., and Vasireddy, G. (2002). Speech-Plans: Generating evaluative responses in spoken dialogue. In *Proceedings of the International Conference on Natural Language Generation (INLG)*. Association for Computational Linguistics.

Wood, G. (2003). *Edison's Eve: A magical history of the quest for mechanical life*. Anchor Books, New York, NY.

Part I

Joint construction

2 Communicative intentions and natural language generation

Nate Blaylock

2.1 Introduction

Natural language generation (NLG) has one significant disadvantage when compared with natural language understanding (NLU): the lack of a clear input format. NLU has as its starting point either a speech signal or text, both of which are well-understood representations. Input for NLG, on the other hand, is quite ill-defined and there is nothing close to consensus in the NLG community. This poses a fundamental challenge in doing research on NLG – there is no agreed-upon starting point.

The issue of input is particularly salient for NLG in interactive systems for one simple reason: interactive systems create their own communicative content. This is due to the role of these systems as dialogue partners, or conversational agents. The necessarily agentive nature of these systems requires them to decide *why* to speak and *what* to convey, as well as *how* to convey it.

Consider some other domains where NLG is used. In machine translation, for example, the system's job is to translate content from one human language to another. Here, the system does not have to create any content at all – that was already done by the human who created the source language text or speech. Work on automatic summarization is another example. Here the system's goal is to take pre-existing content and condense it.

Conversational agents, however, must create content. At a high level, this creation is driven by the need of the agent to communicate something. What the agent wants to accomplish by communicating is referred to as a **communicative intention**. If we can define communicative intentions, they can serve as a standard input representation to NLG for interactive systems.

In this chapter, we will discuss communicative intentions and their use in interactive systems. We then present a model of communicative intentions based on a collaborative problem-solving model of dialogue and show some examples of its use. We then discuss research challenges related to reasoning about high-level communicative intentions for NLG.

2.2 What are communicative intentions?

A **communicative intention** is a communication goal that an agent has for a single turn in the dialogue. It is the answer to the question: What is the agent trying to accomplish by saying this? If we could all plug into one another's brains, we would not need language, but we would still need communication. If that were the case, we could each form communicative intentions and send them directly to the brain of the person we are interacting with. Although this is theoretically possible with artificial agents, humans are not built this way. Instead, we form communicative intentions that we then encode in language (text or speech). Our intended audience (the hearer) has to take this text or speech and try to recover our original communicative intentions.

(2.1)   (An utterance and three possible responses)
Utterance: *Do you have a watch?*

Response 1: *Yes, it's 4:30.*
Response 2: *Yes, it's a Rolex.*
Response 3: *Yes, here it is.* [Takes off watch and hands it to speaker]

Communicative intentions are in no way synonymous with semantic content (meaning), rather, they belong to pragmatics. To illustrate the difference, consider example 2.1, which shows a single utterance paired with three plausible responses for different situations. Here are three different contexts in which the original utterance might have been uttered:

- **Context for Response 1**: The speaker looks at his arm, sees nothing on it, and then approaches the hearer and says *Do you have a watch?* Response 1 seems appropriate here, since it appears that the speaker's intention is to find out what time it is.
- **Context for Response 2**: The speaker and the hearer are arguing about what brand of watch each has. The speaker turns to the hearer and says *Do you have a watch?* Here, Response 2 seems appropriate, as the speaker seems to want to know what brand of watch the hearer has.
- **Context for Response 3**: The hearer is trapped in a burning room with MacGyver.[1] MacGyver looks around the room and tells the hearer he has an idea. He grabs a roll of duct tape and a pencil and then looks at the hearer and says *Do you have a watch?* MacGyver's plan, it appears, is to escape by constructing something using duct tape, a pencil, and a watch. His intention appears to be to obtain the hearer's watch. In this case, Response 3 (including the actual giving of the watch) seems to be the most appropriate response.

The utterance *Do you have a watch?* has the same semantic content in each context above. However, in each case, the speaker has very different goals. Despite this, the (fictitious) hearer is able to successfully decode the speaker's communicative intentions and respond appropriately in each context.[2]

In this example, we did not use the term **communicative intention** in any of the context descriptions. It is difficult to define a communicative intention in layman's terms

---

[1] MacGyver is a television character who typically escapes from dangerous situations by combining everyday items into powerful tools.

[2] The problem of recovering the communicative intentions from an utterance in NLU is called **intention recognition**.

Goal:	Know the time
Non-communicative solutions:	Grab and look at the hearer's watch
	Look at speaker's own cell phone
Communicative intention (CI):	Get the hearer to tell me the time
Possible realizations of CI:	*Do you have a watch?*
	What time is it?
	I will give you $100 if you tell me the time.

Figure 2.1   Example communicative and non-communicative solutions for goal know the time

Goal:	Obtain the hearer's watch
Non-communicative solutions:	Steal the watch
Communicative intention (CI):	Get the hearer to give me the watch
Possible realizations of CI:	*Do you have a watch?*
	Can I have your watch?
	Give me your watch or we're going to die!

Figure 2.2   Example communicative and non-communicative solutions for goal obtain the hearer's watch

(and, as noted above, there is no agreed-upon model of what a communicative intention even is). Figures 2.1 and 2.2 expand on some of the possible goals from this example. Each lists a top-level goal for the speaker from the first and third contexts in the example above (know the time and obtain the hearer's watch, respectively). Note that these goals need not have anything to do with language or communication – they are private goals of the speaker. To illustrate this, Figures 2.1 and 2.2 list a few non-communicative actions the speaker could take, as an individual, to accomplish the goals.

In contrast to those non-communicative actions, each example also has a communicative intention, which is another way in which the speaker can accomplish the goal. Each communicative intention in these examples is of the form *get the hearer to do X*, i.e., communicative intentions do not succeed without the cooperation of the hearer.

Each figure also lists several (very different) possible ways of realizing each communicative intention. Whereas in example 2.1, we had several different communicative intentions that could be realised as the same utterance, here we see that a single communicative intention can be realized in several different ways.

2.3   Communicative intentions in interactive systems

The way in which an interactive system (or dialogue system) models discourse has a direct bearing on how it models communicative intentions. Most interactive systems do not explicitly model communicative intentions at all: in systems that perform a single **fixed task** over a closed domain, all possible discourse states can be mapped in advance and the communicative intentions for both NLU and NLG can be hard-wired in the system. By contrast, some systems perform multiple tasks, or perform a task over an open domain, and these systems typically use **plan-based models**. In this section, we discuss fixed-task and plan-based discourse models and how communicative intentions are handled in each. We argue that these models do not have the power to represent many types of dialogue. Then, in Section 2.4, we present our own model of discourse based on **collaborative problem solving**, which explicitly represents and reasons about communicative intentions.

2.3.1   Fixed-task models

Fixed-task dialogue models make the (explicit or implicit) assumption that the structure of the **domain task** (the non-linguistic task that is the focus of the dialogue) is known (and encoded) at the time the dialogue system is designed. This does not necessarily mean that the structure of the dialogue itself (the **discourse task**) need be fixed. To use the terminology from Section 2.2, the structure of the goals is fixed, but not necessarily the structure of the communicative intentions.

For fixed-task systems, a prototypical dialogue task is that of database lookup for information such as flight, bus, or train schedules (Rudnicky *et al.*, 1999; Lamel *et al.* 2000; Black *et al.*, 2011), local business information (Ehlen and Johnston, 2010), or weather reports (Zue *et al.*, 2000). Here, the structure of the domain task (the database lookup) is fixed, and the system only needs to query the user for certain input slot values, such as departure time or business name, in order to be able to perform the task. (We will see in the section below that this constraint is removed by plan-based models, which allow the user and system to discuss both *information* needed for the task and the *form* of the task itself.) Here, we discuss two types of fixed-task models: **finite state models** and **form-filling models**.

Finite state models

Finite state dialogue models (e.g., Hansen *et al.*, 1996) are the most constrained models we discuss here. In this approach, a dialogue designer must encode not only the domain task, but also all possible dialogue sequences. Typically, a finite state automaton is encoded, with each state representing a system utterance (e.g., a prompt) and each transition arc representing a user utterance. Accepting states typically signify successful completion of the domain task (or at least "successful" completion of the dialogue, e.g., the system said goodbye properly).

The use of finite state models results in very restricted dialogue (see Bohlin *et al.*, 1999 for a good discussion of this issue). First, not only do finite state models require a fixed domain task structure, they also require a fixed, fully enumerated discourse structure. The user is restricted to making utterances that correspond to outgoing transitions from the current state, and more generally, must follow one of the predefined paths through the automaton. In practice, this means that the user must talk about task information in the order that the system designer envisioned.

Second, finite state models suffer from an inherent lack of memory. The system only has knowledge of what state it is in, but not of how it got there. Long-term dependencies, such as an agreed-upon decision, would be difficult if not impossible to represent in a general way.

Communicative intentions, while not explicitly modeled in finite state models, roughly correspond to the nodes and arcs (for system and user intentions, respectively). Generation content is typically encoded in each node in the system, and no explicit reasoning about communicative intentions is required.

Form-filling models

In form-filling dialogue (e.g., Seneff and Polifroni, 2000; Lamel *et al.*, 2000; Chu-Carroll, 2000), a **frame** lists **slots** that represent information needed for the system to perform a (fixed) domain task. Dialogue is modeled as the process of filling in those slots. This results in a much more free exchange at the discourse-task level, as users can now give information (fill slots) in any order they wish within a single form. When the form is completely filled, the system performs the domain task and reports the results. Rudnicky *et al.* (1999) and Bohus and Rudnicky (2003) added the ability to support the filling in of several forms, each representing different parts of the domain task.

Although form-filling does not require a fixed *discourse* structure, the *domain* task structure remains fixed. A dialogue system designer must know the domain task structure in order to design forms that contain the necessary slots for the task.

The disadvantages of form-filling dialogue models are similar to those for finite state models. Both types of model require fixed domain task structures; however, there may be many possible ways to accomplish a goal, some better than others depending on the current state of the world and of the interaction. Pre-specifying possible domain tasks and their structures seriously constrains the ability of an agent to adapt to a changing world, and limits the ability of a dialogue system to cover many desirable domains (such as interaction with an autonomous robot). Communicative intentions are also not explicitly modeled in these models. Systems usually follow a set of fixed rules for generation content, which generally consist of confirming a user input (e.g., *Okay*) and/or asking the value of one of the unfilled slots.

2.3.2   Plan-based models

Plan-based models of dialogue are based on the idea that the purpose of (task-oriented) dialogue is to pursue domain goals – either by planning a course of action (**joint planning**) or by executing such a plan (**joint execution**). In this section, we first discuss the foundations of plan-based dialogue modeling. We then discuss various threads of research using plan-based dialogue models. Finally, we discuss some shortcomings of these approaches.

Foundations – speech acts

Austin (1962), Grice (1969), and Searle (1975) all noted that utterances can cause changes in the world. To give a classic example, the utterance *I pronounce you man and wife,* said by the right person in the right context, causes two people to be married. More subtly, the utterance *John is in the kitchen* may have the effect of causing the hearer to believe that John is in the kitchen. Actions performed by utterances have come to be called **speech acts**.

Cohen, Perrault and Allen (1982) were the first to build a computational theory of speech acts and their relation to agent goals and desires. Allen (1983) concentrated on NLU, using plan recognition and speech act theory to recognize a speaker's intentions. Cohen and Perrault (1979), on the other hand, concentrated on NLG, using artificial intelligence models of planning together with speech acts for NLG based on a speaker's intentions. In particular, they modeled each speech act type as a plan operator with preconditions and effects. Using these, they were able to illustrate the execution of different plans for achieving the same overall goal; for example, getting someone to move either by a REQUEST that they do it, or an INFORM that the speaker wants them to do it. This helps explain some of the variety of utterances presented in Figures 2.1 and 2.2. Almost all the plan-based models presented below use speech acts as basic elements.

Domain-level plans

Carberry (1990) modeled dialogue as joint planning. Her system used a hierarchical plan library that held information about decomposition and parameters of plan schemas in the domain. Based on this library, utterances in a dialogue contribute to the construction of a hierarchical plan by filling in parameters and decomposing subgoals. Carberry's system supported planning in both a bottom-up (talk about actions first) and top-down (talk about goals first) fashion.

Although Carberry's system was able to account for dialogues that build complex plans, it only dealt with utterances that talk directly about actions and goals in the domain. Other dialogue phenomena (such as correction and clarification subdialogues) were not addressed. Also, the system was unable to handle dialogues that support actual plan execution.

In this system, communicative intentions can be seen as contributing to the hierarchical domain plan. As mentioned above, however, utterances not related directly to the plan (such as clarifications) were not modeled.

Meta-plans

In order to support dialogue-level phenomena such as correction and clarification subdialogues, some dialogue models also included a meta-plan level. Litman and Allen (1990) extended Carberry's earlier work to better account for various dialogue phenomena. Although a task-oriented dialogue will focus mainly on the domain task, many utterances serve instead to ensure reliable communication. Litman and Allen added a new layer to the dialogue model to account for these utterances. In this layer, the dialogue focus is on one or more **meta-plans**, domain-independent plans that take other plans as arguments. Litman and Allen's model is able to account for a number of dialogue phenomena, including clarification and correction subdialogues. For example, consider the following dialogue from Litman and Allen (1990):

(2.2)   Teacher: *OK the next thing you do is add one egg to the blender, to the shrimp in the blender.*

> Student: *The whole egg?*
> Teacher: *Yeah, the whole egg. Not the shells.*
> Student: *Gotcha. Done.*

The domain-level plan here is cooking. The student's first utterance (*The whole egg?*), however, is uttered because of confusion about the previous utterance. Instead of replying directly to the teacher's instruction, the student asks a clarification question about one of the objects (the egg) to be used in the plan. Litman and Allen model this as an IDENTIFY-PARAMETER meta-plan, as the egg can be seen as a parameter for the plan. The teacher responds to this question and then the student completes the IDENTIFY-PARAMETER plan, and continues with the domain-level plan.

Litman and Allen model meta-plans using a stack-like structure. Other examples of meta-plans include: CORRECT-PLAN (changes a plan when unexpected events happen at runtime), INTRODUCE-PLAN (shifts the dialogue focus to a new plan), and MODIFY-PLAN (changes some part of a plan).

In this model, communicative intentions can be thought of as either additions to the domain plan (as in Carberry's system, above) or meta-plan actions such as IDENTIFY-PARAMETER.

Other plan levels in dialogue

There have been several efforts to extend the work on meta-plans, creating further levels of plans to support other dialogue phenomena.

Lambert and Carberry (1991) proposed a three-level model of dialogue, consisting of the domain and meta levels of Litman and Allen as well as a level of **discourse plans**, which specifically handles recognition of multi-utterance speech acts. The following utterances, for example, constitute a warning (a discourse plan) only if taken together (Lambert and Carberry, 1991):

(2.3)   U1: *The city of xxx is considering filing for bankruptcy.*
         U2: *One of your mutual funds owns xxx bonds.*

At the same time as Lambert and Carberry, Ramshaw (1991) proposed a different three-level model. Instead of a discourse level, Ramshaw proposed an **exploration** level. The intuition is that some utterances are made simply in the attempt to explore a possible course of action, whereas others are explicitly made to attempt to create or execute a plan. Ramshaw's model allows the system to distinguish between the two cases. The model also allows for other types of communicative intentions, such as comparisons of different plans.

Chu-Carroll and Carberry (2000) extended the three-level model of Lambert and Carberry and added a fourth level, **belief**. They also changed the model so that it distinguished between proposed and accepted plans and beliefs (as in the work of Ramshaw). This extends coverage of the model to include negotiation dialogues, where participants have conflicting views and collaborate to resolve them.

SharedPlans

One of the shortcomings of many of the plan-based models mentioned above is that they do not explicitly model the collaborative nature of dialogue. The SharedPlan formalism (Grosz and Kraus, 1996, 1999) was created in part to model the intentional structure of discourse (Grosz and Sidner, 1986). It describes how agents collaborate together to form a joint plan. The model has four operators, which are used by agents in building SharedPlans:

- **Select_Rec**: An individual agent selects a recipe to be used to attain a given subgoal.
- **Elaborate_Individual**: An individual agent decomposes a recipe into (eventually) completely specified atomic actions.
- **Select_Rec_GR**: Intuitively, the same as **Select_Rec**, only at the multi-agent level.[3] A group of agents select a recipe for a subgoal.
- **Elaborate_Group**: The multi-agent equivalent of **Elaborate_Individual** – a group of agents decompose a recipe.

Using these four operators, a group of agents collaborates until it has completely specified a **full SharedPlan** (which they will presumably execute at some time in the future).

However, SharedPlans only models collaboration for joint *planning* between agents. It does not model the collaboration that occurs when agents are trying to *execute* a joint plan.[4]

SharedPlans models the formulation of joint plans with the four operators previously discussed: **Select_Rec**, **Elaborate_Individual**, **Select_Rec_GR**, and **Elaborate_Group**. Although these operators were sufficient to allow the formalization of group intentions and beliefs about joint plans, they do not provide enough detail for us to model collaboration at an utterance-by-utterance level, which is needed to represent communicative intentions. As an example, consider the **Elaborate_Group** operator, which has the function of decomposing a recipe, instantiating the parameters (including which agent or sub-group will perform which action at what time and which resources will be used), and making sure the rest of the group has similar intentions and beliefs about the plan. An **Elaborate_Group** can (and often does) consist of many individual utterances. In order to build a dialogue system, we need to be able to model the communicative intentions behind a single utterance.

We think that our model (presented in the next section) may be compatible with the SharedPlans formalism and can be seen as specifying the details of the SharedPlan operators.[5]

---

[3] Individual and group operators entail different constraints on individual intentions and beliefs. However, this is not important for understanding the formalism as a model of collaborative planning.

[4] The formalism does specify the needed intentions and beliefs for agents executing joint plans.

[5] This is actually mentioned in Grosz and Kraus (1996) as an area of needed future work.

Table 2.1. Conversation act types (Traum and Hinkelman, 1992).

Discourse level	Act type	Sample acts
Sub UU	**Turn-taking**	Take-turn, Keep-turn, Release-turn, Assign-turn
UU	**Grounding**	Initiate, Continue, Ack, Repair, ReqRepair, ReqAck, Cancel
Discourse Unit	**Core Speech Acts**	Inform, WHQ, YNQ, Accept, Request, Reject, Suggest, Eval, ReqPerm, Offer, Promise
Multiple Discourse Units	**Argumentation**	Elaborate, Summarize Clarify, Q&A, Convince, Find-Plan

2.3.3   Conversation Acts Theory

Traum and Hinkelman (Traum and Hinkelman, 1992) proposed a four-level model of dialogue called **Conversation Acts Theory**. Plan-based dialogue models typically focus on how utterances contribute to the performance of domain tasks. By contrast, Conversation Acts Theory focuses on how utterances contribute to the communicative process. Table 2.1 shows the different types of acts in Conversation Acts Theory.

Core Speech Acts

A major contribution of Conversation Acts Theory is that it takes traditional speech acts and changes them into **Core Speech Acts**, which are multiagent actions requiring efforts from the speaker and hearer to succeed. The span from an utterance that **initiates** a core speech act to the one that completes it is called a **Discourse Unit**.

Grounding acts

In Conversation Acts Theory, speech acts do not succeed in isolation; rather, a speaker's utterances must be **grounded** by the hearer in order to succeed. **Grounding acts** include:

- **Initiate**   The initial part of a Discourse Unit.
- **Continue**   Used when the initiating agent has a turn of several utterances. An utterance that further expands the meaning of the Discourse Unit.
- **Acknowledge**   Signals understanding of the Discourse Unit (although not necessarily agreement, which is at the Core Speech Act level).
- **Repair**   Changes some part of the Discourse Unit.
- **ReqRepair**   A request that the other agent repair the Discourse Unit.
- **ReqAck**   An explicit request for an acknowledgment by the other agent.
- **Cancel**   Declares the Discourse Unit "dead" and ungrounded.

These form part of Traum's computational theory of grounding (Traum, 1994).

Argumentation Acts

Conversation Acts Theory also provides a place for higher-level **Argumentation Acts** that span multiple Discourse Units. As far as we are aware, this level was never well defined. Traum and Hinkelman give examples of possible Argumentation Acts, including rhetorical relations (Mann and Thompson, 1988) and meta-plans (Litman and Allen, 1990).

Turn-taking Acts

At the lowest level in Conversation Acts Theory are Turn-taking Acts. These are concerned with the coordination of speaking turns in a dialogue. An utterance can possibly be composed of several Turn-taking Acts (for example, to take the turn at the start, hold the turn while speaking, and then release it when finished).

One important contribution of Conversation Acts Theory is that a single utterance can encode multiple communicative intentions, with the intentions including acts at different levels of the theory. Stent (2002) used Conversation Acts Theory as the communicative intention input to the NLG subsystem of the TRIPS dialogue system (Allen *et al.*, 2001).

Unlike the other plan-based models we have presented, Conversation Acts Theory is much more centered around dialogue behaviors than around task behaviours. Consequently, the theory only specifies act types, not their content. This is true even at the interface between levels. Also, although this theory improves on speech act theory by modeling speech acts as multiagent actions, it still suffers from some of the issues of speech act theory. In particular, it does not attempt to define a (closed) set of allowable Core Speech Acts. It is in fact unclear whether such a closed set of domain-independent speech acts exists (Clark, 1996; Allen and Core, 1997; Di Eugenio *et al.*, 1997; Alexandersson *et al.*, 1998; Traum, 2000; Bunt *et al.*, 2010).

Finally, as mentioned above, the Argumentation Acts level was never well defined. In Traum and Hinkelman (1992), Argumentation Acts are described vaguely at a level higher than Core Speech Acts, which can be anything from rhetorical relations to operations to change a joint plan.

Our collaborative problem-solving model of dialogue, presented in the next section, builds on the ideas of Conversation Acts Theory and overcomes some of these drawbacks.

2.3.4 Rational behavior models

Several researchers have suggested that dialogue behaviours should be modeled, not reified as speech acts, but simply as an epiphenomenon arising from the rational behavior of agents (Cohen and Levesque, 1990; Levesque *et al.*, 1990; Sadek and De Mori, 1998). These models are predicated on the idea that agents communicate because they are committed to the principles of rational behavior.

As an example, agents, in order to achieve goals, form joint intentions. These joint intentions commit rational agents to certain behaviors, such as helping their fellow agents achieve their part of the plan and letting the other agents know if, for example, they believe the goal is achieved or they decide to drop their intention. These commitments and rationality can be seen as what causes agents to engage in dialogue.

However, rational behavior models have only modeled dialogue supporting joint execution of plans and do not handle dialogue that supports joint planning.[6] Also, although the single-agent levels of formal representations of individual and joint agent rationality are very thorough, the levels of dialogue modeling and agent interaction were never fully developed as far as we are aware.

2.4 Modeling communicative intentions with problem solving

We would like to model a wide range of task-oriented dialogue. Most previous (plan-based) dialogue models for task-oriented dialogue have only supported dialogues about task planning, or dialogues about task execution, but not both. However, the same dialogue can often involve elements of both planning and execution. In fact, task-oriented dialogue can support just about any facet of collaborative activity. This includes not only planning and execution, but also goal selection, plan evaluation, execution monitoring, replanning, and a host of other task-related activities. In addition, conditions often change, requiring dialogue participants to go back and revisit previous decisions, re-evaluate and possibly even abort goals and plans. We have created a dialogue model based on collaborative problem solving, which includes all of these activities. In this section, we briefly present our model. For more details (including a fuller analysis of the examples), see Blaylock (2005) and Blaylock and Allen (2005).

2.4.1 Collaborative problem solving

Problem solving (PS) is the process by which a (single) agent chooses and pursues **objectives** (goals). Specifically, problem solving consists of following three general phases:

- **Determining objectives**: In this phase, an agent manages objectives, deciding to which it is committed, which will drive its current behaviour, etc.
- **Determining and instantiating recipes for objectives**: In this phase, an agent determines and instantiates a hierarchical recipe (i.e., plan) to use to work toward an objective. An agent may either choose a recipe from its recipe library, or it may choose to create a new recipe via planning.
- **Executing recipes and monitoring success**: In this phase, an agent executes a recipe and monitors the execution to check for success.

There are several things to note about this general description. First, we do not impose any strict ordering on the phases above. For example, an agent may begin executing a partially instantiated recipe and do more instantiation later as necessary. An agent may

---

[6] In a way, work on rational behavior models can be seen as a kind of complement to work on SharedPlans – which focuses on planning, but not execution.

also adopt and pursue an objective in order to help it to decide what recipe to use for another objective. Our purpose is not to specify a specific problem-solving strategy or prescriptive model of how an agent *should* perform problem solving. Instead, we want to create a general descriptive model that enables agents with different problem-solving strategies to still communicate. Collaborative problem solving (CPS) follows a similar process to single-agent problem solving. However, here two (or more) agents jointly choose and pursue objectives. The level of collaboration in the problem solving may vary greatly. In some cases, for example, the collaboration may be primarily in the planning phase, but one agent will actually execute the plan alone. In other cases, the collaboration may be active in all stages, including the planning and execution of a joint plan, where both agents execute actions in a coordinated fashion.

2.4.2   Collaborative problem solving state

At the core of the CPS model is the Collaborative Problem Solving (CPS) state, which is an emergent, shared model of the agents' current problem-solving context. It is used to record the objectives the agents are currently jointly committed to, the recipes committed to for those objectives, and the resources (parameters) currently committed to for those objectives. In addition, it also explicitly models which objectives are currently being executed, and which have been successfully completed. (The last two are needed to model explicit dialogue contributions such as *I'm starting to <do some action>* and *I'm done*, respectively.)

Not only does the CPS state track the decisions that have been made by the agents (such as which goal to pursue or how to accomplish it), it also tracks the agents' decision-making process. The state explicitly models each decision point in the model and allows for constraints to be adopted as part of the decision-making process (e.g., *Let's listen to something by the Beatles*, which restricts the parameter of the objective to be a song by the Beatles). Additionally, the state tracks each possible value for the decision point that has been mentioned (e.g., *We could listen to Yesterday or Here Comes the Sun*). For each possible value mentioned, the state tracks any evaluations by the agents of that value in the context of the decision (e.g., *Yesterday is not upbeat enough*).

Our CPS model defines an upper-level ontology of problem-solving objects (objectives, recipes, resources, situations, evaluations, and constraints), which are represented as typed feature structures. Each object type defines a set of slots, which represent the decision points in defining it (such as the recipe to use to achieve an objective). The upper-level ontology is ported to a specific domain through inheritance. For example, in the music player examples above, we may define a new type listen-to-song that inherits from objective and additionally adds a song parameter (which then becomes another decision point).

Changes to the CPS state are realized by the execution of a CPS act. The model defines a number of CPS acts. Each CPS act takes as parameters a CPS object and a decision point in the CPS state. The acts are listed below:

- **focus**: Used to focus problem solving on a particular **object**.
- **defocus**: Removes the focus on a particular **object**.
- **identify**: Used to identify an **object** as a possible option for a decision point.
- **adopt**: Commits the agents to an **object** as the value of a decision point.
- **abandon**: Removes an existing commitment to an **object**.
- **select**: Moves an **objective** into active execution.
- **defer**: Removes an **objective** from active execution (but does not remove a commitment to it).
- **release**: Removes the agents' commitment to an **objective** that they believe has been achieved.

Using these CPS acts, agents change their CPS state while problem solving. However, an agent cannot single-handedly execute CPS acts to make changes to the state; doing so requires the cooperation and coordination of all participating agents. In the model, CPS acts are generated by sets of **interaction acts** (IntActs) – actions that single agents execute in order to negotiate and coordinate changes to the CPS state. An IntAct is a single-agent action that takes a CPS act as an argument. The IntActs are **begin, continue, complete**, and **reject**. An agent beginning a new CPS act proposal performs a **begin**. For successful generation of the CPS act, the proposal is possibly passed back and forth between the agents, being revised with **continues**, until both agents finally agree on it, which is signified by an agent not adding any new information to the proposal but simply accepting it with a **complete**. This generates the proposed CPS act, resulting in a change to the CPS state. At any point in this exchange, either agent can perform a **reject**, which causes the proposed CPS act – and thus the proposed change to the CPS state – to fail.

2.4.3   Grounding

The CPS model as presented so far makes the simplifying assumption that utterances are always correctly heard and interpreted by the hearer (i.e., that the intended (instantiated) interaction acts are correctly recovered). In human communication, mishearing and misunderstanding can be the rule, rather than the exception. Because of this, both speaker and hearer need to *collaboratively* determine the meaning of an utterance through a process called **grounding** (Clark, 1996). To account for grounding in our model, we use the Grounding acts proposed as part of Conversation Acts Theory (Traum and Hinkelman, 1992).[7] In our model, we expand the definition of Grounding acts to allow them to take individual IntActs as arguments.

In our model, an IntAct is not successfully executed until it has been successfully grounded. This is typically after an **acknowledge**.[8] Similarly, the negotiation (IntAct) must be successful (with a **complete**) in order for the CPS act to be successful. Only when a CPS Act is successfully generated does the CPS state change.

---

[7] Our interaction acts and CPS acts could be seen as roughly corresponding to the Core Speech Acts and Argumentation Acts levels in Conversation Acts Theory.
[8] But see Traum (1994) for details.

2.4.4   Communicative intentions

In the full CPS model, we have a Grounding act, which takes as a parameter an IntAct, which in turn takes a CPS Act as a parameter. Loosely, a fully instantiated Grounding act looks something like this:

```
GroundingAct(IntAct(CPSAct(PS Object, Decision Point in CPS State)))
```

We now come back full circle to the question of what communicative intentions are. Our claim is that an instantiated Grounding act *is* a communicative intention. In other words, individual agents communicate in an attempt to successfully negotiate a change to the CPS state. Because (human) agents cannot send Grounding act messages directly to another agent, they encode Grounding acts into language (NLG) and transmit them to the other agent, who then attempts to recover the original Grounding acts (NLU). We now have an explicit model of communicative intentions.

As an example, consider the following utterance with its corresponding interpretation in a typical context (we omit the Grounding acts for the moment to simplify the discussion):

(2.4)   A: *Let's listen to a song.*
$$\text{begin}_1(\text{identify}(\boxed{1}[\text{objective listen to a song}]))$$
$$\text{begin}_2 (\text{adopt-objective}(\boxed{1}))$$
$$\text{begin}_3 (\text{focus}(\boxed{1}))$$

We use the subscript numbers on the IntActs to show their alignment with other acts with the same arguments (in the responses below), and we do not repeat those arguments. We also omit the decision point parameter of the CPS acts to simplify things, and just give a gloss of the problem-solving objects within those acts (for a fuller analysis of this and more examples see Blaylock (2005)). The glosses are preceded by a boxed number, which is then subsequently used to refer to that same gloss.

By assigning these IntActs, we claim that the utterance in example 2.4 has three communicative intentions (corresponding to the three IntActs):

1. To propose that listening to a song be considered as a possible top-level objective.
2. To propose that this objective be adopted as a top-level objective.
3. To propose that problem solving activity be focused on the listen-song objective (e.g., in order to specify a song, to find a recipe to accomplish it, …).

That these are present can be demonstrated by showing possible responses to the utterance in example 2.4 that reject some or all of the proposed CPS acts. Consider the following possible responses to the utterance in example 2.4, tagged with corresponding communicative intentions:

(2.5)   B: *OK.*
$$\text{complete}_1$$
$$\text{complete}_2$$
$$\text{complete}_3$$

This is a prototypical response, which completes all three acts. If this is the response (assuming proper grounding by the original speaker) the CPS state has now been changed to reflect that the focus is on the objective of identifying a song to which to listen.

(2.6)   B: *No.*
          complete$_1$
          reject$_2$
          reject$_3$

This utterance rejects the last two CPS acts (**adopt** and **focus**), but actually completes the first CPS act (**identify**). This means that B is accepting the fact that this identifying a song is a *possible* objective, even though B rejects **committing** to it. The next possible response highlights this contrast:

(2.7)   B: *I don't listen to songs with clients.*
          reject$_1$
          reject$_2$
          reject$_3$

Here all three CPS acts are rejected. The **identify** is rejected by claiming that the proposed class of objectives is situationally impossible, or inappropriate. (It is usually quite hard to reject **identify** acts, which may have something to do with presuppositions about rational behavior from rational agents.)

The next example helps show the existence of the **identify(focus)** act:

(2.8)   B: *OK, but let's talk about where to eat first.*
          complete$_1$
          complete$_2$
          reject$_3$

Here B completes the first two CPS acts, accepting the objective as a possibility and also committing to it. However, the focus move is rejected, and a different focus is proposed (IntAct not shown).

A more complicated example, including Grounding acts, is shown in Figure 2.3.[9] As the example illustrates, the workings of our CPS model can be complex. However, we think this is more due to the natural complexity of dialogue than to the model itself. We are not aware of a simpler model that has the same coverage of naturally occurring dialogues.

2.5   Implications of collaborative problem solving for NLG

We now turn to a discussion of the implications of the CPS model for Natural Language Generation (NLG). As a thought experiment, the reader is encouraged to consider how

---

[9] For more discussion, see Blaylock (2005); Blaylock and Allen (2005).

the utterances in Figure 2.3 might be generated from the communicative intentions and the CPS state at each point in the dialogue. It is clear that the problem is a challenging one, for several reasons:

1. **Representational issues** – mapping from abstract communicative intentions through semantic content to surface forms;
2. **Information packaging** and focusing issues – deciding which intentions can be realized implicitly, which intentions can be realized together, and which intentions should be foregrounded;
3. **Reference issues** – including reference to abstract entities, tasks, and task-related processes;
4. **Presentational issues** – including decisions about which modalities to use to realize different communicative intentions.

First, there are large **representational** gaps between communicative intentions and the surface forms of utterances. We explored this problem earlier in this chapter when we discussed Figures 2.1 and 2.2, where we showed different alternative utterances (containing different semantic content) that can be generated from a single communicative intention. Cohen explored this problem at a level of beliefs, intentions, and speech acts (Cohen and Perrault, 1979), but more research needs to be done to understand the influence of grounding, negotiation, and collaborative problem solving on the NLG problem.

Second, decisions about how to **package and mark information** are key challenges for NLG. Several communicative intentions may be realized through a single utterance. Some are implicit grounding of the previous utterance (e.g., ack_{4-6} in utterance 3.1 in Figure 2.3), but there also seems to be some sort of implicit mechanism being used sometimes at the IntAct level (e.g., $init_{12-13}$ in utterance 4.1). To illustrate this, consider the example in Figure 2.4, which is taken from a well-known expert–apprentice dialogue (Grosz and Sidner, 1986). Here, utterance 18.2 cleverly combines the two intentions of (1) communicating that the apprentice has begun execution of the pulling off the flywheel step of the plan and (2) requesting of the expert a recipe for achieving that objective. There is a presupposition in utterance 18.2 that the apprentice has already started working on getting the wheel off. Contrast that with another possible realization of the same communicative intentions: *I'm starting to work on getting the wheel off. How do I do it?* Decisions about what to say, what to leave out, and what to let be inferred are difficult for NLG algorithms.

In utterance 19.1 in Figure 2.4, the expert identifies a recipe for pulling the flywheel off. This raises another interesting NLG question: how does one **refer** to parts of the CPS state or to things one wants to introduce into the CPS state, such as recipes? In this case, the recipe is referred to by mentioning the use of a resource (the flywheel puller) that participates in it. The generation of descriptions and of referring expressions are classic NLG problems, and when the referents are abstract objects, processes, or events (such as recipes), the task only becomes more difficult (Grosz, 1978).

1.1 U: *okay, the problem is we better ship a boxcar of oranges to Bath by 8 AM.*

$\text{init}_1(\text{begin}_1(\text{identify}(\boxed{1}\,[\text{objective of shipping oranges with constraint [by 8am]}])))$

$\text{init}_2(\text{begin}_2(\text{adopt}(\boxed{1})))$

$\text{init}_3(\text{begin}_3(\text{focus}(\boxed{1})))$

2.1 S: *okay.*

$\text{ack}_{1-3};\ \text{init}_{4-6}(\text{complete}_{1-3})$

3.1 U: *now … umm … so we need to get a boxcar to Corning, where there are oranges.*

ack_{4-6}

$\text{init}_7(\text{begin}_4(\text{identify}(\boxed{2}\,[\text{recipe involving moving boxcar to Corning}])))$

$\text{init}_8(\text{begin}_5(\text{adopt}(\boxed{2})))$

$\text{init}_9(\text{begin}_6(\text{focus}(\boxed{2})))$

3.2 U: *there are oranges at Corning*

$\text{init}_{10}(\text{begin}_7(\text{identify}(\boxed{3}\,[\text{constraint on the situation}])))$

$\text{init}_{11}(\text{begin}_8(\text{adopt}(\boxed{3})))$

3.3 U: *right?*

reqack_{10-11}

4.1 S: *right.*

$\text{ack}_{10-11};\ \text{init}_{12-13}(\text{complete}_{7-8})$

...

(Utterances 5.1 to 13.1 skipped due to space limitations)

...

13.2 U: *or, we could actually move it [Engine E1] to Dansville, to pick up the boxcar there*

$\text{init}_{101}(\text{begin}_{51}(\text{identify}(\boxed{4}\,[\text{objective move E1 to Dansville}])))$

$\text{init}_{102}(\text{begin}_{52}(\text{focus}(\boxed{4})))$

14.1 S: *okay.*

$\text{ack}_{101-102};\ \text{init}_{103-104}(\text{complete}_{51-52})$

15.1 U: *um and hook up the boxcar to the engine, move it from Dansville to Corning, load up some oranges into the boxcar, and then move it on to Bath.*

$\text{ack}_{103-104}$

$\text{init}_{105}(\text{begin}_{53}(\text{identify}(\boxed{5}\,[\text{objective hook the boxcar to engine1}]))$

$\text{init}_{106}(\text{begin}_{54}(\text{adopt}(\boxed{5})))$

$\text{init}_{107}(\text{begin}_{55}(\text{identify}(\boxed{6}\,[\text{before}(\boxed{1}\,,\boxed{2})])))$

$\text{init}_{108}(\text{begin}_{56}(\text{adopt}(\boxed{6}))$

$\text{init}_{109-116}(\text{begin}_{57-64})\,[\text{2 other actions and 2 other ordering constraints}]$

16.1 S: *okay.*

$\text{ack}_{105-116};\ \text{init}_{117-128}(\text{complete}_{53-64})$

17.1 U: *how does THAT sound?*

$\text{ack}_{117-128}$

$\text{init}_{129}(\text{begin}_{65}(\text{identify}(\boxed{7}\,[\text{blank evaluation}])))$

18.1 S: *that gets us to Bath at 7 AM, and (inc) so that's no problem.*

ack_{129}

$\text{init}_{130}(\text{continue}_{65}(\text{identify}(\boxed{7}\,[\text{evaluation: sufficient}])))$

$\text{init}_{131}(\text{begin}_{66}(\text{adopt}(\boxed{7})))$

19.1 U: *good.*

$\text{ack}_{130-131}$

$\text{init}_{132-133}(\text{complete}_{65-66})$

$\text{init}_{134}(\text{begin}_{67}(\text{adopt}(\boxed{4})))$

$\text{init}_{135}(\text{begin}_{68}(\text{defocus}(\boxed{4})))$

20.1 S: *okay.*

$\text{ack}_{132-135};\ \text{init}_{136-137}(\text{complete}_{67-68})$

Figure 2.3 Analysis of a planning dialogue (from Traum and Hinkelman (1992))

18.2   A: *but I'm having trouble getting the wheel off*

$init_{15}(begin_8(select(\boxed{1}[objective pull off wheel])))$

$init_{16}(begin_9(focus(\boxed{1})))$

$init_{17}(begin_{10}(identify(\boxed{2}[blank recipe])))$

$init_{18}(begin_{11}(adopt(\boxed{2})))$

19.1   E: *Use the wheelpuller.*

$ack_{14-18};\ init_{19-21}(complete_{7-9})$

$init_{22}(continue_{10}(identify(\boxed{3}[recipe using wheelpuller])))$

$init_{23}(continue_{11}(adopt(\boxed{3})))$

Figure 2.4   Analysis involving recipe reference from a dialogue (from Grosz and Sidner (1986))

Another issue for NLG in collaborative problem solving is **presentational**: the interaction between communicative intentions and available or desired communication modalities. In developing the SAMMIE system (Becker *et al.*, 2006) for in-car multimodal MP3 player control, which used the CPS model for dialogue management (Blaylock, 2007), we discovered an interesting case that the CPS model was not able to handle. One assumption of the current model is that decisions about communicative modality are below the level of communicative intentions. However, in addition to user utterances like *What Beatles albums are there?*, we also encountered utterances like *Show me Beatles albums*, which was an explicit request for the use of the visual modality. The model may have to be extended to include communication modalities.

2.6   Conclusions and future work

In this chapter, we discussed communicative intentions and their role within interactive systems. We argued for a more inclusive model of task-oriented dialogue, one that can be used both for *planning* and *execution* of tasks, as well as other task-related activities. We briefly introduced a model of communicative intentions in dialogue that is based on collaborative problem solving. The CPS model of dialogue comprises an explicit model of communicative intentions with layers at the grounding, negotiation, and collaborative problem solving levels. We argued that research is needed in natural language generation to enable a dialogue system to generate text or speech based on a very high-level model of communicative intentions like ours, and outlined several potential areas for additional work in this area. As dialogue systems become capable of handling an increasingly rich variety of task-related behaviors, the need for NLG systems that can generate from complex communicative intentions in a complex discourse context will only increase.

Acknowledgments

We would like to thank James Allen and George Ferguson, who were collaborators on earlier versions of this work.

This research was supported by a grant from DARPA (no. #F30602-98-2-0133); a grant from the Department of Energy (no. #P2100A000306); two grants from The National Science Foundation (award #IIS-0328811 and award #E1A-0080124); and the EU-funded TALK Project (no. IST-507802). Any opinions, findings, conclusions, or recommendations expressed in this chapter are those of the author and do not necessarily reflect the views of the above-named organisations.

References

Alexandersson, J., Buschbeck-Wolf, B., Fujinami, T., Kipp, M., Kock, S., Maier, E., Reithinger, N., Schmitz, B., and Siegel, M. (1998). Dialogue acts in VERBMOBIL-2 second edition. Technical Report Verbmobil Report 226, DFKI Saarbrücken, Universität Stuttgart, TU Berlin, Universität des Saarlandes.

Allen, J. F. (1983). Recognizing intentions from natural language utterances. In Brady, M. and Berwick, R., editors, *Computational Models of Discourse*, pages 107–166. MIT Press, Cambridge, MA.

Allen, J. F., Byron, D., Dzikovska, M., Ferguson, G., Galescu, L., and Stent, A. (2001). Towards conversational human-computer interaction. *AI Magazine*, **22**(4):27–37.

Allen, J. F. and Core, M. (1997). Draft of DAMSL: Dialog act markup in several layers. Available from: http://www.cs.rochester.edu/research/cisd/resources/damsl/RevisedManual/. Accessed on 11/24/2013.

Austin, J. L. (1962). *How To Do Things with Words*. Clarendon Press, Oxford, UK.

Becker, T., Blaylock, N., Gerstenberger, C., Kruijff-Korbayová, I., Korthauer, A., Pinkal, M., Pitz, M., Poller, P., and Schehl, J. (2006). Natural and intuitive multimodal dialogue for in-car applications: The Sammie system. In *Proceedings of the European Conference on Artificial Intelligence*, pages 612–616, Riva del Garda, Italy. Italian Association for Artificial Intelligence.

Black, A. W., Burger, S., Conkie, A., Hastie, H., Keizer, S., Lemon, O., Merigaud, N., Parent, G., Schubiner, G., Thomson, B., Williams, J. D., Yu, K., Young, S., and Eskenazi, M. (2011). Spoken dialog challenge 2010: Comparison of live and control test results. In *Proceedings of the SIGdial Conference on Discourse and Dialogue (SIGDIAL)*, pages 2–7, Portland, OR. Association for Computational Linguistics.

Blaylock, N. (2005). *Towards Tractable Agent-based Dialogue*. PhD thesis, Department of Computer Science, University of Rochester.

Blaylock, N. (2007). Towards flexible, domain-independent dialogue management using collaborative problem solving. In *Proceedings of the Workshop on the Semantics and Pragmatics of Dialogue*, pages 91–98, Rovereto, Italy. SemDial.

Blaylock, N. and Allen, J. F. (2005). A collaborative problem-solving model of dialogue. In *Proceedings of the SIGdial Workshop on Discourse and Dialogue (SIGDIAL)*, pages 200–211, Lisbon, Portugal. Association for Computational Linguistics.

Bohlin, P., Bos, J., Larsson, S., Lewin, I., Matheson, C., and Milward, D. (1999). Survey of existing interactive systems. Available from http://www.ling.gu.se/projekt/trindi/publications.html. Accessed on 11/24/2013.

Bohus, D. and Rudnicky, A. I. (2003). RavenClaw: Dialog management using hierarchical task decomposition and an expectation agenda. In *Proceedings of the European Conference on*

Speech Communication and Technology (EUROSPEECH), pages 597–600, Geneva, Switzerland. International Speech Communication Association.

Bunt, H., Alexandersson, J., Carletta, J., Choe, J.-W., Fang, A. C., Hasida, K., Lee, K., Petukhova, V., Popescu-Belis, A., Romary, L., Soria, C., and Traum, D. (2010). Towards an ISO standard for dialogue act annotation. In *Proceedings of the International Conference on Language Resources and Evaluation (LREC)*, Valletta, Malta. European Language Resources Association.

Carberry, S. (1990). *Plan Recognition in Natural Language Dialogue*. MIT Press, Cambridge, MA.

Chu-Carroll, J. (2000). MIMIC: An adaptive mixed initiative spoken dialogue system for information queries. In *Proceedings of the Conference on Applied Natural Language Processing*, pages 97–104, Seattle, WA. Association for Computational Linguistics.

Chu-Carroll, J. and Carberry, S. (2000). Conflict resolution in collaborative planning dialogues. *International Journal of Human-Computer Studies*, **53**(6):969–1015.

Clark, H. (1996). *Using Language*. Cambridge University Press, Cambridge, UK.

Cohen, P. R. and Levesque, H. J. (1990). Rational interaction as the basis for communication. In Cohen, P. R., Morgan, J., and Pollack, M. E., editors, *Intentions in Communication*, pages 221–255. MIT Press, Cambridge, MA.

Cohen, P. R. and Perrault, C. R. (1979). Elements of a plan-based theory of speech acts. *Cognitive Science*, **3**(3):177–212.

Cohen, P. R., Perrault, C. R., and Allen, J. F. (1982). Beyond question answering. In Lehnert, W. G. and Ringle, M., editors, *Strategies for Natural Language Processing*, pages 245–274. Lawrence Erlbaum Associates, Hillsdale, NJ.

Di Eugenio, B., Jordan, P. W., Thomason, R. H., and Moore, J. D. (1997). Reconstructed intentions in collaborative problem solving dialogues. In *Working Notes of the AAAI Fall Symposium on Communicative Action in Humans and Machines*, Boston, MA. Association for the Advancement of Artificial Intelligence.

Ehlen, P. and Johnston, M. (2010). Speak4it: Multimodal interaction for local search. In *Proceedings of the International Conference on Multimodal Interfaces and the Workshop on Machine Learning for Multimodal Interaction (ICMI-MLMI)*, article 10, Beijing, China. Association for Computing Machinery.

Grice, H. P. (1969). Utterer's meaning and intention. *The Philosophical Review*, **78**(2):147–177.

Grosz, B. J. (1978). Focusing in dialog. In *Proceedings of the Workshop on Theoretical Issues in Natural Language Processing*, pages 96–103, Urbana, IL. Association for Computational Linguistics.

Grosz, B. J. and Kraus, S. (1996). Collaborative plans for complex group action. *Artificial Intelligence*, **86**(2):269–357.

Grosz, B. J. and Kraus, S. (1999). The evolution of SharedPlans. In Wooldridge, M. J. and Rao, A., editors, *Foundations and Theories of Rational Agency*, pages 227–262. Kluwer, Dordrecht, The Netherlands.

Grosz, B. J. and Sidner, C. L. (1986). Attention, intentions, and the structure of discourse. *Computational Linguistics*, **12**(3):175–204.

Hansen, B., Novick, D. G., and Sutton, S. (1996). Systematic design of spoken prompts. In *Proceedings of the ACM SIGCHI Conference on Human Factors in Computing Systems (CHI)*, pages 157–164, Vancouver, Canada. Association for Computing Machinery.

Lambert, L. and Carberry, S. (1991). A tripartite plan-based model of dialogue. In *Proceedings of the Annual Meeting of the Association for Computational Linguistics (ACL)*, pages 47–54, Berkeley, CA. Association for Computational Linguistics.

Lamel, L., Rosset, S., Gauvain, J.-L., Bennacef, S., Garnier-Rizet, M., and Prouts, B. (2000). The LIMSI ARISE system. *Speech Communication*, **31**(4):339–354.

Levesque, H. J., Cohen, P. R., and Nunes, J. H. T. (1990). On acting together. In *Proceedings of the Conference on Artificial Intelligence (AAAI)*, pages 94–99, Boston, MA. AAAI Press.

Litman, D. and Allen, J. F. (1990). Discourse processing and commonsense plans. In Cohen, P. R., Morgan, J., and Pollack, M., editors, *Intentions in Communication*, pages 365–388. MIT Press, Cambridge, MA.

Mann, W. C. and Thompson, S. A. (1988). Rhetorical structure theory: Toward a functional theory of text organization. *Text*, **8**(3):243–281.

Ramshaw, L. A. (1991). A three-level model for plan exploration. In *Proceedings of the Annual Meeting of the Association for Computational Linguistics (ACL)*, pages 39–46, Berkeley, CA. Association for Computational Linguistics.

Rudnicky, A. I., Thayer, E., Constantinides, P., Tchou, C., Shern, R., Lenzo, K., Xu, W., and Oh, A. (1999). Creating natural dialogs in the Carnegie Mellon Communicator system. In *Proceedings of the European Conference on Speech Communication and Technology (EUROSPEECH)*, pages 1531–1534, Budapest, Hungary. International Speech Communication Association.

Sadek, D. and De Mori, R. (1998). Dialogue systems. In De Mori, R., editor, *Spoken Dialogues with Computers*, pages 523–561. Academic Press, Orlando, FL.

Searle, J. (1975). Indirect speech acts. In Cole, P. and Morgan, J. L., editors, *Speech Acts*, pages 59–82. Academic Press, New York, NY.

Seneff, S. and Polifroni, J. (2000). Dialogue management in the Mercury flight reservation system. In *Proceedings of the ANLP/NAACL Workshop on Conversational Systems*, pages 11–16, Seattle, WA. Association for Computational Linguistics.

Stent, A. J. (2002). A conversation acts model for generating spoken dialogue contributions. *Computer Speech and Language*, **16**(3–4):313–352.

Traum, D. and Hinkelman, E. (1992). Conversation acts in task-oriented spoken dialogue. *Computational Intelligence*, **8**(3):575–599.

Traum, D. R. (1994). *A Computational Theory of Grounding in Natural Language Conversation*. PhD thesis, Department of Computer Science, University of Rochester.

Traum, D. R. (2000). 20 questions for dialogue act taxonomies. *Journal of Semantics*, **17**(1):7–30.

Zue, V., Seneff, S., Glass, J., Polifroni, J., Pao, C., Hazen, T. J., and Hetherington, L. (2000). JUPITER: A telephone-based conversational interface for weather information. *IEEE Transactions on Speech and Audio Processing*, **8**(1):85–96.

3 Pursuing and demonstrating understanding in dialogue

David DeVault and Matthew Stone

3.1 Introduction

The appeal of natural language dialogue as an interface modality is its ability to support open-ended mixed-initiative interaction. Many systems offer rich and extensive capabilities, but must support novice or infrequent users. It is unreasonable to expect untrained users to know the actions they need in advance, or to be able to specify their goals using a regimented scheme of commands or menu options. Dialogue allows the user to talk through their needs with the system and arrive collaboratively at a feasible solution. Dialogue, in short, becomes more useful to users as the interaction becomes more potentially problematic.

However, the flexibility of dialogue comes at a cost in system engineering. We cannot expect the user's model of the task and domain to align with the system's. Consequently, the system cannot count on a fixed schema to enable it to understand the user. It must be prepared for incorrect or incomplete analyses of users' utterances, and must be able to put together users' needs across extended interactions. Conversely, the system must be prepared for users that misunderstand it, or fail to understand it.

This chapter provides an overview of the concepts, models, and research challenges involved in this process of pursuing and demonstrating understanding in dialogue. We start in Section 3.2 from analyses of human–human conversation. People are no different from systems: they, too, face potentially problematic interactions involving misunderstandings. In response, they avail themselves of a wide range of discourse moves and interactive strategies, suggesting that they approach communication itself as a collaborative process wherein all parties establish agreement, to their mutual satisfaction, on the distinctions that matter for their discussion and on the expressions through which to identify those distinctions. In the literature, this process is often described as **grounding** communication, or identifying contributions well enough so that they become part of the **common ground** of the conversation (Clark and Marshall, 1981; Clark and Schaefer, 1989; Clark and Wilkes-Gibbs, 1990; Clark, 1996).

For a computer system, grounding can naturally be understood in terms of problem solving. When a system encounters an utterance whose interpretation is incomplete, ambiguous, unlikely, or unreliable, it has to figure out how to refine and confirm that interpretation without derailing the interaction. When a system gets evidence that one of its own utterances may not have been understood correctly, it has to figure out how to revise and reframe its contributions to keep the conversation on track. Solving such problems means coming up with ways of building on partial understandings of previous contributions, while formulating utterances with a reasonable expectation that they will be understood correctly, and will move the conversation forward. We call this reasoning process **contribution tracking**. It is a core requirement for natural language generation in dialogue.

In Section 3.3, we describe a preliminary version of contribution tracking in a prototype dialogue system, COREF (DeVault and Stone, 2006, 2007, 2009; DeVault *et al.*, 2005; DeVault, 2008), and offer a new assessment of the qualitative capabilities that contribution tracking gives the current version of COREF. We emphasize how contribution tracking influences all the steps of planning and acting in dialogue systems, but particularly the process of natural language generation. Our analysis suggests that dialogue system designers must be wary of idealizations often adopted in natural language generation research. For example, it may not be possible to specify a definite context for generation, the system may not be able to formulate an utterance that the user will be guaranteed to understand, and the system may have a communicative goal and content to convey that overlaps in important ways with the communicative goals and content of previously formulated utterances.

Relaxing these idealizations remains a major challenge for generation and grounding in dialogue. In particular, in Section 3.4, we use our characterization of contribution tracking as problem solving to analyze the capabilities of grounding models in current conversational systems. Many applied frameworks restrict grounding actions to simple acknowledgments, confirmation utterances, and clarification requests. This narrow focus lets systems streamline the reasoning required to generate grounding utterances, through approaches that abstract away from aspects of the system's uncertainty about the conversational state and its consequences for system choices.

On the other hand, as we survey in Section 3.5, a wide range of emerging applications will require a much more sophisticated understanding of the grounding work that systems and users are doing. A conversational agent that generates communicative gestures and facial expressions will need to model how non-verbal actions help to signal understanding or lack of understanding. A collaborative robot that carries out physical activities jointly with a human partner will need to model how real-world actions give evidence of conversational participants' interpretations of utterances. A virtual human, with naturalistic mechanisms of attention, cognition, and emotion, will need to be able to realize aspects of its internal state, including its understanding of what is happening in the conversation, in the work it is obviously doing to participate in the conversation. In our view, these emerging applications for dialogue technology give a lasting importance to general accounts of grounding as problem solving – and offer an exciting range of practical test cases for generating new kinds of grounding phenomena.

3.2 Background

Avoiding misunderstanding in problematic situations is a joint effort. If the addressee gives no feedback about what he understands, there is no way that the speaker can confirm that she was understood as she intended. Conversely, unless the speaker acts on the feedback the addressee provides, the addressee cannot correct an incomplete or faulty understanding. In human–human conversations, conversational participants work jointly to stay in sync. Understanding of what people do in the face of difficulties can provide a starting point for achieving similar grounding in dialogue systems.

3.2.1   Grounding behaviors

As Clark (1996, Ch. 5) reminds us, successful understanding, for humans and machines, involves recognizing what the speaker is doing across a hierarchy of levels. At the lowest level, recognizing words, people sometimes face substantial difficulty – though perhaps not so severe as systems with automatic speech recognition. In such situations, conversational participants often use confirmation strategies to make sure they get the information right. Example 3.1 illustrates:

> (3.1)   June   *ah, what ((are you)) now, *where**
> Darryl   **yes* forty-nine Skipton Place*
> June   *forty-one*
> Darryl   *nine. nine*
> June   *forty-nine, Skipton Place,*
> Darryl   *W one.*
> London–Lund Corpus (9.2a.979) cited in Clark (1996, p. 236)

June's confirmation strategy involves repeating what Darryl says; this enables Darryl to catch and correct June's mishearing of *forty-nine* as *forty-one*. Grounding at the recognition level is facilitated by coordination: Darryl speaks in instalments that can be echoed and corrected easily, while June echoes as expected. The strategy allows specifications, confirmations, and corrections to come fluidly and elliptically. It's not just June that's taking grounding into account in managing the dialogue here; it's also Darryl.

Understanding at the next-higher level involves grammatical analysis of utterances. A representative task here is the resolution of word-sense ambiguities. Systems famously face a vocabulary problem because users are so variable in the meanings they assign to words in unfamiliar situations (Furnas *et al.*, 1987). So do people, like conversational participants A and B in example 3.2.

> (3.2)   B   *k– who evaluates the property —*
> A   *u:h whoever you asked,. the surveyor for the build-
> ing society*
> B   *no, I meant who decides what price it'll go on the
> market —*
> A   *(– snorts), whatever people will pay —*
> London–Lund Corpus (4.2.298) cited in Clark (1996, p. 234)

In the complex process of valuing real estate, property is evaluated in one sense by the seller and sales agent to fix an offering price, and in another sense by an appraisal or survey carried out before the buyer can get a mortgage, which in the UK prototypically comes from a building society. In example 3.2, A's first answer clearly assumes one construal of B's term *evaluate*. Even though A offers no overt confirmation or acknowledgment of B's meaning, the response allows B to recognize A's construal and to reframe the original question more precisely. In this case, grounding is accomplished through explicit meta-level language that takes meaning itself as the topic of conversation in grounding episodes.

The highest level of understanding concerns the relationship of conversational participants' meanings to the ongoing task and situation. Problematic reference, as in example 3.3, illustrates the difficulties both people and systems face in this process, and the dynamics through which people achieve grounding.

> (3.3)   A   *Okay, the next one is the rabbit.*
>   B   *Uh —*
>   A   *That's asleep, you know, it looks like it's got ears*
>   *and a head pointing down?*
>   B   *Okay*
>   Clark and Wilkes-Gibbs (1990) cited in Clark (1993, p. 127)

Here B offers a simple signal of non-understanding. At this point it is up to A to develop the initial description further. A produces an **expansion**, in Clark and Wilkes-Gibbs's (1990) terminology. A ignores B's possible invitation to address the explicit topic of what A means, and simply provides a syntactic and semantic continuation of the initial description.

In our characterization of grounding so far, we have seen that an addressee's contribution to grounding can include confirmation, relevant followup utterances, and signals of non-understanding. We close with two other examples that underscore the range of grounding moves in human–human conversation. In example 3.4, B responds to an unclear description by offering an alternative description that seems clearer. A accepts and adopts B's reformulation.

> (3.4)   A   *Okay, and the next one is the person that looks like*
>   *they're carrying something and it's sticking out to*
>   *the left. It looks like a hat that's upside down.*
>   B   *The guy that's pointing to the left again?*
>   A   *Yeah, pointing to the left, that's it! (laughs)*
>   B   *Okay*
>   Clark and Wilkes-Gibbs (1990) cited in Clark (1993, p. 129)

Such cases show the benefits of joint effort in grounding. Finally, we should not forget simple cases like example 3.5:

> (3.5)   Burton   *how was the wedding —*
>   Anna   *oh it was really good, it was uh it was a lovely day*
>   Burton   *yes*

Anna   *and. it was a super place,. to have it. of course*
Burton   *yes —*
Anna   *and we went and sat on sat in an orchard, at Grantchester, and had a huge tea *afterwards (laughs —)**
Burton   **(laughs —)**
London–Lund corpus (7.3l.1441) cited in (Clark, 1996, p. 237)

Grounding is necessary even in unproblematic interactions, and it often takes the form of straightforward acknowledgments, like Burton's in example 3.5.

3.2.2   Grounding as a collaborative process

Clearly, grounding in human–human conversation is a complex, wide-ranging skill. Its effects are pervasive, multifaceted, and varied. Implementing analogous skills in dialogue systems requires an overarching framework that reveals the fundamental commonalities behind people's grounding strategies and links them to mechanisms that plausibly underlie them.

Accounts of grounding in cognitive science start from Grice's influential account of conversation as rational collaboration (Grice, 1975). Grice proposes that conversation is governed by a central Cooperative Principle: all conversational participants are expected do their part in the conversation, by making appropriate contributions. This includes showing how they understand previous utterances in the conversation and creating follow up utterances to ensure understanding (Clark and Schaefer, 1989; Brennan, 1990; Clark and Wilkes-Gibbs, 1990). Clark (1996) characterizes this in terms of the concept of **closure**, or having good evidence that communicative actions have succeeded. Traum and Allen (1994) describe grounding in even stronger terms. They argue that conversational participants have an **obligation** to provide evidence of their understanding and address the issues others raise, above and beyond the real-world collaborative interests they share. Both kinds of approach suggest an architecture for conversation where conversational participants regularly assess what they understand and estimate what their addressees understand. Accordingly, conversational participants bring goals, preferences, or obligations for mutual understanding that they pursue at the same time as they track task-related goals for sharing information and achieving real-world results.

The diversity of moves that accomplish grounding, as surveyed in Section 3.2.1, imposes further constraints on a theory of grounding. Grounding is, in Brennan's phrase (1990), a matter of seeking and providing **evidence** for understanding. This evidence can take a variety of forms (see Clark, 1996, pp. 223ff). It is clear that repetitions (as in example 3.1), reformulations (as in example 3.4), and assertions of understanding (as in example 3.5), provide evidence about the addressee's level of understanding. But followup utterances (as in example 3.2), and assertions of non-understanding (as in example 3.3), do the same. In fact, followup utterances can provide quite good evidence about an addressee's understanding (or lack thereof), regardless of whether or not the addressee intends the utterance to do so. Conversely, assertions of understanding that reveal only the addressee's own judgment may be quite unreliable.

These cases show that conversational participants face extended periods of transient uncertainty during grounding. During these periods, they must assess the evidence they have about grounding, and trade off the costs of further clarification against the risks of persistent misunderstanding as they select their next contributions to the conversation (Horvitz and Paek, 2001). The resulting dynamics of interaction can be approximated by models of grounding that require that each contribution be acknowledged and accepted before it can be admitted into an incrementally updated representation of common ground and before the conversation can move forward (Traum, 1994; Matheson *et al.*, 2000). But as we discuss in Section 3.4.1, this type of model has a limited ability to resolve uncertainty using evidence from multiple utterances; conversational participants' reasoning may be more nuanced than these models allow.

Conversation is structured hierarchically, with short segments that address focused subtasks nested within longer segments that address larger tasks (Grosz and Sidner, 1986). This structure accords with the difficulty we have when we approach tasks in the wrong order or have to revisit issues we thought were resolved. We also have distinctive knowledge about how to negotiate as part of a collaborative task-oriented activity such as conversation. Collaborative negotiation typically involves a distinctive inventory of possible contributions to the activity: negotiators can make proposals, revise them, accept them conditionally or unconditionally, and so forth (Sidner, 1994; Carberry and Lambert, 1999; Di Eugenio *et al.*, 2000). Collaborative negotiation is frequently modeled through additional layers of discourse structure, which explicitly represent the ordinary contributions of utterances as the object of meta-level discussion (Carberry and Lambert, 1999; Allen *et al.*, 2002). Models along these lines naturally describe conversational participants' collaborative efforts to agree on linguistic utterances and their interpretations (Heeman and Hirst, 1995). This is an important consideration in when and how to ground.

Another such consideration is the context dependence of utterance interpretation. Many utterances, including those typically relevant for grounding, express a specific relationship to the information presented and the open questions raised in the prior discourse (Hobbs *et al.*, 1993; Kehler, 2001; Asher and Lascarides, 2003). Many also involve implicit references to salient entities from the discourse context. Both kinds of contextual link must be resolved as part of understanding the utterance. It is largely through these links that utterances provide evidence about the speaker's acceptance and understanding of prior contributions to the conversation (Lascarides and Asher, 2009; Stone and Lascarides, 2010). Thus, these links explain efficient utterances that ground implicitly, as well as less efficient constructions that are designed to avoid grounding incorrectly.

Discourse theory also highlights the specific grammatical and interactive resources that make it easy for conversational participants to provide evidence of their understanding. For example, the rules of grammar sometimes permit interpreting fragmentary utterances together with previous utterances, as in the successive expansions of example 3.3. See Gregoromichelaki *et al.* (2011) for a wide range of further cases. Other types of fragment, such as the reprise fragments of example 3.1, seem to carry semantic constraints that closely tie their interpretation to those of prior utterances (Ginzburg and Cooper, 2004; Purver, 2004). Both cases create interpretive connections that readily signal grounding.

To sum up, all the moves we make in conversation, including grounding moves, respect the distinctive structure and process of human collaborative activity. Approaches to grounding can benefit from detailed models of discourse interpretation, including both general constraints and specific syntactic, semantic, and interactive resources that are available for providing evidence of understanding.

3.2.3   Grounding as problem solving

The examples of Section 3.2.1 and the theoretical accounts of Section 3.2.2 portray grounding strategies as flexible responses to a speaker's information state and goals, given the affordances of grammar and collaboration. As a framework for realizing such responses in dialogue systems, we advocate a characterization of grounding as problem solving.

Problem solving is a general perspective on flexible intelligent behaviour (Newell, 1982; Russell and Norvig, 1995). A problem solving system is endowed with general knowledge about the actions available to it and their possible effects, and with goals or preferences that it must strive to make true through the actions it chooses. The system approaches a new situation by distinguishing key features of the situation: those that the system needs to change, on the one hand, and those that define the system's opportunities to act, on the other. This representation constitutes a problem for the system. The system solves such problems by reasoning creatively. It systematically explores possibilities for action and predicts their results, until it improvises a program of action that it can use to further its goals to a satisfactory degree in the current situation.

To treat grounding as problem solving is to design a conversational agent with knowledge about possible utterances that includes the contributions that utterances can make to the evolving conversation and the evidence that utterances offer about their speaker's understanding. In the case of grounding, this knowledge must describe general models of collaborative discourse, along with particularly relevant grammatical constructions and negotiation moves. The conversational agent must then track its own understanding and its estimate of the understanding of its conversational participants, and aim to reduce the uncertainty in these understandings to a satisfactory level. In particular, a specific pattern of ambiguity in a specific conversational situation is a trigger for producing a new utterance whose interpretation highlights the potentially problematic nature of the interaction and initiates a possible strategy to resolve it.

Methodologically, our appeal to problem solving plays two roles. First, it serves to link system design to empirical and theoretical results about grounding, by emphasizing the knowledge that is realized in systems rather than the specific algorithms or processing through which systems deploy that knowledge. Second, it provides a transparent explanation of the design of certain kinds of grounding systems: namely, those that navigate through a large and heterogeneous space of possible utterances to generate creative utterances for new situations. This suits our purpose in documenting and analyzing our COREF system in Section 3.3.

Problem solving is a theoretical perspective rather than a specific technique. Dialogue strategies are often engineered using specific models, such as Partially-Observable Markov Decision Processes (POMDPs; Williams and Young 2007), which clarify specific aspects of the system's behavior. Mathematically, a POMDP starts from a precise probabilistic definition of what any utterance can contribute and the evidence a system gets about these contributions moment by moment. It derives an overall strategy that chooses utterances with an eye to long-term success. This accounts for the system's reasoning in managing uncertainty about the interaction and in making quantitative tradeoffs between gathering information and advancing task goals. By contrast, the specific choices made in natural language generation depend on methods that let us compute what new utterances can do, creatively, across an open-ended generative space. Where POMDP models assume that this computation can be accomplished straightforwardly, a problem-solving perspective lets us clarify the contributions of techniques for action representation, discourse modeling, and the effective management of search. Of course, the information a POMDP encodes also needs to be present in a problem-solving model to handle the decisions that are emphasized in POMDP research. This is why the choice of representation is in part a matter of theoretical perspective.

Our approach to grounding elaborates the model of NLG as problem solving from the SPUD generator (Stone *et al.*, 2003). SPUD solves problems of contributing specific new information to conversational participants against the backdrop of a determinate common ground. SPUD's linguistic knowledge takes the form of a lexicalized grammar with entries characterized in syntactic, semantic, and pragmatic terms. A model of interpretation predicts how an utterance with a given semantics can link up to the context to convey relevant information. A solution to an NLG problem is a syntactically complete derivation tree whose meaning, as resolved unambiguously in context, contributes all the new information to the conversation, without suggesting anything false.

SPUD's problem-solving approach offers an integrated account of a range of generation effects, including the aggregation of related information into complex sentences, the planning of referring expressions, and the orchestration of lexical and syntactic choices. Importantly, it tracks the creative, open-ended ways in which these effects overlap in complex utterances. Moreover, the problem-solving approach emphasizes the role for declarative techniques in system design. The knowledge SPUD uses can be derived from diverse data sources, such as grammars designed for language understanding, linguistic analyses of target corpora, and machine learning over attested or desired uses of utterances in context.

SPUD's generation model must be extended in a number of ways to handle the kinds of grounding phenomena illustrated in Section 3.2.1. It must be able to handle the many different kinds of contribution that utterances can make to conversation, besides just providing new information. It must be able to predict not just the contributions that utterances make directly to the conversation, but also the indirect effects that utterances can have on conversational participants' information, especially their assessments of grounding. Finally, it must take into account possible uncertainties in the context, to calculate what interpretations an utterance could have or whether the addressee will understand it as intended. In our work with the COREF system, we have developed extensions to SPUD to handle these issues. In Section 3.3, we outline our approach and illustrate its role in enabling our system to exhibit a wide range of grounding strategies.

3.3   An NLG model for flexible grounding

COREF participates in a two-agent object identification game that we adapted from the experiments of Clark and Wilkes-Gibbs (1990) and Brennan and Clark (1996). Our game plays out in a special-purpose graphical user interface, shown in Figure 3.1, which can support either human–human or human–agent interactions. The objective is for the two players to work together to create a specific configuration of objects, or a "scene," adding objects into the scene one at a time. The players participate from physically separated locations so that communication can only occur through the interface. Each has their own version of the interface, which displays the same set of candidate objects but in differently shuffled spatial locations. The shuffling prevents the use of spatial expressions such as "the object at the top left".[1]

As in the experiments of Clark and Wilkes-Gibbs (1990) and Brennan and Clark (1996), one of the players, who plays the role of **director**, instructs the other player, who plays the role of **matcher**, which object goes next. As the game proceeds, the next object is automatically determined by the interface and privately indicated to the director using a blue arrow. The director's job is then to get the matcher to click on (their version of) this object.

To achieve agreement about a target object, the two players can talk back and forth using written English, in an instant-messaging style modality. Each player's interface provides a real-time indication that their partner is "Active" while the other is composing an utterance in their text box, but the interface does not show in real time what characters are being typed. Thus, it is not possible for a player to view or interpret an utterance by their partner incrementally (as it is typed). Once the Enter key is pressed, the utterance appears on the screens of both players as part of a scrollable history field in the interface, which provides full access to all the previous utterances in the dialogue.

When the matcher clicks on an object they believe is the target, their version of that object is moved into their scene. The director has no visible indication that the matcher has done this. However, the director must click the Continue (next object) button

---

[1] Note that in a human–human game, there are literally two versions of the graphical interface on separate computers. In a human–agent interaction, the agent uses a software interface that provides the same information that the graphical interface would provide to a human participant. For example, the agent is not aware of the spatial locations of objects on the screen of its human partner.

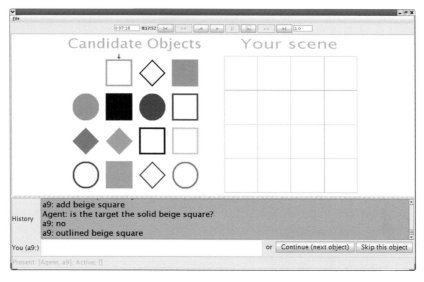

Figure 3.1 A human user plays an object identification game with COREF. The figure shows the perspective of the user (a9). The user is playing the role of director, and trying to identify the square at upper left (indicated by the arrow) to COREF

in order to move the current target into the director's scene, and move on to the next target object. This means that the players need to discuss not just what the target object is, but also whether the matcher has added it into their scene, so that they can coordinate on the right moment to move on to the next object. If the coordination succeeds perfectly, then the director and matcher will create the same scene in their separate interfaces.

We have run a series of evaluation experiments in which COREF interacts with human subjects (DeVault and Stone, 2007; DeVault, 2008). Here, we illustrate qualitative aspects of COREF's grounding behavior by describing the reasoning that COREF uses to generate questions in three key interactions from this data set. In example 3.6, COREF plays the matcher, and begins by trying to guess the next object to add by producing *is the target pink?* (the display features pink, light brown, and dark brown objects).

(3.6)   Agent   *is the target pink?*
         a18   *no*
         a18   *it is brown*
         Agent   *do you mean dark brown?*
         a18   *yes*

COREF's subsequent clarification question *do you mean dark brown?* attempts to disambiguate the color characterizing the target object.

In example 3.7, COREF is the director.

(3.7)   Agent   *the blue circle*
          s1   *ok*
          Agent   *did you add it?*
          s1   *yes*

COREF interprets the user's *ok* as ambiguous: maybe the user has added the blue circle to the display; maybe they are simply signaling understanding. COREF's *did you add it?* is designed to move the conversation forward in either case.

Finally, COREF, playing the matcher in example 3.8, finds two objects that might plausibly be described as *beige diamonds*.

(3.8)   s2   *the object is the beige diamond*
          Agent   *do you mean the solid beige diamond?*
          s2   *no*
          s2   *i mean the empty beige diamond*
          s2   *did you add it?*
          Agent   *done*

COREF's clarification question *do you mean the solid beige diamond?* attempts to distinguish between two target objects based on their fill.

A uniform problem-solving model leads to the production of all of these questions in COREF. The model is distinguished by its flexible mapping from utterances to task actions (Section 3.3.1), its explicit representation of the effects of uncertainty on interpretation (Section 3.3.2), and its specification of generation problems in terms of a pattern of acceptable interpretations each of which moves the conversation forward despite uncertainty (Section 3.3.3).

3.3.1   Utterances and contributions

Our modeling challenge in COREF is to characterize the different contributions that utterances can make. Our approach is to use an architecture of **dialogue acts**. We follow Bunt (1994, 1996, 2000) in defining a dialogue act as a "functional unit used by a speaker to change the context." In particular, each dialogue act comprises a **semantic content** and a **communicative function**. The semantic content is the information the speaker is introducing into the context; for example, some proposition p. The communicative function is the intended role for that information in the context; for example, inform, yn-question, or correct (Bunt, 2000). Together, the communicative function and semantic content determine an update function that maps the previous context to the updated context that results from the dialogue act (Larsson and Traum, 2000).

Researchers commonly hypothesize dialogue acts that specifically govern grounding. For example, Bunt (1994) offers a top-level distinction between **dialogue control acts** and **task-oriented acts**, and then subdivides dialogue control acts into **feedback acts**, **discourse structuring acts**, and **interaction management acts**. Traum and Hinkelman (1992) distinguish four top-level types of **conversation act**: **turn-taking acts**, **grounding acts**, **core speech acts**, and **argumentation acts**. The DAMSL annotation scheme

(Core and Allen, 1997) captures grounding by distinguishing between the **backward** and **forward communicative functions** of **communicative acts**.

By contrast, we have designed COREF with a special-purpose set of about twenty action types. Our inventory endows COREF with detailed knowledge about reference and other structured collaborations. Grounding is not implemented as an explicit goal for COREF, nor is grounded material given a special representational status; rather, grounding emerges as a side effect of the reasoning and strategies that motivate the agent's action choices.

Our action set builds on stack-based models of clarification sub-dialogue (Ginzburg and Cooper, 2004; Purver, 2004), collaborative negotiation models (Sidner, 1994; Carberry and Lambert, 1999; Di Eugenio *et al.*, 2000), and the use of collaborative discourse theory to characterize user interface tasks (Rich *et al.*, 2001). For example, one action type, pushCollabRef$[D, M, T]$, lets the director D initiate collaborative reference with matcher M to a target T. This action type is tacit: it corresponds with the mental decision by D to collaboratively identify target T to M. Its effects are to push a new task onto a stack of active tasks in the context and set up an initially empty constraint network recording the information that the conversational participants have used to describe T. Once this task is underway, the director can perform a dialogue act, addcr$[T, C]$, whose effects add the constraint C to the constraint network for T.

The identification by the matcher that the target T is some entity R is captured by a tacit mental action setVarValue$[T, R]$. Its effect is to add the proposition varValue(T, R) to the context. After identifying R as a target object for the scene, the matcher can take the tacit action addToScene$[R]$. Although the matcher needs to click the object in their display in order to add it to their scene, this is invisible to the director. The effect of the matcher clicking the object is to physically move the object into the scene part of the matcher's experiment interface. The effect of agent A taking action addToScene$[R]$ is that the proposition inScene(R, A) is added to the list of propositions in the context.

In addition, COREF's interpretation process causes certain additional updates to the dialogue state whenever an observable action occurs. These additional updates allow COREF to keep track of its uncertainty in interpretation; they also push a ManageAmbiguity subtask, which makes it coherent for the observer x of action A to perform followup actions to deal with uncertainty.

We associate utterances with dialogue acts by modeling utterance interpretation as an intention-recognition problem (Stone, 2004). We assume that each utterance by a speaker is intended to **generate** (Goldman, 1970; Pollack, 1986) one or more dialogue acts. We also assume that utterances are often intended to reveal the occurrence of certain tacit dialogue acts indirectly, through a kind of conversational implicature we have called **enlightened update** (Thomason *et al.*, 2006). To recognize the speaker's intention(s) requires integrating three sources of information: the logical structure of the ongoing collaborative activity, the current state of the dialogue, and grammatical constraints.

Our model of dialogue structure takes the form of a function $N(A, s)$ that captures the set of alternative actions that agent A can coherently perform next in dialogue state s. This function captures similar regularities to **adjacency pair** models, to **dialogue grammars** that distinguish coherent sequences of dialogue acts from incoherent ones, to state machines with dialogue acts attached to the state transitions, and to plan-based models of dialogue coherence; see (Cohen, 1997) for a review of these techniques. This function is populated in two ways in COREF. Some of its results are determined directly from the topmost task on the stack of active tasks in the dialogue context: handbuilt graphs outline all possible paths for deploying COREF actions to pursue any task. Other results define actions that can happen coherently at a number of points in a dialogue. These actions include operations that push subtasks, including describing the current state of the dialogue and resolving a yes–no or wh– question, and operations that manage dialogue flow, such as abandoning the topmost dialogue subtask. COREF uses its model of dialogue structure as constraints on the possible intentions of conversational participants at any point in the conversation. The actual constraints are derived in two stages to account for the possibility of tacit actions. First, given a representation s that captures a possible state of the dialogue after the last observable action, COREF builds a **horizon graph** (DeVault, 2008) capturing the coherent ways the speaker might have tacitly acted to update the dialogue state before performing their observable action. For instance, in example 3.7, the matcher may have tacitly clicked and added an object to their scene prior to uttering *ok*. For efficiency, we use heuristics to discard tacit actions that *could* be performed coherently, according to COREF's task models, but which an interpreter would never actually hypothesize in a particular situation, given the overt action that the agent has chosen.

Second, COREF uses the horizon graph to solve constraints associated with the observed action. In this stage, COREF instantiates free parameters associated with the action to contextually relevant values. For utterances, the relevant constraints are identified using a grammar that determines both presupposed constraints (which must hold in the context) and dialogue acts (which must be coherent in the context). The constraints associated with an utterance are determined by performing a bottom-up chart parse of the utterance and joining the presuppositions and dialogue acts associated with each edge in the chart.

Our architecture of dialogue acts and interpretive inference allows us to reason about the contributions of question-asking moves in collaborative reference at a fine granularity. The questions of examples 3.6–3.8, for example, differ in whether the question is intended to add a constraint about a target referent – a property or object, whether referenced in a previous utterance or designated as the next thing to add to the scene – or whether it is intended to establish the occurrence of a specific event. The utterances also differ in the tacit moves that relate them to the ongoing discourse: they may push a reminder task, begin a clarification sub-dialogue amplifying on a previous utterance, or simply introduce a specific issue whose resolution contributes to the current task. Each of these interpretations is obtained by the process we have sketched here: constraint-satisfaction inference that matches the grammatical structure of a candidate utterance against a context-specific inventory of coherent contributions to the dialogue.

3.3.2   Modeling uncertainty in interpretation

Examples 3.6–3.8 involve user utterances for which the interpretation models of Section 3.3.1 offer multiple plausible analyses. In example 3.6 the presence of light brown and dark brown objects on the display translates into two possible ways that COREF could understand the user's reference to a color *brown*. In example 3.7, the state of the task offers two different dialogue acts, each of which constitutes a possible resolution for *ok*. In example 3.8, COREF sees two possible referents for *the beige diamond*.

When a conversational participant uses such ambiguous utterances, COREF models the evolving state of the conversation as uncertain. In particular, COREF tracks the dialogue acts possibly generated by the ambiguous utterance by spawning a new thread of interpretation for each reading. By spawning different threads of interpretation capturing the user's alternative contributions, COREF can continue to assign coherent, principled interpretations to the user's ongoing utterances, and continue to participate in the dialogue. COREF uses probabilistic reasoning to reallocate probability mass among the different threads of interpretation over time, while retaining a principled approach to linguistic interpretation in context within each thread. This updating allows COREF to model the clarification sub-dialogues in examples 3.6 and 3.8 as introducing and then resolving transient uncertainty. Thus, in example 3.6, the possible *light brown* interpretation creates a thread of interpretation in which the target has been constrained to have a light brown color, because of the user's contribution in making the utterance *it is brown*. When, however, the user answers *yes* to the question *do you mean dark brown?*, this thread of interpretation becomes unlikely. Accordingly, COREF assigns high probability to the other thread, which, COREF now realizes, has always tracked the correct understanding of the user's utterance *it is brown*. COREF's clarifications can resolve relevant ambiguities without necessarily pinpointing one correct thread of interpretation. With the *did you add it?* of example 3.7, COREF ensures that the user clicks to add the next object to the scene. But because COREF recognizes that an effective reminder could prompt the user to both add the object and *then* answer yes, COREF never discovers whether the user's original *ok* was intended as a simple acknowledgment or also as a confirmation of an addToScene action.

Since the present chapter is focused on generation, we simply note that COREF's context for generation is often a set of alternative threads. We detail COREF's interpretation and dialogue management under uncertainty, in particular our efforts to learn probability models from interacting with users, more fully in DeVault and Stone (2007), DeVault (2008), and DeVault and Stone (2009).

3.3.3   Generating under uncertainty

In COREF, the task of generation is to formulate an utterance that will make a clear positive contribution to the dialogue no matter what the true interpretation thread turns out to be. COREF's utterances are represented along with specific links to the context and implicatures that spell out exactly how the utterance contributes to the ongoing task.

COREF's uncertainty may therefore make it impossible to formulate an utterance with one definite interpretation.

For example, after *ok* in example 3.7, our model predicts that *did you add it?* will carry different implicatures depending on what the user intended to acknowledge. If the user meant only to show that they understood COREF's utterance, then *did you add it?* will implicate that COREF has given enough information so that the user should have identified the target. This implicature is not present if the user meant that they added the object to the scene – the user already knows and has acted on this. Accordingly, in addition to a set of dialogue acts to generate that serve as explicit communicative goals, COREF's generation problems are specified in terms of a set of **acceptable contributions** that represent reasonable ways to move the conversation forward. COREF aims for an utterance that is likely to achieve the specified communicative goals, and will make an acceptable contribution in any case.

COREF's model of context also shapes the problem of disambiguating its utterances. Disambiguation requires COREF to take into account both its own uncertainty and that of its conversational participant. On COREF's side, each of the possible interpretation threads for the dialogue so far represents a possible source of ambiguity. For example, if reference resolution predicts that an anaphoric expression would be resolved one way in one context but another way in another context, COREF needs to represent and track this potential ambiguity. In addition, COREF must track the distinctive ambiguities that the user faces in identifying the implicatures that COREF is tacitly making. As in interpretation, these implicatures are determined by a horizon graph that maps out coherent ways to continue the conversation. COREF is designed to anticipate these ambiguities in a streamlined way during generation, without searching over implicatures and corresponding states of the conversation. COREF constructs a special model of constraint interpretation for generation, which we call a **pseudo-context**. A pseudo-context C behaves as follows: For each state s that represents one possible interpretation thread of the conversation, and each accessible state s_i that is on the horizon graph of s, any resolution of the utterance's presuppositions and dialogue acts in s_i also counts as resolution of the utterance's presuppositions and dialogue acts in C. The effect of this behavior is to increase the amount of ambiguity that is visible to the generator beyond what would be present in a single context, to reflect ambiguities arising throughout the possible horizon. To completely disambiguate its implicatures, the generator should identify an intended horizon state in the possible horizon graph (Thomason *et al.*, 2006), so it would perhaps be ideal to use this reasoning explicitly in generation. However, we have not yet assessed how important this reasoning could be to COREF's behavior in practice.

In COREF, generation occurs in two stages. The first stage starts from a set of communicative goals and a set of acceptable contributions, and a pseudo-context for resolving meaning. The communicative goals are selected by the dialogue manager, targeting both the system's uncertainty and the specific dialogue state that COREF thinks is most likely. The set of acceptable contributions is determined by COREF's handbuilt dialogue policy. COREF searches through its lexicalized grammar, building a candidate utterance, word-by-word, until it finds a complete derivation that achieves the specified communicative goals unambiguously in the extended pseudo-context. This is essentially the same search algorithm as SPUD, but using a richer model of interpretation.

In the second generation stage, COREF interprets the candidate utterance as though it were actually uttered. This process evaluates the candidate utterance on a context-by-context basis; no pseudo-context is involved. This stage therefore offers an accurate assessment of any remaining implicatures and ambiguities in the candidate utterance. If all of the actions that appear in all the interpretations found are acceptable to COREF, then COREF accepts the candidate utterance. In the case that multiple interpretations are supported by a candidate utterance, if COREF accepts the utterance, we view COREF as **willing to be interpreted as making any of those contributions**. This scenario is one in which COREF makes an underspecified contribution. If one or more of the interpretations for a candidate utterance is unacceptable to COREF, it reconsiders its dialogue policy by formulating a different communicative goal or different implicatures. This process repeats until an acceptable candidate utterance is found, or until all communicative options are exhausted.

3.3.4   Examples

Across its dialogues with human participants in our experimental evaluations, COREF asked users 559 questions of 90 distinct sentence types. Examples 3.6–3.8 illustrate the different questioning moves COREF used. The variation in COREF's utterances arises from problem solving to achieve specific communicative goals in specific contexts. For example, COREF uses different vocabulary to describe properties in the display depending on the salient contrasts, uses different referring expressions to describe objects depending on the alternative objects, and varies in its use of full referring expressions versus pronouns depending on the dialogue history. Analyses of the dialogue logs reveal that these questions are usually effective in moving the conversation forward, for example by resolving COREF's uncertainty in cases of ambiguity. See DeVault (2008) for full details about these experimental results.

As we have seen, the generation model in COREF differs from that in SPUD in several ways. COREF models a full range of dialogue acts, communicated both explicitly and by implicature. Also, COREF tracks uncertainty about the context, and is correspondingly more flexible in how it assesses utterance interpretation and what counts as a successful output for generation. To show how these innovations are crucial to grounding in COREF, we consider in detail the reasoning that COREF uses to produce *did you add it?* in example 3.7. COREF starts from an uncertain context. There are two possible threads of interpretation at this point in the dialogue: one (the most probable in COREF's model of interpretation) where the user has just acknowledged COREF's description, and another where the user has just asserted that they have put the next target in place. COREF sets its communicative goal based on the coherent options in the most probable interpretation. We briefly survey these options, as determined by COREF's horizon graph. Even though COREF is uncertain what the user's utterance of *ok* meant, COREF is willing to tacitly decline the opportunity to clarify by performing a tacitNop action. Moving forward, COREF's rules for action acceptability allow it to implicate, on the grounds that its prior utterance *the blue circle* added a constraint that uniquely identified the target object, that the user must have identified the target object. COREF's domain model captures this as action `s1 : setVarValue[t3, e2_2]`. Here `s1` is COREF's symbol for the user, `t3` is COREF's symbol for the target of the current collaborative reference episode, and `e2_2` is COREF's symbol for the next object to add to the scene. The effect of this tacit action is to add to the context the proposition `varValue(t3, e2_2)`. COREF's action acceptability rules further allow the agent to push a `Remind` task, which allows COREF to coherently ask whether `s1` has added the object to their scene. COREF's policy is to try to ask for clarification whenever the task state is ambiguous, so COREF's dialogue manager proposes to generate the dialogue act `COREF : askYNQ[past[s1, addToScene[e2_2]]]`.

In this situation, in order to proceed with generation, COREF anticipates interpretations in both interpretation threads by constructing an expanded pseudo-state that maps out a range of possibilities. These include the interpretations just surveyed, as well as corresponding interpretations in the horizon graph for the other thread of interpretation. Search through the grammar identifies *did you add it?* as an unambiguous utterance expressing the goal dialogue act in this pseudo-state. COREF then explicitly assesses how the user will interpret an utterance of *did you add it?*. This involves recognizing the specific implicatures the utterance will generate in context, using the horizon graph. COREF predicts that *did you add it?* supports two interpretations, one from each of the current possible dialogue states. If the user's previous *ok* was a simple acknowledgment, the utterance supports the interpretation in (3.9):

(3.9) ⟨ `COREF:  tacitNop[[s1 does say[ok]]]`,
      `s1:     setVarValue[t3, e2_2]`,
      `COREF:  pushRemind[COREF, s1, past,`
      `        refuseTaskAction, addToScene[e2_2]]`,
      `COREF:  askYNQ[past[s1, addToScene[e2_2]]]`,
      `COREF:  setPrag[inFocus(Y), inFocus(e2_2)]⟩`

This is the specific set of contributions that COREF makes with this utterance in this specific context: COREF is declining to manage any perceived uncertainty regarding `s1`'s utterance of *ok*, implicating that `s1` has identified the target object, implicating the start of a reminder subtask task, overtly asking a yes–no question as part of that reminder task, and putting object `e2_2` into focus.

If the user's previous *ok* was an assertion that the user had put the next target in place, COREF associates *did you add it?* with the interpretation in (3.10):

(3.10) ⟨ `COREF:  tacitNop[[s1 does say[ok]]]`,
       `COREF:  pushRemind[COREF, s1, past,`
       `        refuseTaskAction, addToScene[e2_2]]`,
       `COREF:  askYNQ[past[s1, addToScene[e2_2]]]`,
       `COREF:  setPrag[inFocus(Y), inFocus(e2_2)]⟩`

This interpretation is similar, except now COREF would not be implicating that the user has identified the target object.

Because COREF finds both of these contributions acceptable, in the corresponding dialogue states, COREF's generator returns action sequence ⟨say[did you add it?]⟩, along with the anticipated user interpretations. COREF accepts the utterance because it is willing to be recognized as making either contribution. In this way, COREF models itself as having extended two threads of coherent interpretation which it spawned upon perceiving the ambiguity in the user's utterance of *ok*. We view COREF as working collaboratively to move the conversation forward, but making an underspecified contribution.

3.4   Alternative approaches

Our work showcases the possibility of achieving grounding in conversational systems through general problem solving. In Section 3.5, we argue that such approaches will be particularly important for embodied conversational agents, in emerging applications such as human–robot interaction. In these domains, many different kinds of action can give evidence about conversational participants' understanding of each another, and a problem-solving model may be necessary to capture the complexity and diversity of the reasoning involved.

It is important not to underestimate the difficulty of realizing general problem solving in a working system. A problem-solving system needs to reason from general and accurate knowledge, which can be extremely challenging to specify by hand or learn from data. A problem-solving system also needs powerful search mechanisms to explore the consequences of its knowledge and generate solutions; these mechanisms usually require a combination of careful engineering and substantial computational resources. Moreover, the complex inference that problem-solving systems perform makes their results unpredictable, exacerbates the errors associated with incorrect knowledge, and makes it difficult to offer guarantees about system behavior and performance. COREF has a number of limitations of this sort. For example, in our experimental evaluations, one unfortunate interaction between COREF's construction of pseudo-contexts and its characterization of acceptable interpretations led to its occasional use of *do you mean it?* as a clarification question.

For all these reasons, applied spoken dialogue systems typically do not generate grounding utterances using general problem-solving models. Alternative implementation frameworks can lead to insights in areas including empirical models of understanding, robust handling of uncertainty, and the ability to make quantitative tradeoffs in generation. We briefly survey some of these frameworks in this section. In particular, we consider abstract models of dialogue in terms of qualitative grounding moves, which minimize the need for reasoning about uncertainty in conversation (Section 3.4.1); models of user state that use probabilistic inference to restrict the need for open-ended problem solving for generation under uncertainty (Section 3.4.2); and feature engineering to avoid the need for deep generalizations about how systems should address their own or their users' uncertainty (Section 3.4.3).

3.4.1   Incremental common ground

Historically, linguists and philosophers have characterized conversational state qualitatively in terms of a notion of **mutual knowledge** or **common ground** (Stalnaker, 1974, 1978; Clark and Marshall, 1981). This is a body of information that conversational participants know they can rely on in formulating utterances. A simple way to adapt natural language generation techniques to dialogue, and handle grounding, is through rules that approximate the incremental evolution of common ground in conversation.

Purver (2004) offers a particularly influential model of common ground in problematic dialogue. He proposes that utterances, by default, update the common ground to reflect their contributions. However, utterances like clarification questions, which show that the previous discourse may not have been understood well enough for the purposes of the conversation, trigger "downdates" that erase previous contributions from the common ground. (By contrast, earlier models by Traum (1994) and Matheson *et al.* (2000) assign contributions a "pending" status until they are acknowledged and accepted by all conversational participants.) Incremental common ground models are limited in their ability to assimilate information that accrues across multiple utterances, particularly when misunderstandings are discovered late, as in example 3.2. Such cases call for probabilistic models and Bayesian reasoning (DeVault and Stone, 2006; Stone and Lascarides, 2010). However, incremental common ground models enable elegant natural language generation implementations, because they can describe communicative context and communicative goals in familiar qualitative terms. Stent (2002) describes in detail how grounding acts can be realized as fluid utterances in incremental common ground models.

3.4.2   Probabilistic inference

A different direction in grounding research focuses on the accurate representation of uncertainty in the system's understanding of the common ground. In noisy settings, including most speech applications, effective dialogue management depends on using all available evidence about the state of the conversation. Spoken dialogue systems research has gone a long way toward capturing the evidence about user utterances that is actually provided by speech recognition systems, and understanding the strategies that let systems exploit this evidence to maximize their understanding (Horvitz and Paek, 2001; Roy *et al.*, 2000; Williams and Young, 2006, 2007).

This research adopts decision-theoretic frameworks for optimizing choice in the face of uncertainty, such as the POMDPs reviewed in Section 3.2.3. For these frameworks to be computationally feasible, the underlying model of the conversational state must be streamlined, so researchers typically focus on accurately modeling the user's private state or goal. For example, in the human–robot dialogue application of Roy *et al.* (2000), the POMDP captures uncertainty about the command the user is trying to give to the robot. In the POMDP-based call center application of Williams (2008), the state is the directory listing (name and listing type) to which the caller wishes to be connected. Richer models of conversational state can be accommodated only with shortcuts in solution strategies. In particular, as we will see in Section 3.4.3, most decision-theoretic models of dialogue management do not actually represent the system's moment-to-moment uncertainty about the state of the conversation, and so they cannot learn to ask clarification questions indirectly.

Dialogue systems optimized using decision-theoretic frameworks also streamline the generation process, generally drawing from a small fixed inventory of utterances that have been hand-authored by a system designer to effectively convey specific messages. For grounding, for example, the system might select from general prompts, acknowledgments, explicit confirmation questions, and implicit confirmations that verbalize the system's understanding of a prior utterance (word-for-word) while moving forward in the dialogue. Optimization allows the system to tune its strategy and complete tasks even in the face of frequent errors in speech recognition. Similar optimization techniques can orchestrate the choice of high-level generation strategies in dialogue (Lemon, 2011). However, with only coarse tools for describing utterance structure and semantics, these frameworks offer limited insight into generating the kinds of coordinated, context-sensitive linguistic expressions that can potentially realize grounding actions.

3.4.3 Correlating conversational success with grounding features

Another type of approach to dialogue management optimizes tradeoffs about how to ground based directly on dialogue outcome. This type of approach uses Markov decision processes (Levin and Pieraccini, 1997; Levin *et al.*, 1998). It simplifies the dialogue model to include only directly observable features in the dialogue state. Rather than represent the user's utterance as a hidden variable, it simply represents the recognition result as actually delivered by the speech recognizer. The dialogue model captures the possibility of recognition errors probabilistically: the system sometimes gets a bad outcome in a dialogue when it acts on incorrectly recognized utterances. Rather than reason explicitly about the effects of clarification on the system's understanding, the dialogue model is instead designed to distinguish those dialogue states in which the system has obtained confirmation of a recognized dialogue state parameter from those in which it has not. The usefulness of grounding actions is revealed by differences in rates of dialogue failure across these different kinds of states.

This type of approach is very appealing because it supports powerful empirical methods for estimating model parameters and powerful computational techniques for optimizing dialogue strategies. The optimization metric can be tuned to the factors that actually govern dialogue success in specific applications (Walker, 2000). Models of dialogue can be learned accurately from small amounts of dialogue data using bootstrapping and simulation (Rieser and Lemon, 2011). Also, function approximation and state abstraction techniques make it possible to compute good strategies for complex dialogue models (Henderson *et al.*, 2008). However, because the models ultimately describe the effects of system decisions directly through observable features of the dialogue, all grounding depends on the insights of the system builders in describing the dialogue state through the right features. For example, Tetreault and Litman (2006) explicitly study which features to include in a dialogue state representation based on the impact those features have on learned dialogue policies.

3.5   Future challenges

One way to think of the streamlined grounding models that characterize many current systems is that they provide empirically based protocols that succeed in practice in overcoming specific kinds of communication problems. These protocols work because they help to align a system's behavior with the true state of the user. Problem-solving models, by contrast, aim to endow the system with more open-ended abilities to explicitly reason about understanding. We believe that such abilities will become increasingly important as dialogue systems begin to handle richer interactions.

3.5.1   Explicit multimodal grounding

One important emerging direction for dialogue research is the design of embodied conversational agents (Cassell, 2000) that contribute to conversation using the affordances of a physical robot or graphical character, including gestures, gaze, facial expressions, and modulations of position and posture of the body as a whole. By using complex communicative actions that pair spoken utterances with actions in other modalities, embodied conversational agents can communicate their understanding in increasingly flexible and natural ways.

For example, Nakano et al. (2003) describe an embodied conversational agent that gives directions with reference to a tabletop map. The system detects and adapts to the non-verbal grounding cues that followers spontaneously provide in human–human conversation. For example, Nakano et al. (2003) found that when listeners could follow instructions, they would nod and continue to direct their attention to the map, but when something was unclear, listeners would gaze up toward the direction-giver and wait attentively for clarification. The system interpreted these cues as grounding actions using an incremental common ground model, and varied its generation strategies accordingly. Nakano et al. (2003) found that their system elicited from its conversational participants many of the same qualitative dynamics they found in human–human conversations.

Multimodal grounding requires a framework for generating and interpreting multimodal communicative acts. A common strategy is "fission," or distributing a planned set of dialogue acts across different modalities to match patterns that are derived from analyses of effective interactions in a specific domain. SmartKom (Wahlster et al., 2001) is an influential early example of multimodal fission for dialogue generation. By contrast, problem-solving models of multimodal generation, such as those of Cassell et al. (2000) and Kopp et al. (2004), reason about the affordances and interdependencies of body and speech to creatively explore the space of possible multimodal utterances and synthesize utterances that link specific behaviors to specific functions opportunistically and flexibly.

One potential advantage of a problem-solving model is the ability to reason indirectly about grounding. This is important if, as theoretical analyses suggest (Lascarides and

Stone, 2009), gesture parallels language in its context-sensitivity, and so affords similar indirect evidence about grounding. Lascarides and Stone (2009) analyze a fragment of conversation in which one speaker explains Newton's Law of Gravity to his conversational participant. The speaker explains the logic of the equation in part through a series of gestures that depict the Galilean experiment of dropping a series of weights in tandem. His addressee demonstrates that she understands his explanation by gesturing her own Galilean experiment, which tracks and eventually anticipates the speaker's own. The evidence each gesture gives of understanding, just like the evidence spoken words give in cases like example 3.4, is inseparable from the interpretation the gesture gets by linking up with the context and contributing content to it.

Researchers have not yet attempted to link problem-solving models of grounding, such as those explored in Section 3.3, with sophisticated descriptions of the interpretation of the form and meaning of multimodal utterances. We believe that doing so will lead to a much broader and more natural range of grounding functions for non-verbal behaviors in embodied conversational agents.

3.5.2   Implicit multimodal grounding

Explicit communicative contributions are not the only evidence embodied agents give of their understanding. When virtual humans (Swartout *et al.*, 2006) realize embodied behaviors through computational architectures that focus information processing and trigger emotional responses in human-like ways, these behaviors can make the agent's conversational strategies and judgments more legible to human conversational participants. This is an indirect kind of grounding.

When people speak in problematic situations, their utterances reveal their uncertainty in recognizable ways (Brennan and Williams, 1995; Swerts and Krahmer, 2005; Stone and Oh, 2008). People are slower to respond in cases of uncertainty. They also make different appraisals of the process of the conversation and their contributions to it when they are uncertain. Uncertainty may feel difficult, for example, or may lead to a contribution that feels unsatisfactory. These appraisals shape the affect of the conversational participants and so influence their facial expressions. To the extent that virtual humans exhibit the same cognitive and affective dynamics, their uncertainty will also be recognizable. See Stone and Oh (2008) for a case study.

In fact, Sengers (1999) has argued that agents that aim to be understood must not only exhibit the right cognitive and affective dynamics – they must work actively to reveal these dynamics to their audience. Sengers focused on clarifying the goals and decisions of animated characters, by dramatizing how they see, react to, and engage with events in their virtual environments. It is an open problem to integrate these techniques with approaches to grounding. A problem-solving approach like the one in COREF offers a natural and attractive way to do this, since it promises to describe the evidence that agents provide about understanding through their communication in probabilistic terms that are compatible with other evidence that might come from agents' attention, processing, or emotion.

3.5.3   Grounding through task action

A final source of evidence about grounding comes from the real-world activity that accompanies dialogue in task-oriented interactions. Understanding what conversational participants have *done* is continuous with reasoning about what they have said and what they have understood. For example, suppose a speaker has given an instruction to an addressee. The action the addressee performs to carry out the instruction gives very good evidence about how the addressee understood the speaker. It is natural to associate instrumental actions with grounding functions, especially in settings such as human–robot interaction where conversation is used to coordinate embodied activities among physically co-present conversational participants.

A starting point for modeling such functions might be to model task actions as generating grounding acts in certain cases. For example, carrying out an expected action might constitute a specific form of acknowledgment. Such models could encode useful strategies for carrying out efficient dialogues that avoid misunderstandings. However, the study of human–human dialogue again suggests that more general problem-solving models will be necessary to reproduce humans' use of physical action in grounding.

For example, consider the results of Clark and Krych (2004). They analyzed dyadic conversations in which a director leads a matcher through the assembly of a Lego structure. As you might expect, directors' instructions often include instalments (as in example 3.1) and expansions (as in example 3.3) formulated in real time in response to feedback from the matcher. This feedback, however, often takes the form of physical action: Matchers pose blocks tentatively and use other strategies, including accompanying verbal and non-verbal communicative action, to mark their actions as provisional. Some of the most effective teamwork features tight coordination where directors offer short fragments to repeatedly correct proposed matcher actions. Clark and Krych (2004) cite one example where the director's iterative critique of four successive poses – over just four seconds – frees the conversational participants from having to agree on a precise description of the complex spatial configuration of a difficult-to-describe piece.

COREF's problem-solving model already interprets task actions (such as `Continue (next object)`) using the same intention-recognition framework it uses to interpret utterances. That means that COREF expects these actions to meet the constraints established by the conversational participants in the preceding dialogue as well as the natural organization of the ongoing activity. Thus, as with an utterance, COREF can simultaneously enrich its understanding of the action and resolve uncertainty in the context by reconciling its observations and its interpretive constraints.

COREF, however, is a long way from the fluidity found in human–human dialogues like Clark and Krych's (2004). We suspect that modeling the grounding function of provisional or even incorrect actions requires extending the kinds of models in COREF with a generative account of the relationship between actions and instructions that factors in the underspecification, ambiguities, and errors common to natural language descriptions (see Stone and Lascarides, 2010). The negotiation involved, meanwhile, requires richer accounts of the use of fragmentary utterances to link up with and coordinate ongoing real-world action. These phenomena again highlight the long-term research opportunities and challenges of problem-solving models of grounding.

3.6 Conclusions

A fundamental problem in deploying natural language generation in dialogue systems involves enriching models of language use. Dialogue systems must be able to handle problematic interactions, and that means that they cannot simply exploit static models of form and meaning. They must be able to negotiate their contributions to conversation flexibly and creatively, despite missteps and uncertainties, across extended interactions. Natural language generation, as a field, is just beginning to engage meaningfully with these requirements and the challenges they bring. Many important problems are open, and this chapter has given a corresponding emphasis to exploratory and open-ended research.

In particular, our focus has been on problem-solving models: general accounts that help us operationalize the generation of productive contributions in problematic situations. Problem solving provides a tool for systematizing the grounding behaviors that people use in human–human conversation, for understanding the knowledge of activity, context, and language that underpins grounding behaviors, and for mapping the possible interactions between language, embodied communication, and task action in keeping conversation on track.

Of course, we can build on systematic thinking about grounding in dialogue using a wide range of implementations. Simpler and more constrained frameworks often provide the most efficient and robust realization of the insights of more general models. Indeed, for the moment, problem-solving models may be most important as a bridge between descriptive accounts of human–human conversation and the strategies we choose to realize in practical dialogue systems. Descriptive analyses characterize grounding in common-sense terms: people in problematic dialogues offer evidence of understanding, act collaboratively, negotiate, reach agreement. Computational models – and their limits – remind us how subtle and sophisticated this common-sense talk really is. The concepts involved tap directly into our most powerful principles of social cognition, principles for which science offers only the barest sketch. Linguistic meaning ranks among the triumphs of our abilities to relate to one another. In understanding, systematizing and implementing grounding in conversational agents, we deepen and transform our understanding of those abilities, and the abilities themselves.

Acknowledgments

This research was funded by the NSF under grant numbers HLC 0308121 and HSD 0624191.

References

Allen, J. F., Blaylock, N., and Ferguson, G. (2002). A problem solving model for collaborative agents. In *Proceedings of the International Joint Conference on Autonomous Agents and Multiagent Systems (AAMAS)*, pages 774–781, Bologna, Italy. International Foundation for Autonomous Agents and Multiagent Systems.

Asher, N. and Lascarides, A. (2003). *Logics of Conversation*. Cambridge University Press, Cambridge, UK.

Brennan, S. E. (1990). *Seeking and Providing Evidence for Mutual Understanding*. PhD thesis, Department of Psychology, Stanford University.

Brennan, S. E. and Clark, H. H. (1996). Conceptual pacts and lexical choice in conversation. *Journal of Experimental Psychology*, **22**(6):1482–1493.

Brennan, S. E. and Williams, M. (1995). The feeling of another's knowing: Prosody and filled pauses as cues to listeners about the metacognitive states of speakers. *Journal of Memory and Language*, **34**(3):383–398.

Bunt, H. (1994). Context and dialogue control. *THINK Quarterly*, **3**:19–31.

Bunt, H. (1996). Interaction management functions and context representation requirements. In *Proceedings of the Twente Workshop on Language Technology*, pages 187–198, University of Twente. University of Twente.

Bunt, H. (2000). Dialogue pragmatics and context specification. In Bunt, H. and Black, W., editors, *Abduction, Belief and Context in Dialogue. Studies in Computational Pragmatics*, pages 81–150. John Benjamins, Amsterdam, The Netherlands.

Carberry, S. and Lambert, L. (1999). A process model for recognizing communicative acts and modeling negotiation subdialogues. *Computational Linguistics*, **25**(1):1–53.

Cassell, J. (2000). Embodied conversational interface agents. *Communications of the ACM*, **43**(4):70–78.

Cassell, J., Stone, M., and Yan, H. (2000). Coordination and context-dependence in the generation of embodied conversation. In *Proceedings of the International Conference on Natural Language Generation (INLG)*, pages 171–178, Mitzpe Ramon, Israel. Association for Computational Linguistics.

Clark, H. (1996). *Using Language*. Cambridge University Press, Cambridge, UK.

Clark, H. H. (1993). *Arenas of Language Use*. University of Chicago Press, Chicago, IL.

Clark, H. H. and Krych, M. (2004). Speaking while monitoring addressees for understanding. *Journal of Memory and Language*, **50**(1):62–81.

Clark, H. H. and Marshall, C. R. (1981). Definite reference and mutual knowledge. In Joshi, A., Webber, B., and Sag, I., editors, *Elements of Discourse Understanding*, pages 10–63. Cambridge University Press, Cambridge, UK.

Clark, H. H. and Schaefer, E. F. (1989). Contributing to discourse. *Cognitive Science*, **13**(2): 259–294.

Clark, H. H. and Wilkes-Gibbs, D. (1990). Referring as a collaborative process. In Cohen, P. R., Morgan, J., and Pollack, M. E., editors, *Intentions in Communication*, pages 463–493. MIT Press, Cambridge, MA.

Cohen, P. R. (1997). Dialogue modeling. In Cole, R., Mariani, J., Uszkoreit, H., Varile, G. B., Zaenen, A., and Zampolli, A., editors, *Survey of the State of the Art in Human Language Technology (Studies in Natural Language Processing)*, pages 204–210. Cambridge University Press, Cambridge, UK.

Core, M. G. and Allen, J. F. (1997). Coding dialogues with the DAMSL annotation scheme. In *Working Notes of the AAAI Fall Symposium on Communicative Action in Humans and Machines*, Boston, MA. AAAI Press.

DeVault, D. (2008). *Contribution Tracking: Participating in Task-Oriented Dialogue under Uncertainty*. PhD thesis, Department of Computer Science, Rutgers, The State University of New Jersey, New Brunswick, NJ.

DeVault, D., Kariaeva, N., Kothari, A., Oved, I., and Stone, M. (2005). An information-state approach to collaborative reference. In *Proceedings of the Annual Meeting of the Association for Computational Linguistics (ACL)*, pages 1–4, Ann Arbor, MI. Association for Computational Linguistics.

DeVault, D. and Stone, M. (2006). Scorekeeping in an uncertain language game. In *Proceedings of the Workshop on the Semantics and Pragmatics of Dialogue (brandial)*, pages 139–146, Potsdam, Germany. SemDial.

DeVault, D. and Stone, M. (2007). Managing ambiguities across utterances in dialogue. In *Proceedings of the Workshop on the Semantics and Pragmatics of Dialogue (DECALOG)*, pages 49–56, Rovereto, Italy. SemDial.

DeVault, D. and Stone, M. (2009). Learning to interpret utterances using dialogue history. In *Proceedings of the Conference of the European Chapter of the Association for Computational Linguistics (EACL)*, pages 184–192, Athens, Greece. Association for Computational Linguistics.

Di Eugenio, B., Jordan, P. W., Thomason, R. H., and Moore, J. D. (2000). The agreement process: An empirical investigation of human–human computer-mediated collaborative dialogue. *International Journal of Human-Computer Studies*, **53**(6):1017–1076.

Furnas, G. W., Landauer, T. K., Gomez, L. M., and Dumais, S. T. (1987). The vocabulary problem in human–system communications. *Communications of the ACM*, **30**(11):964–971.

Ginzburg, J. and Cooper, R. (2004). Clarification, ellipsis and the nature of contextual updates in dialogue. *Linguistics and Philosophy*, **27**(3):297–365.

Goldman, A. (1970). *A Theory of Human Action*. Prentice Hall, Upper Saddle River, NJ.

Gregoromichelaki, E., Kempson, R., Purver, M., Mills, G. J., Cann, R., Meyer-Viol, W., and Healey, P. G. (2011). Incrementality and intention-recognition in utterance processing. *Dialogue and Discourse*, **2**(1):199–233.

Grice, H. P. (1975). Logic and conversations. In Cole, P. and Morgan, J. L., editors, *Syntax and Semantics III: Speech Acts*, pages 41–58. Academic Press, New York, NY.

Grosz, B. J. and Sidner, C. L. (1986). Attention, intentions, and the structure of discourse. *Computational Linguistics*, **12**(3):175–204.

Heeman, P. A. and Hirst, G. (1995). Collaborating on referring expressions. *Computational Linguistics*, **21**(3):351–383.

Henderson, J., Lemon, O., and Georgila, K. (2008). Hybrid reinforcement/supervised learning of dialogue policies from fixed datasets. *Computational Linguistics*, **34**(4):487–513.

Hobbs, J. R., Stickel, M., Appelt, D., and Martin, P. (1993). Interpretation as abduction. *Artificial Intelligence*, **63**(1–2):69–142.

Horvitz, E. and Paek, T. (2001). Harnessing models of users' goals to mediate clarification dialog in spoken language systems. In *Proceedings of the International Conference on User Modeling*, pages 3–13, Sonthofen, Germany. Springer.

Kehler, A. (2001). *Coherence, Reference and the Theory of Grammar*. CSLI Publications, Stanford, CA.

Kopp, S., Tepper, P., and Cassell, J. (2004). Towards integrated microplanning of language and iconic gesture for multimodal output. In *Proceedings of the International Conference on Multimodal Interfaces (ICMI)*, pages 97–104, State College, PA. Association for Computing Machinery.

Larsson, S. and Traum, D. (2000). Information state and dialogue management in the TRINDI dialogue move engine toolkit. *Natural Language Engineering*, **6**(3–4):323–340.

Lascarides, A. and Asher, N. (2009). Agreement, disputes and commitments in dialogue. *Journal of Semantics*, **26**(2):109–158.

Lascarides, A. and Stone, M. (2009). Discourse coherence and gesture interpretation. *Gesture*, **9**(2):147–180.

Lemon, O. (2011). Learning what to say and how to say it: Joint optimisation of spoken dialogue management and natural language generation. *Computer Speech & Language*, **25**(2):210–221.

Levin, E. and Pieraccini, R. (1997). A stochastic model of computer–human interaction for learning dialogue strategies. In *Proceedings of the European Conference on Speech Communication and Technology (EUROSPEECH)*, pages 1883–1886, Rhodes, Greece. International Speech Communication Association.

Levin, E., Pieraccini, R., and Eckert, W. (1998). Using Markov decision process for learning dialogue strategies. In *Proceedings of the IEEE International Conference on Acoustics, Speech, and Signal Processing (ICASSP)*, volume 1, pages 201–204, Seattle, WA. Institute of Electrical and Electronics Engineers.

Matheson, C., Poesio, M., and Traum, D. (2000). Modelling grounding and discourse obligations using update rules. In *Proceedings of the Conference of the North American Chapter of the Association for Computational Linguistics (NAACL)*, pages 1–8, Seattle, WA. Association for Computational Linguistics.

Nakano, Y. I., Reinstein, G., Stocky, T., and Cassell, J. (2003). Towards a model of face-to-face grounding. In *Proceedings of the Annual Meeting of the Association for Computational Linguistics (ACL)*, pages 553–561, Sapporo, Japan. Association for Computational Linguistics.

Newell, A. (1982). The knowledge level. *Artificial Intelligence*, **18**:87–127.

Pollack, M. (1986). A model of plan inference that distinguishes between the beliefs of actors and observers. In *Proceedings of the Annual Meeting of the Association for Computational Linguistics (ACL)*, pages 207–214, New York, NY. Association for Computational Linguistics.

Purver, M. (2004). *The Theory and Use of Clarification Requests in Dialogue*. PhD thesis, Department of Computer Science, King's College, University of London.

Rich, C., Sidner, C. L., and Lesh, N. (2001). COLLAGEN: Applying collaborative discourse theory to human–computer interaction. *Artificial Intelligence Magazine*, **22**(4):15–25.

Rieser, V. and Lemon, O. (2011). Learning and evaluation of dialogue strategies for new applications: Empirical methods for optimization from small data sets. *Computational Linguistics*, **37**(1):153–196.

Roy, N., Pineau, J., and Thrun, S. (2000). Spoken dialog management for robots. In *Proceedings of the Annual Meeting of the Association for Computational Linguistics (ACL)*, pages 93–100, Hong Kong. Association for Computational Linguistics.

Russell, S. and Norvig, P. (1995). *Artificial Intelligence: A Modern Approach*. Prentice Hall, Upper Saddle River, NJ.

Sengers, P. (1999). Designing comprehensible agents. In *Proceedings of the International Joint Conference on Artificial Intelligence (IJCAI)*, pages 1227–1232, Stockholm, Sweden. International Joint Conference on Artificial Intelligence.

Sidner, C. L. (1994). Negotiation in collaborative activity: A discourse analysis. *Knowledge Based Systems*, **7**(4):265–267.

Stalnaker, R. (1974). Pragmatic presuppositions. In Munitz, M. K. and Unger, P. K., editors, *Semantics and Philosophy*, pages 197–213. New York University Press, New York, NY.

Stalnaker, R. (1978). Assertion. In Cole, P., editor, *Syntax and Semantics*, volume 9, pages 315–332. Academic Press, New York, NY.

Stent, A. J. (2002). A conversation acts model for generating spoken dialogue contributions. *Computer Speech and Language*, **16**(3–4):313–352.

Stone, M. (2004). Communicative intentions and conversational processes in human–human and human–computer dialogue. In Trueswell, J. C. and Tanenhaus, M. K., editors, *Approaches to Studying World-Situated Language Use: Bridging the Language-as-Product and Language-as-Action Traditions*, pages 39–70. MIT Press, Cambridge, MA.

Stone, M., Doran, C., Webber, B., Bleam, T., and Palmer, M. (2003). Microplanning with communicative intentions: The SPUD systems. *Computational Intelligence*, **19**(4):314–381.

Stone, M. and Lascarides, A. (2010). Coherence and rationality in dialogue. In *Proceedings of the Workshop on the Semantics and Pragmatics of Dialogue (SEMDIAL)*, pages 51–58, Poznán, Poland. SemDial.

Stone, M. and Oh, I. (2008). Modeling facial expression of uncertainty in conversational animation. In Wachsmuth, I. and Knoblich, G., editors, *Modeling Communication with Robots and Virtual Humans*, pages 57–76. Springer, Heidelberg, Germany.

Swartout, W., Gratch, J., Hill, R. W., Hovy, E., Marsella, S., Rickel, J., and Traum, D. (2006). Toward virtual humans. *AI Magazine*, **27**(2):96–108.

Swerts, M. and Krahmer, E. (2005). Audiovisual prosody and feeling of knowing. *Journal of Memory and Language*, **53**(1):81–94.

Tetreault, J. and Litman, D. (2006). Using reinforcement learning to build a better model of dialogue state. In *Proceedings of the Conference of the European Chapter of the Association for Computational Linguistics (EACL)*, pages 289–296, Trento, Italy. Association for Computational Linguistics.

Thomason, R. H., Stone, M., and DeVault, D. (2006). Enlightened update: A computational architecture for presupposition and other pragmatic phenomena. For the Ohio State Pragmatics Initiative, 2006, available at http://www.research.rutgers.edu/~ddevault/. Accessed on 11/24/2013.

Traum, D. and Allen, J. F. (1994). Discourse obligations in dialogue processing. In *Proceedings of the Annual Meeting of the Association for Computational Linguistics (ACL)*, pages 1–8, Las Cruces, NM. Association for Computational Linguistics.

Traum, D. and Hinkelman, E. (1992). Conversation acts in task-oriented spoken dialogue. *Computational Intelligence*, **8**(3):575–599.

Traum, D. R. (1994). *A Computational Theory of Grounding in Natural Language Conversation*. PhD thesis, Department of Computer Science, University of Rochester.

Wahlster, W., Reithinger, N., and Blocher, A. (2001). SmartKom: Multimodal communication with a life-like character. In *Proceedings of the European Conference on Speech Communication and Technology (EUROSPEECH)*, pages 1547–1550, Aalborg, Denmark. International Speech Communication Association.

Walker, M. A. (2000). An application of reinforcement learning to dialogue strategy selection in a spoken dialogue system for email. *Journal of Artificial Intelligence Research*, **12**:387–416.

Williams, J. and Young, S. (2006). Scaling POMDPs for dialog management with composite summary point-based value iteration (CSPBVI). In *Proceedings of the AAAI Workshop on Statistical and Empirical Approaches for Spoken Dialogue Systems*. AAAI Press.

Williams, J. D. (2008). Demonstration of a POMDP voice dialer. In *Proceedings of the Annual Meeting of the Association for Computational Linguistics: Human Language Technologies (ACL-HLT)*, pages 1–4, Columbus, OH. Association for Computational Linguistics.

Williams, J. D. and Young, S. (2007). Partially observable Markov decision processes for spoken dialog systems. *Computer Speech and Language*, **21**(2):393–422.

4 Dialogue and compound contributions

Matthew Purver, Julian Hough, and Eleni Gregoromichelaki

4.1 Introduction

In this chapter, we examine the phenomenon of **compound contributions** (CCs) and the implications for NLG in interactive systems. Compound contributions are contributions in dialogue that continue or complete an earlier contribution, thus resulting in a single syntactic or semantic unit built across multiple contributions provided by one or more speakers (example 4.1). The term as used here therefore includes more specific cases that have been referred to as **expansions**, **(collaborative) completions**, and **split** or **shared utterances**.

(4.1) (Friends of the Earth club meeting:)
    A: *So what is that? Is that er ... booklet or something?*
    B: *It's a* [[*book*]]
    C: [[*Book*]]
    B: *Just ...* [[*talking about al— you know alternative*]]
    D: [[*On erm ... renewable yeah*]]
    B: *energy really I think*
    A: *Yeah*   (*BNC D97 2038-2044*)[1]

As we discuss in Section 4.2, a dialogue agent that processes or takes part in CCs must change between understanding (NLU) and generation (NLG), possibly many times, while keeping track of the representation being constructed. This imposes some strong requirements on the nature of NLG in interactive systems, in terms both of NLG incrementality and of NLG–NLU interdependency. As we explain in Section 4.3, current approaches to NLG, while exhibiting incrementality to substantial degrees for various reasons, do not yet entirely satisfy these requirements. In Sections 4.4 and 4.5 we outline one possible approach to NLG that is compatible with CCs; it uses the **Dynamic Syntax** (DS) grammatical framework with **Type Theory with Records** (TTR). We explain how DS-TTR might be incorporated into an interactive system. In Section 4.6 we outline how this approach can be used in principle without recourse to recognizing or modeling interlocutors' intentions, and how it is compatible with emerging empirical evidence about alignment between dialogue participants.

---

[1] Examples labeled *BNC* are taken from the British National Corpus (Burnard, 2000).

4.2   Compound contributions

4.2.1   Introduction

Interlocutors very straightforwardly shift between the roles of listener and speaker, without necessarily waiting for sentences to end. This results in the phenomenon we describe here as **compound contributions** (CCs): syntactic or semantic units made up of multiple dialogue contributions, possibly provided by multiple speakers. In (4.1) above, B begins a sentence; C interrupts (or takes advantage of a pause) to clarify a constituent, but B then continues to extend his or her initial contribution seamlessly. Again, D interrupts to offer a correction; B can presumably process this, and continues with a final segment – which may be completing either B's original or D's corrected version of the contribution. We can see C's and D's contributions, as well as B's continuations and completions, as examples of the same general CC phenomenon: in each case, the new contribution is continuing (optionally including editing or repair) an antecedent that may well be incomplete, and may well have been produced by another speaker.

CCs can take many forms. Conversation analysis research has paid attention to some of these, in particular noting the distinction between expansions (contributions that add additional material to an already complete antecedent (4.2)) and completions (contributions that complete an incomplete antecedent (4.3)). A range of characteristic structural patterns for these phenomena, and the corresponding speaker transition points, have been observed: expansions often involve optional adjuncts such as sentence relatives (4.2); and completions often involve patterns such as IF–THEN (4.3) or occur opportunistically after pauses (4.4) (see Lerner, 1991, 1996, 2004; Ono and Thompson, 1993; Rühlemann and McCarthy, 2007, among others).

(4.2)   A: *profit for the group is a hundred and ninety thousand pounds.*
        B: *Which is superb.*   (BNC FUK 2460-2461)

(4.3)   A: *Before that then if they were ill*
        B: *They get nothing.*   (BNC H5H 110-111)

(4.4)   A: *Well I do know last week thet=uh Al was certainly very <pause 0.5>*
        B: *pissed off*   (Lerner, 1996, p. 260)

Clearly, examples such as these impose interesting requirements for NLG. Agent B must generate a contribution that takes into account the possibly incomplete contribution from agent A, both in syntactic terms (continuing in a grammatical fashion) and in semantic terms (continuing in a coherent and/or plausible way). And corpus studies (Skuplik, 1999; Szczepek Reed, 2000; Purver *et al.*, 2009) suggest that these are not isolated examples: CCs are common in both task-oriented and general open-domain dialogue, with around 3% of contributions in dialogue continuing some other speaker's material (Howes *et al.*, 2011).

4.2.2   Data

The regular patterns of (4.2)–(4.4) already show that agents can continue or extend utterances across speaker and/or turn boundaries, but these patterns are by no means the only possibilities. In this section, we review some other possible CC forms, using data from Purver *et al.* (2009) and Gregoromichelaki *et al.* (2011), and note the rather strict requirements they impose.

Incrementality

In dialogue, participants ground each other's contributions (Allen *et al.*, 2001) through backchannels like *yeah, mhm*, etc. This is very often done at incremental points within a sentence: the initial listener shifts briefly to become the speaker and produce a grounding utterance (with the initial speaker briefly becoming a listener to notice and process it), and roles then revert to the original:

(4.5) A: *Push it, when you want it, just push it* [pause] *up there*
       B: *Yeah.*
       A: *so it comes out.*   (*BNC KSR 30-32*)

(4.6) A: *So if you start at the centre* [pause] *and draw a line and mark off seventy two degrees,*
       B: *Mm.*
       A: *and then mark off another seventy two degrees and another seventy two degrees and another seventy two degrees and join the ends,*
       B: *Yeah.*
       A: *you'll end up with a regular pentagon.*   (*BNC KND 160-164*)

We see these as examples of CCs in that the overall sentential content is spread across multiple speaker turns (although here, all by the same speaker). NLG processes must therefore be interruptible at the speaker transition points, and able to resume later.

In addition, the speaker must also be able to process and understand the grounding contribution – at least, to decide whether it gives positive or negative feedback. In (4.5)–(4.6) above, one might argue that this requires little in the way of understanding; however, the grounding contribution may provide or require extra information which must be processed in the context of the partial contribution produced so far:

(4.7) A: *And er they X-rayed me, and took a urine sample, took a blood sample. Er, the doctor*
       B: *Chorlton?*
       A: *Chorlton, mhm, he examined me, erm, he, he said now they were on about a slide* [unclear] *on my heart.*   (*BNC KPY 1005-1008*)

(4.8) (Friends of the Earth club meeting (repeat of (4.1) above):)
       A: *So what is that? Is that er ... booklet or something?*
       B: *It's a* [[*book*]]
       C: [[*Book*]]
       B: *Just ...* [[*talking about al– you know alternative*]]
       D: [[*On erm ... renewable yeah*]]
       B: *energy really I think*
       A: *Yeah*   (*BNC D97 2038-2044*)

Contributions by B in (4.7) and by C and D in (4.8) clarify, repair, or extend the partial utterances, with the clarification apparently becoming absorbed into the final, collaboratively derived, content. Processing these contributions must require not only suspending the initial speaker's NLG process, but providing the partial representations it provides to his or her NLU processes, allowing for understanding and evaluation of the requested confirmation or correction. As NLG then continues, it must do so from the newly clarified or corrected representation, including content from all contributions so far.

Note that (4.7) shows that the speaker transition point can come even in the middle of an emergent clause; and the transitions around D's contribution in (4.8) occur within a noun phrase (between adjective and noun). Indeed, transitions within any kind of syntactic constituent seem to be possible, with little or no constraint on the possible position (Howes *et al.*, 2011; Gregoromichelaki *et al.*, 2011), suggesting that incremental processing must be just that – operating on a strictly word-by-word basis:

(4.9)   A: [...] *whereas qualitative is* [pause] *you know what the actual variations*
         B: *entails*
         A: *entails. you know what the actual quality of the variations are.*
                                                        (*BNC G4V 114-117*)

(4.10)   A: *We need to put your name down. Even if that wasn't a*
         B: *A proper* [[*conversation*]]
         A: [[*a grunt*]].   (*BNC KDF 25-27*)

(4.11)   A: *All the machinery was*
         B: [[*All steam.*]][2]
         A: [[*operated*]] *by steam*   (*BNC H5G 177-179*)

(4.12)   A: *I've got a scribble behind it, oh annual report I'd get that from.*
         B: *Right.*
         A: *And the total number of* [[*sixth form students in a division.*]]
         B: [[*Sixth form students in a division.*]] *Right.*
                                                        (*BNC H5D 123-127*)

Syntactic dependencies

However, despite this apparent flexibility in transition point, syntactic dependencies seem to be preserved across the speaker transition:

(4.13)   (With smoke coming from the kitchen:)
         A: *I'm afraid I burnt the kitchen ceiling*
         B: *But have you*
         A: *burned myself? Fortunately not.*

(4.14)   A: *Do you know whether every waitress handed in*
         B: *her tax forms?*
         A: *or even any payslips?*

---

[2] The [[brackets indicate overlapping speech between two subsequent utterances – i.e., here A's "operated" overlaps B's "All steam."

The negative polarity item *any* in A's final contribution (4.14) is licensed by the context set up in the initial partial antecedent; similarly the scope of A's quantifier *every* must include B's anaphoric *her*. In (4.13), A's reflexive pronoun depends on B's initial subject. The grammar, then, seems to be crucially involved in licensing CCs. However, this grammar cannot be one that licenses strings: the complete sentence gained by joining together B's and A's CC in (4.13), *have you burned myself*, is not a grammatical string. Rather, semantic and contextual representations must be involved with the syntactic characterizations utilized to underpin the co-construction but not inducing a separate representation level.

Semantics and intentionality

In many CC examples, the respondent appears to have guessed what they think was intended by the original speaker. These have been called *collaborative completions* (Poesio and Rieser, 2010):

(4.15)   (Conversation from A and B, to C:)
         A: *We're going to ...*
         B: *Bristol, where Jo lives.*

(4.16)   A: *Are you left or*
         B: *Right-handed.*

Such examples must require that semantic representations are being created incrementally, making the partial meaning available at speaker transition in order to allow the continuing agent to make a correct guess. The process of inferring original intentions may be based on an agent's understanding of the extra-linguistic context or domain (see Poesio and Rieser (2010) and discussion below). But this is not the only possibility: as (4.17)–(4.18) show, such completions by no means need to be what the original speaker actually had in mind:

(4.17)   Morse: *in any case the question was*
         Suspect: *a VERY good question inspector*   (*Morse*, BBC Radio 7)

(4.18)   Daughter: *Oh here dad, a good way to get those corners out*
         Dad: *is to stick yer finger inside.*
         Daughter: *well, that's one way.*   (from Lerner, 1991)

In fact, such continuations can be completely the opposite of what the original speaker might have intended, as in what we will call **hostile continuations** or **devious suggestions**, which are nevertheless collaboratively constructed from a grammatical point of view:

(4.19)   (A and B arguing:)
         A: *In fact what this shows is*
         B: *that you are an idiot*

(4.20)   (A mother, B son:)
         A: *This afternoon first you'll do your homework, then wash the dishes
         and then*
         B: *you'll give me £10?*

Note, though, that even such examples show that syntactic matching is preserved, and suggest the availability of semantic representations in order to produce a continuation that is coherent (even if not calculated on the basis of attributed intentions).

4.2.3   Incremental interpretation vs. incremental representation

It is clear, then, that a linguistic system that is to account for the data provided by CCs must be incremental in some sense: at apparently any point in a sentence, partial representations must be provided from the comprehension (NLU) to the production (NLG) facility, and vice versa. These processes must therefore be producing suitable representations **incrementally**; they must also be able to exchange them, requiring the quality of **reversibility**, in that representations available in interpretation should be available for generation too (see Neumann, 1998, and below).

A pertinent question, then, is to what degree incrementality is required, and at which levels. In terms of interpretation, Milward (1991) points out the difference between a linguistic system's capacity for **strong incremental interpretation** and its ability to access and produce **incremental representations**. Strong incremental interpretation is defined as a system's ability to extract the maximal amount of information possible from an unfinished utterance as it is being produced, particularly the semantic dependencies of the informational content (e.g., a representation such as $\lambda x. like'(john', x)$ should be available after parsing *John likes*). Incremental representation, on the other hand, is defined as a representation being available for each substring of an utterance, but not necessarily including the dependencies between these substrings (e.g., a representation such as *john'* being available after consuming *John* and then $john' * \lambda y. \lambda x. like'(y, x)$ being available after consuming *likes* as the following word).

Systems may exhibit only one of these different types of incrementality. This is perhaps most clear for the case of a system producing incremental representations but not yielding strict incremental interpretation – that is to say, a system that incrementally produces representations $\lambda y. \lambda x. like'(y, x)$ and *john'*, but does not carry out functional application to give the maximal possible semantic information $\lambda x. like'(john', x)$. But the converse is also possible: another system might make the maximal interpretation for a partial utterance available incrementally, but if this is built up by adding to a semantic representation without maintaining lexical information – for example, by the incremental updating of Discourse Representation Structures (DRS; for details see Kamp and Reyle, 1993) – it may not be possible to determine which word or sequence of words was responsible for which part of the semantic representation, and therefore the procedural or construction elements of the context may be irretrievable.

The evidence reviewed above, however, suggests that a successful model of CCs would need to incorporate both strong incremental interpretation *and* incremental representation, for each word uttered sequentially in a dialogue. Representations for substrings and their contributions are required for clarification and confirmation behavior (4.7–4.8); and partial sentential meanings including semantic dependencies must be available for coherent, helpful (or otherwise) continuations to be suggested (4.17–4.20).

4.2.4   CCs and intentions

However, incremental comprehension cannot be based primarily on guessing speaker intentions or recognizing known discourse plans: for instance, it is not clear that in (4.17)–(4.20) the addressee has to have guessed the original speaker's (propositional) intention/plan before offering a continuation.[3] Moreover, speaker plans need not necessarily be fully formed before production: the assumption of fully formed propositional intentions guiding production will predict that all the cases above where the continuation is not as expected, as in (4.17)–(4.20), would have to involve some kind of revision or backtracking on the part of the original speaker. But this is not a necessary assumption: as long as the speaker is licensed to operate with partial structures, he or she can start an utterance without a fully formed intention/plan as to how it will develop (as the psycholinguistic models in any case suggest), relying on feedback from the hearer to shape the utterance (Goodwin, 1979). The importance of feedback in co-constructing meaning in communication has already been documented at the propositional level (the level of speech acts) within Conversational Analysis (CA; see, e.g., Schegloff, 2007). However, it seems here that the same processes can operate sub-propositionally, but only relative to grammar models that allow the incremental, sub-sentential integration of cross-speaker productions.

4.2.5   CCs and coordination

Importantly, phenomena such as (4.1)–(4.20) are not dysfunctional uses of language, unsuccessful acts of communication, performance issues involving repair, or deviant uses. If one were to set them aside as such, one would be left without an account of how people manage to understand what each other has said in these cases. In fact, it is now well documented that such "miscommunication" not only provides vital insights as to how language and communication operate (Schegloff, 1979), but also facilitates dialogue coordination: as Healey (2008) shows, the local processes involved in the detection and resolution of misalignments during interaction lead to significantly positive effects on measures of successful interactional outcomes (see also Brennan and Schober, 2001); and, as Saxton (1997) shows, in addition, such mechanisms, in the form of negative evidence and embedded repairs, crucially mediate language acquisition (see also Goodwin, 1981, pp. 170–171). Therefore, miscommunication and the specialized repair procedures made available by the structured linguistic and interactional resources

---

[3] These are cases not addressed by DeVault *et al.* (2009), who otherwise offer a method for getting full interpretation as early as possible. Lascarides and Asher (2009) and Asher and Lascarides (2008) also define a model of dialogue that partly sidesteps many of the issues raised in intention recognition. But, in adopting the essentially suprasentential remit of Segmented Discourse Representation Theory (SDRT), their model does not address the step-by-step incrementality required to model split-utterance phenomena.

available to interlocutors are the sole means that can guarantee intersubjectivity and coordination.

4.2.6   Implications for NLG

To summarize, the data presented above show that CCs impose some strong requirements on NLG (and indeed on NLU):

- Full word-by-word **incrementality**: NLG and NLU processes must both be able to begin and end at any syntactic point in a sentence (including within syntactic or semantic constituents).
- Strong **incremental interpretation**: an agent must be able to produce and access meaning representations for partial sentences on a word-by-word basis, to be able to determine a coherent, plausible, or collaborative continuation.
- **Incremental representation**: an agent must be able to access the lexical, syntactic, and semantic information contributed by the constituent parts processed so far, to process or account for clarifications and confirmations.
- **Incremental context**: agents must incrementally add to and read from context on a word-by-word basis, to account for cross-speaker anaphora and ellipsis and for changing references to participants.
- **Reversibility**, or perhaps better **interchangeability**: the partial representations (meaning and form) built by NLU at the point of speaker transition must be suitable for use by NLG as a starting point, and vice versa, preserving syntactic and semantic constraints across the boundary.
- **Extensibility**: the representations of meaning and form must be extendable, to allow the incorporation of extensions (adjuncts, clarifications, etc.) even to complete antecedents.

As we shall see in the next section, previous and current work on incremental NLG has produced models that address many of these requirements, but not all; in subsequent sections, we then outline a possible approach that does.

4.3   Previous work

In this section, we review existing research in incremental production and CCs, both from a psycholinguistic and from a computational perspective.

4.3.1   Psycholinguistic research

The incrementality of online linguistic processing is now uncontroversial. Standard psycholinguistic models assume that language comprehension operates incrementally, with partial interpretations being built more or less on a word-by-word basis (see, e.g., Sturt and Crocker, 1996). Language production has also been argued to be incremental (Kempen and Hoenkamp, 1987; Levelt, 1989; Ferreira, 1996), with evidence also coming from self-repairs and various types of speech errors (Levelt, 1983; van Wijk and Kempen, 1987).

Guhe (2007) further argues for the incremental conceptualization of observed events in a visual scene. He uses this domain to propose a model of the incremental generation of preverbal messages, which in turn guides down-stream semantic and syntactic formulation. In the interleaving of planning, conceptual structuring of the message, syntactic structure construction, and articulation, incremental models assume that information is processed as it becomes available, operating on minimal amounts of characteristic input to each phase of generation, reflecting the introspective observation that the end of a sentence is not planned when one starts to utter its beginning (Guhe *et al.*, 2000). The evidence from CCs described above supports these processing claims, along with providing additional evidence for the ease of switching roles between incremental parsing and incremental generation during dialogue.

4.3.2   Incrementality in NLG

Early work on incremental NLG was motivated not only by the emerging psychological evidence but also by attempts to improve the user experience in natural language interfaces: systems that did not need to compile complete sentence plans before beginning surface realization could provide decreased response times. Levelt's (1989) concepts of the **conceptualization** and **formulation** stages of language production lead to a more concrete computational distinction between the **tactical** and **strategic** stages of generation (Thompson, 1977), with the incremental passing of units between these becoming important. Parallel and distributed processing across modules stands in contrast to traditional pipeline approaches to NLG (see, e.g., Reiter and Dale, 2000), a shortcoming of generation architectures outlined perspicuously by De Smedt *et al.* (1996).

With formalisms such as Functional Unification Grammar (FUG; Kay, 1985) and Tree Adjoining Grammar (TAG; Joshi, 1985), researchers began to address incrementality explicitly. Kempen and Hoenkamp (1987) made the first notable attempt to implement an incremental generator, introducing their Incremental Procedural Grammar (IPG) model. Schematically, IPG was driven by parallel processes whereby a team of syntactic modules worked together on small parts of a sentence under construction, with the sole communication channel as a stack object (with different constituents loaded onto it), rather than the modules being controlled by a central constructing agent. This approach was consistent with emerging psycholinguistic theories that tree formation was simultaneously conceptually and lexically guided, and that production did not take place in a serial manner; it was capable of generating elliptical answers to questions and also some basic self-repairs.

De Smedt (1990) took incrementality a stage further, showing how developing the syntactic component of the formulation phase in detail could support cognitive claims, shedding light on lexical selection and memory limitations (De Smedt, 1991). De Smedt's Incremental Parallel Formulator (IPF) contained a further functional decomposition between grammatical and phonological encoding, meaning that syntactic processes determining surface form elements like word order and inflection could begin before the entire input for a sentence had been received.

Early incremental systems allowed input to be underspecified in the strategic component of the generator before the tactical component began realizing an utterance, paving the way for shorter response times in dialogue systems but without implementational evidence of such capability. It is worth noting the analogous situation in psycholinguistics: models including the functional decomposition of production stages, as described above, were influential in the autonomous processing camp of psycholinguistics; however, they did not extend to explaining the role of incremental linguistic processing in interaction.

4.3.3   Interleaving parsing and generation

In moving toward the requirements of an interactive system capable of dealing with CCs, notable work on interleaving generation with parsing in an incremental fashion came from Neumann (1994), Neumann and van Noord (1994), and Neumann (1998), who showed how the two processes could be connected using a reversible grammar. The psychological motivation came mainly from Levelt's (1989) concept of a feedback loop to parsing during generation for self-monitoring. The representations used by the parser and generator were explicitly reversible, based around **items**, pairs of logical forms (LFs – in this case, HPSG-like attribute–value matrices) and the corresponding strings.

Processing too was reversible, following the proposal by Shieber (1988), and implemented as a Uniform Tabular Algorithm (UTA), a data-driven selection function that was a generalization of the Earley deduction scheme. The UTA had a uniform indexing mechanism for items and an agenda-based control that allowed item sharing between parsing and generation: partial results computed in one direction could be computed in the other. Items would have either the LF or the string specified but not both: the parser would take items with instantiated string variables but with uninstantiated LFs, and vice versa for the generator. This model therefore fulfilled some of the conditions required for CCs (reversibility and a degree of incrementality) but not all, as it was intended to parse its own utterances for on-line ambiguity checking (self-monitoring), rather than for interactivity and simultaneous interpretation of user input.

4.3.4   Incremental NLG for dialogue

Recent work on incremental dialogue systems, driven by evidence that incremental systems are more efficient and pleasant to use than their non-incremental counterparts (Aist *et al.*, 2007), has brought the challenges for interactive NLG to the fore. In particular, Schlangen and Skantze's (2009, 2011) proposal for an abstract incremental architecture for dialogue, the **Incremental Unit** (IU) framework, has given rise to several interactive systems, including some with interesting NLG capabilities.

In Schlangen and Skantze's architecture, modules comprise a **left buffer** for input increments, a **processor**, and a **right buffer** for output increments. It is the **adding,**

commitment to, and **revoking** of IUs in a module's right buffer and the effect of doing so on another module's left buffer that determines system behavior. Multiple competing IU hypotheses may be present in input or output buffers, and dependencies between them (e.g., the dependency of inferred semantic information from lexical information) are represented by **groundedIn**[4] relations between IUs.

The fact that all modules are defined in this way allows incremental behavior throughout a dialogue system, and this has been exploited to create systems capable of some CC types, including the generation and interpretation of mid-utterance backchannels (Skantze and Schlangen, 2009) and interruptions (Buß *et al.*, 2010). However, most of these systems have focused on the incremental management of the dialogue, rather than on NLU and NLG themselves or on their interdependence. As a result, they tend to use canned text output for NLG; consequently they lack interchangeability, and are therefore not suited for more complex CC phenomena.

Skantze and Hjalmarsson (2010), however, describe a model and system (Jindigo) that incorporates incremental NLG (although still using canned text rather than a more flexible approach). Jindigo can begin response generation before the end of a user utterance: as **word** hypotheses become available from incoming speech input, these are sent in real time to the NLU module, which in turn incrementally outputs **concept** hypotheses to the dialogue manager. This incrementally generates a **speech plan** for the speech synthesizer, which in turn can produce verbal output composed of speech **segments** divided into individual words. This incremental division allows Jindigo to begin speech output before speech plans are complete (*well, let's see* ...). It also provides a mechanism for self-repair in the face of changing speech plans during generation, when input concepts are revised or revoked. By cross-checking the speech plan currently being synthesized against the new speech plan, together with a record of the words so far output, the optimal word/unit position can be determined from which the repair can be integrated. Depending on the progress of the synthesizer through the current speech plan, this repair may be either **covert** (before synthesis) or **overt** (after synthesis), on both the segment and word levels. However, the use of different representations in NLU and NLG, together with the use of atomic semantic representations for entire multiword segments in NLG, means that our criteria of interchangeability and incremental semantic interpretation are not met, and a full treatment of CCs is still lacking.

4.3.5   Computational and formal approaches

Skantze and Hjalmarsson (2010) and Buß *et al.* (2010), as mentioned above, provide models that can handle some forms of compound contributions: mid-utterance backchannels, interruptions, and (some) clarifications and confirmations. A few recent

---

[4] **groundedIn** links are transitive dependency relations between IUs specified by the system designer that may be exploited by modules. For instance, a word hypothesis IU may be grounded in a particular automatic speech recognition (ASR) result, and only added to the word hypothesizer's output graph once that part of the ASR graph is **committed**. See Skantze and Schlangen (2009) and Skantze and Hjalmarsson (2010) for more details.

computational implementations and formal models focus specifically on more complex aspects of CCs.

DeVault *et al.* (2009, 2011) present a framework for predicting and suggesting completions for partial utterances: given partial speech recognition (ASR) hypotheses, their domain-specific classification-based approach can robustly predict the completion of an utterance begun by a user in real time. Given a corpus that pairs ASR output features from user utterances with the corresponding hand-annotated final semantic frames, they train a maximum-entropy classifier to predict frames from a given partial ASR result. They achieve high precision in the correct selection of semantic frames, and provide some indication of possible transition points by using another classifier trained to estimate the point in the incoming utterance at which the probability of the semantic frame currently selected being correct is unlikely to improve with further ASR results.

While their focus is on incremental interpretation rather than generation, this provides a practical model for part of the process involved in a CC: the jump from partial NLU hypotheses to a suggested completion. DeVault *et al.* (2009, 2011) provide a basic NLG strategy for such completions by their system: by finding training utterances that match the predicted semantics the partial input seen so far, the selection of the remainder of the utterance can be produced as the generator's completion. However, while such a model produces incremental semantic interpretations, its lack of syntactic information and its restriction to a finite set of semantic frames known in the domain prevent it from being a full model for CCs: such a model must be more flexible and able to account for syntactic constraints across speaker transitions.

Poesio and Rieser (2010), in contrast, describe a grammar-based approach that incorporates syntactic, semantic, and pragmatic information via a lexicalized tree-adjoining grammar (TAG) paired with the PTT model for incremental dialogue interpretation (Poesio and Traum, 1997). They provide a full account of the incremental interpretation process, incorporating lexical, syntactic, and semantic information and meeting the criteria of incremental interpretation and representation. Beyond this, they also provide a detailed account of how a suggested collaborative completion might be derived using inferential processes and the recognition of plans at the utterance level: by matching the partial representation at speaker transition against a repository of known plans in the relevant domain, an agent can determine the components of these plans that have not yet been made explicit and make a plan to generate them. Importantly, the plans being recognized are at the level of speech planning: the desired continuation is determined by recognizing the phrases and words observed so far as being part of a plan that makes sense in the domain and the current context.

This model therefore meets many of the criteria we defined: both interpretation and representation are incremental, with semantic and syntactic information being present; the use of PTT suggests that linguistic context can be incorporated suitably. However, while reversibility might be incorporated by the choice of suitable parsing and generation frameworks, this is not made explicit; and the extensibility of the representations seems limited by TAG's approach to adjunction (extension via syntactic adjuncts seems easy to treat in this approach, but more general extension is less clear). The use of TAG also seems to restrict the grammar to licensing grammatical strings, problematic for some CCs (see Section 4.2.2).

4.3.6 Summary

Previous work provides models for NLG that are incremental at the word-by-word level, and that can run in parallel with incremental parsing of user contributions, with some form of reversible representation. These models variously provide incremental syntactic construction during generation (Kempen and Hoenkamp, 1987; De Smedt, 1990) and incremental changing of the inputs to generation (Guhe, 2007; Skantze and Hjalmarsson, 2010). However, they do not generally explain how meaning is built up strictly incrementally – how partial structures in generation can be related to maximal semantic content on a word-by-word basis. On the other hand, approaches specifically targeted at collaborative contributions and the required incremental modeling either lack strong incremental representation, so that the parts of the utterance responsible for parts of the meaning representation cannot be determined (DeVault *et al.*, 2009, 2011), or lack reversibility or extensibility while relying on licensing strings rather than meaning representations (Poesio and Rieser, 2010). In addition, little attention has been paid to the availability of linguistic context to NLG, and its sharing with NLU, on an incremental basis. An incremental approach is needed that not only has the qualities of reversibility and extensibility but also the ability to generate incremental semantic interpretations and lexically anchored representations.

4.4 Dynamic Syntax (DS) and Type Theory with Records (TTR)

The approaches outlined so far all lack one or more of the criteria for a successful treatment of CCs. In this section, we describe an incremental grammar formalism and show how it can be extended to meet all these criteria, including strong incremental interpretation, incremental representation, and reversibility.

4.4.1 Dynamic Syntax

One formalism with potential to satisfy the criteria for handling CCs described above is Dynamic Syntax (DS; Cann *et al.* 2005; Kempson *et al.* 2001; *inter alia*). DS is an action-based and semantically oriented incremental grammar formalism that dispenses with an independent level of syntax, instead expressing grammaticality via constraints on the word-by-word monotonic growth of semantic structures. In its original form, these structures are **trees**, with nodes corresponding to terms in the lambda calculus; these nodes are annotated with labels expressing their semantic type and formula, and beta-reduction determines the type and formula at a mother node from those at its daughters (4.21):

(4.21) $Ty(t), \diamond, arrive'(john')$

$Ty(e), john'$ $Ty(e \rightarrow t), \lambda x. arrive'(x)$

The DS lexicon comprises **lexical actions** associated with words, and also a set of globally applicable **computational actions**. Both of these are defined as monotonic tree update operations, and take the form of IF–THEN action structures. In traditional DS notation, the lexical action corresponding to the word *John* has the preconditions and update operations in (4.22). Trees are updated by these actions during the parsing process as words are consumed from the input string.

(4.22) *John*:
  IF $?Ty(e)$
  THEN $put(Ty(e))$
          $put(Fo(john'))$
  ELSE   abort

(4.23) $?Ty(t)$ $\longrightarrow$ $?Ty(t)$

$?Ty(e), \diamond$ $?Ty(e \rightarrow t)$ $?Ty(e), \diamond, john'$ $?Ty(e \rightarrow t)$

DS parsing begins with an **axiom tree** (a single requirement for a truth value, $?Ty(t)$), and at any point, a tree can be **partial**, with nodes annotated with requirements for future development (written with a ? prefix) and a **pointer** (written $\diamond$) marking the node to be developed next (4.23). Actions can then satisfy and/or add requirements. Lexical actions generally satisfy a requirement for their semantic type, but may also add requirements for items expected to follow (e.g., a transitive verb may add a requirement for an object of type $Ty(e)$). Computational actions represent generally available strategies such as removing requirements that are already satisfied, and applying beta-reduction. (4.23) shows the application of the action for *John* defined in (4.22). Grammaticality of a word sequence is defined as satisfaction of all requirements (**tree completeness**) leading to a complete semantic formula of type $Ty(t)$ at the root node, thus situating grammaticality as **parseability**. The left-hand side of Figure 4.1 shows a sketch of a parse for the simple sentence *John likes Mary*: transitions represent the application of lexical actions together with some sequence of computational actions, monotonically constructing partial trees until a complete tree is yielded.

Generation by parsing

Tactical generation in DS can be defined in terms of the parsing process and a subsumption check against a goal tree – a complete and fully specified DS tree such as (4.21) which represents the semantic formula to be expressed (Otsuka and Purver, 2003; Purver and Otsuka, 2003; Purver and Kempson, 2004a). The generation process uses exactly the same tree and action definitions as the parsing process, applied in the same way: trees are extended incrementally as words are added to the output string, and the process is constrained by checking for compatibility with the goal tree. Compatibility

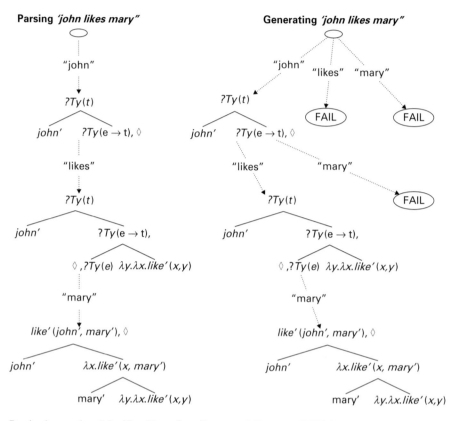

Figure 4.1 Parsing/generating *John likes Mary*, from Purver and Kempson (2004a)

is defined in terms of tree **subsumption**: a tree subsumes a goal tree if it contains no nodes or node annotations absent in the goal tree.[5]

Generation thus follows a "parse-and-test" procedure: lexical actions are chosen from the lexicon and applied; and after each successful application, a subsumption check removes unsuitable candidates from the parse state – see the right-hand side of Figure 4.1 for a sketch of the process. From an NLG perspective, lexicalization and linearization (or in psycholinguistic terms, formulation and word ordering) are thus combined into one process: each word in the lexicon is tested for its applicability at each point of possible tree extension, and if accepted by the generator it is both selected and realized in the output string in one single action. As with Neumann's (1998) framework, DS inherently has the quality of **reversibility**, as the input for generating a string is the semantic tree that would be derived from parsing that string.

---

[5] More correctly, if it contains no nodes that do not subsume some node in the goal tree; as node labels and addresses may be underspecified, the distinction is important for many syntactic phenomena, but we will ignore it here – see Purver and Kempson (2004a).

Context in Dynamic Syntax

Access to some model of linguistic context is required for processing discourse phenomena such as anaphora and ellipsis. For DS, being an incremental framework, this context is taken to include not only the end product of parsing or generating a sentence (the semantic tree and corresponding string), but information about the dynamics of the parsing process itself – the lexical and computational action sequence used to build the tree. Strict readings of anaphora and verb phrase ellipsis (VPE) are obtained by copying semantic formulae (as node annotations) from context: anaphoric elements such as pronouns and elliptical auxiliaries annotate trees with typed metavariables, and computational rules allow the substitution of a contextual value to update those metavariables. Sloppy readings are obtained by re-running a sequence of actions from context: a previous action sequence triggered by a suitable semantic type requirement (and resulting in a formula of that type) can be re-used, again providing a complete semantic formula for a node temporarily annotated with just a metavariable – see (Purver *et al.*, 2006; Kempson *et al.*, 2011) for details.

As defined in Purver and Kempson (2004b) and Purver *et al.* (2006), one possible model for such a context can be expressed in terms of triples $\langle T, W, A \rangle$ of a tree T, a word-sequence W, and the sequence of actions A, both lexical and computational, that are employed to construct the tree. In parsing, the parser state P at any point is characterized as a set of these triples; in generation, the generator state G consists of a goal tree T_G and a set of possible parser states paired with their hypothesized partial strings S. This definition of a generator state in terms of parse states ensures equal access to context for NLU and NLG, as required, with each able to use dynamically a full representation of the linguistic context produced so far.

Suitability for compound contributions

In the basic definitions of the formalism, DS fulfills some of our criteria to provide a model of CCs. It is inherently **incremental**: as each word is parsed or generated, it monotonically updates a set of partial trees. The definition of generation in terms of parsing means that it is also ensured to be **reversible**, with representations being naturally interchangeable between parsing and generation processes at any point. And context is defined to be incrementally and equally available to both parsing and generation.

Correspondingly, Purver and Kempson (2004b) outline a DS model for CCs, showing how the shift from NLU (hearer) to NLG (speaker) can be achieved at any point in a sentence, with context and grammatical constraints transferred seamlessly. The parser state at transition P_T (a set of triples $\langle T, W, A \rangle$) serves as the starting point for the generator state G_T, which becomes $(T_G, \{S_T, P_T\})$, where S_T is the partial string heard so far and T_G is whatever goal tree the generating agent has derived. The standard generation process can then begin, testing lexical and computational actions on the tree under construction in P_T, and successful applications will result in words extending S_T such that the tree is extended while subsuming T_G. The transition from speaker to hearer can be modeled in a directly parallel way, without the need to produce a goal tree, due to the interchangeability of generation and parse states and the context they contain. Gargett *et al.* (2009) also show how such a model can handle mid-sentence clarification and confirmation.

However, it is not clear that the DS model on its own fulfills all the conditions set out in Section 4.2.6. It produces representations that express semantic, syntactic, and lexical information on an incremental basis; but in order to fulfill our criterion of strong incremental interpretation we require these to be in a form suitable for reasoning about possible meanings and continuations, and for determining the contribution of words and phrases. It is not clear how an agent can reason from a partial semantic tree (without a semantic formula annotating its top node) to a goal for generation, especially if this goal must itself be in the form of a tree – and Purver and Kempson (2004a) give no account of how T_G can be derived. The account therefore lacks a way to account for how appropriate completions can be decided, and remains entirely tactical rather than strategic. Given this, there is also a question about the criterion of **extensibility**: while partial DS tree structures are certainly extendable, the lack of a clearly defined semantic interpretation at each stage means it is unclear whether extensibility applies in a semantic sense.

The criterion for incremental representation is also not entirely met. The model of context includes the contributions of the words and phrases seen, as it includes lexical and computational action sequences; but it is not clear how to retrieve from context the correspondence between a word and its contribution (information needed to resolve anaphora and clarifications). While word–action–formula correspondences are present for individual parse path hypotheses (individual $\langle T, W, A \rangle$ triples), there is no straightforward way to retrieve all action or formula hypotheses for a given word when the context contains a set of such triples with no explicit links between them.

4.4.2   Meeting the criteria

However, recent extensions to DS do provide a model that fulfills these missing criteria. The use of Type Theory with Records (TTR) for semantic representation permits incremental interpretation and full extensibility; and the use of a graph-based model of context permits incremental representation.

Type Theory with Records
Recent work in DS has started to explore the use of Type Theory with Records (TTR; Betarte and Tasistro, 1998; Cooper, 2005; Ginzburg, 2012) to extend the DS formalism, replacing the atomic semantic type and epsilon calculus formula node labels with more complex **record types**, and thus providing a more structured semantic representation. Purver *et al.* (2010) provide a sketch of one way to achieve this and show how it can be used to incorporate pragmatic information such as illocutionary force and participant reference (thus, among other things, giving a full account of the grammaticality of examples such as (4.13)). Purver *et al.* (2011) introduce a slightly different variant using a Davidsonian event-based representation, and this is shown in (4.24) below. The semantic formula annotation of a node is now a TTR record type: a sequence of fields consisting of pairs of **labels** with **types**, written $[x : e]$ for a label x of type e. The identity

of predicates and arguments is expressed by use of **manifest** types written $[x_{=john} : e]$ where *john* is a singleton subtype of *e*. The standard DS type label $Ty()$ is now taken to refer to the final (i.e., lowest) field of the corresponding record type. Tree representations otherwise remain as before, with functional application of functor nodes to argument nodes giving the overall TTR record type of the mother node.

(4.24)   *John arrives* $\mapsto$

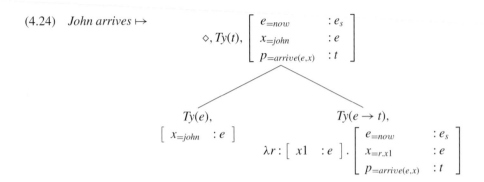

As well as providing the structure needed for representation of pragmatic information, the use of TTR allows us to provide a semantic representation at the root node of a tree, with this becoming more fully specified (via TTR **subtyping**) as more information becomes available. In TTR, a record type *x* is a subtype of a record type *y* if *x* contains at least all fields present in *y*, modulo renaming of labels, with *x* possibly also containing other fields not present in *y*. As Hough (2011) shows, this allows a version of DS in which root nodes are annotated with the maximal semantic content currently inferable given the labels present at the daughters; as more words are parsed (or generated) and the daughter nodes become more fully specified, the root node content is updated to a subtype of its previous type.

As Figure 4.2 shows, this provides a semantic representation (the TTR representation at the root node) that is incrementally updated to show the maximal semantic information known – precisely meeting our criterion of strong incremental **interpretation**. After the word *John*, we have information about an entity of manifest type *john* (a subtype of *e*), and know there will be some overall sentential predicate of type *t*, but do not yet know anything about the predicate or the argument role that *john* plays in it. As more words are added, this information is specified and the TTR subtype becomes more specific. This information was of course already present in partial DS tree structures, but implicit; this approach allows it to be explicitly represented, as required in CC generation (see Section 4.4.2).

The use of TTR also permits semantic **extensibility**, giving a straightforward analysis of continuations as extensions of an existing semantic representation. Adding fields to a record type results in a more fully specified record type that is still a subtype of the original. There is no requirement that the extension be via a complete syntactic constituent (e.g., an adjunct), as the semantic representation is available fully incrementally.

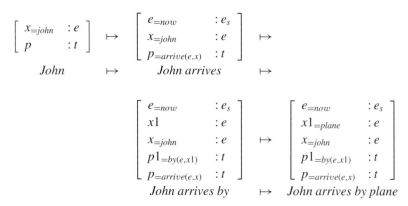

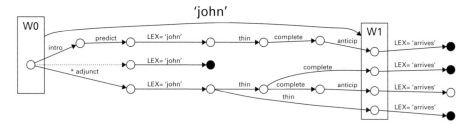

Figure 4.2   Incremental interpretation via TTR subtypes

Figure 4.3   DS context as a DAG, consisting of a parse DAG (circular nodes = trees, solid edges = lexical(bold) and computational actions) **groundedIn** the corresponding word DAG (rectangular nodes = tree sets, dotted edges = word hypotheses) with word hypothesis *john* spanning tree sets W0 and W1

Parsing and generation context as a graph

A further modification provides the required incremental **representation**. Rather than seeing linguistic context as centered around a set of essentially unrelated action sequences, an alternative model is to characterize it as a Directed Acyclic Graph (DAG). Sato (2011) shows how a DAG with DS **actions** for edges and (partial) **trees** for nodes allows a compact model of the dynamic parsing process; and Purver *et al.* (2011) extend this to integrate it with a word hypothesis graph (or "word lattice") as obtained from a standard speech recognizer.

This results in a model of context as shown in Figure 4.3, a hierarchical model with DAGs at two levels. At the action level, the parse graph DAG (shown in the lower half of Figure 4.3 with solid edges and circular nodes) contains detailed information about the actions (both lexical and computational) used in the parsing or generation process: edges corresponding to these actions connect nodes representing the partial trees built by them, and a path through the DAG corresponds to the action sequence for any given tree. At the word level, the word hypothesis DAG (shown at the top of Figure 4.3 with dotted edges and rectangular nodes) connects the words to these action sequences: edges in this DAG correspond to words, and nodes correspond to sets of parse DAG nodes (and therefore sets of hypothesized trees). Note that many possible word hypotheses may be present for NLU in a spoken dialogue system, as multiple ASR hypotheses may be available; in NLG, many possible competing word candidates may be being considered at any point. In both cases, this can be represented by alternative word DAG edges.

For any partial tree, the context (the words, actions, and preceding partial trees involved in producing it) is now available from the paths back to the root in the word and parse DAGs. Moreover, the sets of trees and actions associated with any word or word subsequence are now directly available as that part of the parse DAG spanned by the required word DAG edges. This, of course, means that the contribution of any word or phrase can be directly obtained, fulfilling the criterion of **incremental representation**. It also provides a compact and efficient representation for multiple competing hypotheses, compatible with DAG representations commonly used in interactive systems, including the incremental dialogue system Jindigo (Skantze and Hjalmarsson 2010; see Section 4.3.4 and below). Importantly, the DS definition of generation in terms of parsing still means this model will be equally available to both parsing and generation, and used in the same way by both. The criteria of **interchangeability** and equal access to **incremental context** are therefore still assured.

4.5   Generating compound contributions

Given this suitable framework for parsing and generation, we show how it can be used to provide a possible solution to the challenge posed by CCs for the NLG process in dialogue, in line with the requirements described at the end of Section 4.2. We describe how an incremental dialogue system can handle the phenomenon through use of the incremental parsing and generation models of Dynamic Syntax (DS) combined with semantic construction of TTR record types.

4.5.1   The DyLan dialogue system

DyLan[6] is a prototype incremental dialogue system based around Jindigo (Skantze and Hjalmarsson, 2010) and incorporating an implementation of DS-TTR to provide the NLU (Eshghi et al., 2011; Purver et al., 2011) and NLG (Hough, 2011) modules. Following Jindigo, it uses Schlangen and Skantze's (2009) abstract model of incremental processing: each module is defined in terms of its **incremental unit** (IU) inputs and outputs. For the NLU module, input IUs are updates to a word hypothesis DAG, as produced by a speech recognizer; and output IUs are updates to the context DAG as described in Section 4.4.2 above, including the latest semantic representations as TTR record types annotating the root nodes of the latest hypothesized trees. For the NLG module, input IUs must be representations of the desired semantics (the goal for the current generation process), while output IUs are again updates to the context DAG,

---

[6] DyLan stands for "DYnamics of LANguage."

including the latest word sequence for output. The context DAG is shared by parser and generator: both modules have access to the same data structure, allowing both NLU and NLG processes to add incrementally to the trees and context currently under construction at any point.

Goal concepts and incremental TTR construction

The original DS generation process required a semantic **goal tree** as input, but the strong incremental semantic interpretation property of the extended TTR model simplifies this requirement. As semantic interpretations (TTR record types) are available for any partial tree under construction, the generation process can now be driven by a **goal concept** in the form of a TTR record type, rather than a full goal tree. This reduces the complexity of the input, making the system more compatible with standard information state update and issue-based dialogue managers (e.g., Larsson, 2002). Goal concepts and output strings for the generator now take a form exactly like the partial strings and maximal semantic types for the parser shown in Figure 4.2.

The DyLan parsing system constructs a record type for each path-final tree in the parse DAG as each input word is received, allowing maximal semantic information for **partial** as well as complete trees to be calculated; this is implemented via a simple algorithm that places underspecified metavariables on nodes that lack TTR record types, and then continues with right corner-led beta-reduction as for a complete tree (see Hough, 2011 for details). As words in the lexicon are tested for generation, the generator checks for a **supertype** (subsumption) relation between path-final record types and the current goal record type, proceeding on a word-by-word basis. Parse paths that construct record types not in a supertype relation to the goal may be abandoned, and when a **type match** (i.e., subsumption in both directions) with the goal concept is found, generation is successful and the process can halt.

4.5.2 Parsing and generation co-constructing a shared data structure

The use of TTR record types in NLG removes the need for grammar-specific parameters (a real need when creating goal trees) and means that little modification is required for an off-the-shelf dialogue manager to give the system a handle on CCs. Domain knowledge can also be expressed via a small ontology of domain-specific record types. Born out of a long-standing use of frames for generating stereotypical dialogues in given situations (Lehnert, 1978) the idea of conversation **genres** (Ginzburg, 2012) can be employed here: domain concepts can be assumed to be of a given conversational TTR record type, as in the simple travel domain example below in Figure 4.4; this can contain underspecified fields (the $x1, x2, x3$ values) for information that varies by user and context, as well as fully specified manifest fields.

The interchangeability of representations between NLU and NLG means that the construction of a data structure such as that in Figure 4.4 can become a collaborative process between dialogue participants, permitting a range of varied user input behavior and flexible system responses. As with Purver's *et al.* (2006) original model for CCs, the use of the same representations by NLU and NLG guarantees the ability both to begin

$$\begin{bmatrix} e & : e_s \\ x3 & : e \\ x2 & : e \\ x1 & : e \\ x_{=user} & : e \\ p3_{=by(e,x3)} & : t \\ p2_{=from(e,x2)} & : t \\ p1_{=to(e,x1)} & : t \\ p_{=go(e,x)} & : t \end{bmatrix}$$

Figure 4.4   A TTR record type representing a simple travel domain concept

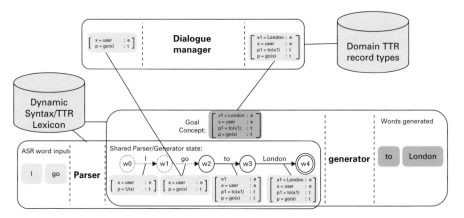

Figure 4.5   Completion of a compound contribution using incremental DS-TTR record type construction, with parser and generator constructing a shared parse state

parsing from the end-point of any generation process (even mid-utterance) and to begin generation from the end-point of any parsing process. Both NLU and NLG models are now characterized entirely by the parse context DAG, with the addition for generation of a TTR goal concept. The transition from generation to parsing now becomes almost trivial: the parsing process can continue from the final node(s) of the generation DAG, with parsing actions extending the trees available in the final node set as normal.

The transition from parsing to generation also requires no change of representation, with the DAG produced by parsing acting as the initial structure for generation (see Figure 4.5); now, though, we also require the addition of a goal concept to drive the generation process. But given the full incremental interpretation provided by the use of record types throughout, we can now also see how a generator might produce such a goal at a speaker transition.

4.5.3   Speaker transition points

The same record types are now used throughout the system: as the concepts for generating system plans, as the goal concepts in NLG, and for matching user input against known concepts in suggesting continuations. In interpretation mode, the emerging conversational record type in context can be incrementally checked against any known domain concept record types; if generation of a continuation is now required, this can be used to select the next generation goal on the basis of any matching knowledge in the system's knowledge base or information state. A suitable generation goal can be any subtype of the current top record type in the parse DAG at speaker transition; a match against a known concept (from domain or conversational context) can provide this.

Given the formal tools of DS-TTR parsing and generation to license CCs, we can therefore equip a system with the ability to generate them quite simply. Possible system transition points trigger the alternation between modules in their co-construction of the shared parse/generator DAG; in DyLan, this is provided by a simplistic dialogue manager with high-level methods without reference to syntax or lexical semantics. We may employ a single method, `continueContribution`, which simply reacts to a silence threshold message from the ASR module, halts the parser, and selects an appropriate goal TTR record type from its concept ontology – achieving this by searching for the first concept that stands in a subtype relation to the record type under construction. The selected record type then acts as the new goal concept, allowing generation to begin immediately. While speech act information can be delivered to the synthesizer for the purposes of prosody alteration, in terms of semantics, no additional information about the utterance is required for a continuation. This stands in contrast to Rieser's (2010) account, which requires inference about speaker intentions; of course, this is not to say that such processes might not be useful, or required for certain situations to fully capture the dialogue data. Importantly, though, we can provide a model for the underlying mechanisms, and for the suggestion of simple continuations on the basis of domain knowledge, without such inference.

To ensure coherence between the different utterances making up a CC, the stipulation that a goal record type selected upon user silence be a subtype of the record type under construction facilitates the joint construction of a completed record type (see Figure 4.5). However, this does not mean the system must have a complete domain concept as a goal, as the selected goal may be an underspecified supertype of a domain concept. This allows contributions even when the system does not possess full information, but knows something about the continuation required, as in (4.25). Here, generation can begin at the first speaker transition if a suitable goal concept can be obtained (e.g., as in Figure 4.4, giving the information that a mode of transport is required but not the reference itself) – the word *from* can therefore be generated. At this point generation will then stop as the utterance covers all information in this goal concept (i.e., leaving no unvocalized goal information left to drive further generation), at which point parsing can take over again if user input is available:

(4.25)   USER: *I'm going to London*
         SYS: *from …*
         USER: *Paris.*

If the user interrupts or extends a system utterance, Jindigo's continuously updating buffers allow notification of input from the ASR module to be sent to the dialogue manager quickly[7] and the switch to parsing mode may take place. The dialogue manager's method `haltGeneration` stops the NLG and transfers the parse DAG construction role back to the NLU. Upon the generation or recognition of each word, a **strong incremental representation** can be extracted for the utterance so far, as each word is parsed semantically and syntactically, with incremental self-monitoring (Levelt, 1989) coming for free in that the utterance's string does not need to be passed back to a parsing module. This is not possible in string-based approaches (e.g., DeVault *et al.* 2009).

Extension contributions (e.g., adjuncts or prepositional phrases in the limited travel domain below, such as *to Paris* and *on Monday*, but in principle any extension) can be dealt with straightforwardly in both parsing and generation, as they introduce subtypes of the record type under construction. The user over-answering a prompt to extend a CC as in (4.26) is therefore handled straightforwardly, as the goal concept during generation may be overridden by the user's input if it is commensurate with the record type constructed up to the speaker transition (i.e., stands in a subtype relation to it). In this sense, a continuation from the user can be "unexpected" but not destructive to the continual build up of meaning, or the maintenance of the parse DAG.

(4.26)   USER: *I want to go . . .*
         SYS: *to . . .*
         USER: *Paris from London*

The system is therefore capable of taking part in arbitrary speaker transitions, including multiple transitions during one co-constructed utterance, and generating any part of a contribution whose parse path will lead to constructing a domain concept record type. At any word position, and *a fortiori*, at any syntactic position, in the utterance, the module responsible for building this path may change depending on the user's behavior, consistent with the psycholinguistic observations summarized in Section 4.2.

4.6   Conclusions and implications for NLG systems

In this chapter, we have outlined the phenomenon of compound contributions (CCs), detailed some of the many forms they can take, and explained why they are of interest to NLG researchers. CCs provide a stringent set of criteria for NLG itself and for NLG/NLU interaction. As set out in Section 4.2.6, these criteria entail full word-by-word incrementality in terms of representation, interpretation, and context access, while requiring full interchangeability of representations between NLU and NLG. While previous research has produced NLG and dialogue models that provide incrementality in many ways, none of them fulfills all of these criteria.

---

[7] A demonstration of the system's capability for rapid turn-taking is shown in Skantze and Schlangen (2009).

In particular, we have seen that the use of standard string-licensing grammars is problematic: contextual pronominal reference changes with speaker transitions, resulting in successful, acceptable utterances which would have to be characterized as ungrammatical if considered merely as surface strings. We have also seen that neither lexical, syntactic, nor semantic incremental processing is sufficient on its own; a fully CC-capable system must produce incremental representations of meaning, structure, and lexical content *together*. However, by extending Dynamic Syntax to incorporate a structured, type-theoretic semantic representation and a graph-based context model, we can provide a model that meets all the criteria for handling CCs, and use this within a prototype interactive system.

Speaker intentions

One feature of note is that our framework allows us to model CCs without *necessarily* relying on speaker intention recognition. While intention recognition may well play a role in many CC cases, and may be a strategy available to hearers in many situations, our model does not rely on it as primary, instead allowing parsing and generation of CCs based only on an agent's internal knowledge and context. Such a model is compatible with existing approaches to interactive systems based on, e.g., information state update rather than higher-order reasoning about one's interlocutor.

Alignment

The main focus of this chapter has been on providing a model that can license the grammatical features of the CC phenomena in question: one that is capable of generating (and parsing) the phenomena in principle. However, NLG systems in real interactive settings need to look beyond this, to features that characterize the naturalness or human-like qualities of the discourse and give us a way to choose between possible alternative formulations. One such feature is **alignment** – the tendency of human interlocutors to produce similar words and/or structures to each other (Pickering and Garrod, 2004). Giving a full account of alignment in the DS-TTR framework is a matter for future research (for one thing, requiring a general model of preferences in DS parsing dynamics – for an initial model see Sato, 2011), and we see this as an area of interest for NLG research. We note here, though, that the graphical model of context does provide an interesting basis for such research. Taking the context DAG as a basis for lexical action choice – with a basic strategy being to re-use actions in context by DAG search before searching through the NLG lexicon – provides an initial platform for an explanation of alignment. More recently used words would tend to be re-used, and the sharing and co-construction of the context model between parsing and generation explains how this happens between hearing and speaking (and vice versa). Interestingly, however, this model would predict that syntactic alignment should arise mainly from lexical alignment – through re-use of lexical action sequences – rather than being driven as an individual process. Recent empirical data suggest that this may indeed be so, with syntactic alignment in corpora perhaps explainable as due to lexical repetition (Healey *et al.*, 2010).

Coordination and repair

It is also worth noting briefly here that speaker transition in CCs is often associated not merely with a smooth transition, but with self-repair phenomena such as repetition and reformulation, as well as the other-repair phenomena such as mid-utterance clarification that we have already described. A full model must account for this, and we look forward to NLG research that incorporates self- and other-repair together into an incremental model of speaker change and CCs. One proposal for how to go about this within the framework described here, using backtracking along the context graph to model repetition and reformulation, is currently being investigated (Hough, 2011).

Acknowledgments

This research was carried out under the *Dynamics of Conversational Dialogue* project, funded by the UK ESRC (RES-062-23-0962), with additional support through the *Robust Incremental Semantic Resources for Dialogue* project, funded by the EPSRC (EP/J010383/1). We also thank Ruth Kempson, Chris Howes, Arash Eshghi, Pat Healey, Wilfried Meyer-Viol, and Graham White for many useful discussions.

References

Aist, G., Allen, J. F., Campana, E., Gomez Gallo, C., Stoness, S., Swift, M., and Tanenhaus, M. K. (2007). Incremental dialogue system faster than, preferred to its nonincremental counterpart. In *Proceedings of the Annual Conference of the Cognitive Science Society (CogSci)*, pages 761–766, Nashville, TN. Cognitive Science Society.

Allen, J. F., Byron, D., Dzikovska, M., Ferguson, G., Galescu, L., and Stent, A. (2001). Towards conversational human–computer interaction. *AI Magazine*, **22**(4):27–37.

Asher, N. and Lascarides, A. (2008). Commitments, beliefs and intentions in dialogue. In *Proceedings of the Workshop on the Semantics and Pragmatics of Dialogue (LONDIAL)*, pages 29–36, London, UK. SemDial.

Betarte, G. and Tasistro, A. (1998). Extension of Martin–Löf type theory with record types and subtyping. In *25 Years of Constructive Type Theory*, pages 21–40. Oxford University Press, Oxford, UK.

Brennan, S. E. and Schober, M. (2001). How listeners compensate for disfluencies in spontaneous speech. *Journal of Memory and Language*, **44**(2):274–296.

Burnard, L. (2000). Reference Guide for the British National Corpus (World Edition). Available from: http://www.natcorp.ox.ac.uk/archive/worldURG/urg.pdf. Accessed on 11/24/2013.

Buß, O., Baumann, T., and Schlangen, D. (2010). Collaborating on utterances with a spoken dialogue system using an ISU-based approach to incremental dialogue management. In *Proceedings of the SIGdial Conference on Discourse and Dialogue (SIGDIAL)*, pages 233–236, Tokyo, Japan. Association for Computational Linguistics.

Cann, R., Kaplan, T., and Kempson, R. (2005). Data at the grammar–pragmatics interface: The case of resumptive pronouns in English. *Lingua*, **115**(11):1551–1577.

Cooper, R. (2005). Records and record types in semantic theory. *Journal of Logic and Computation*, **15**(2):99–112.

De Smedt, K. (1990). IPF: An incremental parallel formulator. In *Current Research in Natural Language Generation*, pages 167–192. Academic Press, San Diego, CA.

De Smedt, K. (1991). Revisions during generation using non-destructive unification. In *Proceedings of the European Workshop on Natural Language Generation (EWNLG)*, pages 63–70, Judenstein/Innsbruck, Austria. Association for Computational Linguistics.

De Smedt, K., Horacek, H., and Zock, M. (1996). Some problems with current architectures in natural language generation. In Adorni, G. and Zock, M., editors, *Trends in Natural Language Generation: An Artificial Intelligence Perspective*, pages 17–46. Springer LNCS, Berlin, Germany.

DeVault, D., Sagae, K., and Traum, D. (2009). Can I finish?: Learning when to respond to incremental interpretation results in interactive dialogue. In *Proceedings of the SIGdial Conference on Discourse and Dialogue (SIGDIAL)*, pages 11–20, London, UK. Association for Computational Linguistics.

DeVault, D., Sagae, K., and Traum, D. (2011). Incremental interpretation and prediction of utterance meaning for interactive dialogue. *Dialogue and Discourse*, 2(1):143–170.

Eshghi, A., Purver, M., and Hough, J. (2011). DyLan: Parser for dynamic syntax. Technical Report EECSRR-11-05, School of Electronic Engineering and Computer Science, Queen Mary University of London.

Ferreira, V. S. (1996). Is it better to give than to donate? Syntactic flexibility in language production. *Journal of Memory and Language*, 35(5):724–755.

Gargett, A., Gregoromichelaki, E., Kempson, R., Purver, M., and Sato, Y. (2009). Grammar resources for modelling dialogue dynamically. *Cognitive Neurodynamics*, 3(4):347–363.

Ginzburg, J. (2012). *The Interactive Stance: Meaning for Conversations*. Oxford University Press, Oxford, UK.

Goodwin, C. (1979). The interactive construction of a sentence in natural conversation. In Psathas, G., editor, *Everyday Language: Studies in Ethnomethodology*, pages 97–121. Irvington, New York, NY.

Goodwin, C. (1981). *Conversational Organization: Interaction between Speakers and Hearers*. Academic Press, New York, NY.

Gregoromichelaki, E., Kempson, R., Purver, M., Mills, G. J., Cann, R., Meyer-Viol, W., and Healey, P. G. (2011). Incrementality and intention-recognition in utterance processing. *Dialogue and Discourse*, 2(1):199–233.

Guhe, M. (2007). *Incremental Conceptualization for Language Production*. Lawrence Erlbaum Associates, Mahwah, NJ.

Guhe, M., Habel, C., and Tappe, H. (2000). Incremental event conceptualization and natural language generation in monitoring environments. In *Proceedings of the International Conference on Natural Language Generation (INLG)*, pages 85–92, Mitzpe Ramon, Israel. Association for Computational Linguistics.

Healey, P. G. (2008). Interactive misalignment: The role of repair in the development of group sub-languages. In Cooper, R. and Kempson, R., editors, *Language in Flux: Dialogue Coordination, Language Variation, Change and Evolution*. College Publications, London, UK.

Healey, P. G., Purver, M., and Howes, C. (2010). Structural divergence in dialogue. In *Abstracts of the Meeting of the Society for Text & Discourse*, Chicago, IL. Society for Text & Discourse.

Hough, J. (2011). Incremental semantics driven natural language generation with self-repairing capability. In *Proceedings of the International Conference on Recent Advances in Natural Language Processing (RANLP)*, pages 79–84, Hissar, Bulgaria. Association for Computational Linguistics.

Howes, C., Purver, M., Healey, P. G., Mills, G. J., and Gregoromichelaki, E. (2011). On incrementality in dialogue: Evidence from compound contributions. *Dialogue and Discourse*, **2**(1):279–311.

Joshi, A. (1985). Tree adjoining grammars: How much context-sensitivity is required to provide reasonable structural descriptions? In Dowty, D., Karttunen, L., and Zwicky, A. M., editors, *Natural Language Parsing: Psychological, Computational, and Theoretical Perspectives*, pages 206–250. Cambridge University Press, Cambridge, UK.

Kamp, H. and Reyle, U. (1993). *From Discourse to Logic: Introduction to Modeltheoretic Semantics of Natural Language, Formal Logic and Discourse Representation Theory*. Kluwer Academic Publishers, Dordrecht, The Netherlands.

Kay, M. (1985). Parsing in functional unification grammar. In Dowty, D., Karttunen, L., and Zwicky, A. M., editors, *Natural Language Parsing: Psychological, Computational and Theoretical Perspectives*, pages 251–278. Cambridge University Press, Cambridge, UK.

Kempen, G. and Hoenkamp, E. (1987). An incremental procedural grammar for sentence formulation. *Cognitive Science*, **11**(2):201–258.

Kempson, R., Cann, R., Eshghi, A., Gregoromichelaki, E., and Purver, M. (2011). Ellipsis. In Lappin, S. and Fox, C., editors, *The Handbook of Contemporary Semantic Theory*. Wiley, New York, NY, 2nd edition.

Kempson, R., Meyer-Viol, W., and Gabbay, D. (2001). *Dynamic Syntax: The Flow of Language Understanding*. Blackwell, Oxford, UK.

Larsson, S. (2002). *Issue-based Dialogue Management*. PhD thesis, Department of Linguistics, Göteborg University.

Lascarides, A. and Asher, N. (2009). Agreement, disputes and commitments in dialogue. *Journal of Semantics*, **26**(2):109–158.

Lehnert, W. G. (1978). *The Process of Question Answering: A Computer Simulation of Cognition*. Lawrence Erlbaum Associates, Mahwah, NJ.

Lerner, G. H. (1991). On the syntax of sentences-in-progress. *Language in Society*, **20**(3): 441–458.

Lerner, G. H. (1996). On the "semi-permeable" character of grammatical units in conversation: Conditional entry into the turn space of another speaker. In Ochs, E., Schegloff, E. A., and Thompson, S. A., editors, *Interaction and Grammar*, pages 238–276. Cambridge University Press, Cambridge, UK.

Lerner, G. H. (2004). Collaborative turn sequences. In Lerner, G. H., editor, *Conversation Analysis: Studies from the First Generation*, pages 225–256. John Benjamins, Amsterdam, The Netherlands.

Levelt, W. J. M. (1983). Monitoring and self-repair in speech. *Cognition*, **14**(1):41–104.

Levelt, W. J. M. (1989). *Speaking: From Intention to Articulation*. MIT Press, Cambridge, MA.

Milward, D. (1991). *Axiomatic Grammar, Non-Constituent Coordination, Incremental Interpretation*. PhD thesis, University of Cambridge.

Neumann, G. (1994). *A Uniform Computational Model for Natural Language Parsing and Generation*. PhD thesis, Universitaat des Saarlandes, Saarbrücken.

Neumann, G. (1998). Interleaving natural language parsing and generation through uniform processing. *Artificial Intelligence*, **99**(1):121–163.

Neumann, G. and van Noord, G. (1994). Reversibility and self-monitoring in natural language generation. In Strzalkowski, T., editor, *Reversible Grammar in Natural Language Processing*, pages 59–96. Kluwer, Dordrecht, The Netherlands.

Ono, T. and Thompson, S. A. (1993). What can conversation tell us about syntax? In Davis, P., editor, *Alternative Linguistics: Descriptive and Theoretical Modes*, pages 213–271. John Benjamins, Amsterdam, The Netherlands.

Otsuka, M. and Purver, M. (2003). Incremental generation by incremental parsing: Tactical generation in dynamic syntax. In *Proceedings of the CLUK Colloquium*, pages 93–100, Edinburgh, Scotland. The UK Special Interest Group for Computational Linguistics.

Pickering, M. and Garrod, S. (2004). Toward a mechanistic psychology of dialogue. *Behavioral and Brain Sciences*, **27**(2):169–226.

Poesio, M. and Rieser, H. (2010). Completions, coordination, and alignment in dialogue. *Dialogue and Discourse*, **1**:1–89.

Poesio, M. and Traum, D. R. (1997). Conversational actions and discourse situations. *Computational Intelligence*, **13**(3):309–347.

Purver, M., Cann, R., and Kempson, R. (2006). Grammars as parsers: Meeting the dialogue challenge. *Research on Language and Computation*, **4**(2–3):289–326.

Purver, M., Eshghi, A., and Hough, J. (2011). Incremental semantic construction in a dialogue system. In *Proceedings of the International Conference on Computational Semantics*, pages 365–369, Oxford, UK. Association for Computational Linguistics.

Purver, M., Gregoromichelaki, E., Meyer-Viol, W., and Cann, R. (2010). Splitting the 'I's and crossing the 'you's: Context, speech acts and grammar. In *Proceedings of the Workshop on the Semantics and Pragmatics of Dialogue (PozDial)*, pages 43–50, Poznán, Poland. SemDial.

Purver, M., Howes, C., Healey, P. G., and Gregoromichelaki, E. (2009). Split utterances in dialogue: A corpus study. In *Proceedings of the SIGdial Conference on Discourse and Dialogue (SIGDIAL)*, pages 262–271, London, UK. Association for Computational Linguistics.

Purver, M. and Kempson, R. (2004a). Context-based incremental generation for dialogue. In *Proceedings of the International Conference on Natural Language Generation (INLG)*, pages 151–160, Brockenhurst, UK. Springer.

Purver, M. and Kempson, R. (2004b). Incrementality, alignment and shared utterances. In *Proceedings of the Workshop on the Semantics and Pragmatics of Dialogue (Catalog)*, pages 85–92, Barcelona, Spain. SemDial.

Purver, M. and Otsuka, M. (2003). Incremental generation by incremental parsing: Tactical generation in dynamic syntax. In *Proceedings of the European Workshop on Natural Language Generation (ENLG)*, pages 79–86, Budapest, Hungary. Association for Computational Linguistics.

Reiter, E. and Dale, R. (2000). *Building Natural Language Generation Systems*. Cambridge University Press, Cambridge, UK.

Rühlemann, C. and McCarthy, M. (2007). *Conversation in Context: A Corpus-Driven Approach*. Continuum, London, UK.

Sato, Y. (2011). Local ambiguity, search strategies and parsing in dynamic syntax. In Gregoromichelaki, E., Kempson, R., and Howes, C., editors, *The Dynamics of Lexical Interfaces*. CSLI Publications, Stanford, CA.

Saxton, M. (1997). The contrast theory of negative input. *Journal of Child Language*, **24**(1): 139–161.

Schegloff, E. (1979). The relevance of repair to syntax-for-conversation. In Givon, T., editor, *Discourse and Syntax*, pages 261–286. Academic Press, New York, NY.

Schegloff, E. (2007). *Sequence Organization in Interaction: A Primer in Conversation Analysis I*. Cambridge University Press, Cambridge, UK.

Schlangen, D. and Skantze, G. (2009). A general, abstract model of incremental dialogue processing. In *Proceedings of the Conference of the European Chapter of the Association for Computational Linguistics (EACL)*, pages 710–718, Athens, Greece. Association for Computational Linguistics.

Schlangen, D. and Skantze, G. (2011). A general, abstract model of incremental dialogue processing. *Dialogue and Discourse*, **2**(1):83–111.

Shieber, S. M. (1988). A uniform architecture for parsing and generation. In *Proceedings of the International Conference on Computational Linguistics (COLING)*, pages 614–619, Budapest, Czech Republic. International Committee on Computational Linguistics.

Skantze, G. and Hjalmarsson, A. (2010). Towards incremental speech generation in dialogue systems. In *Proceedings of the SIGdial Conference on Discourse and Dialogue (SIGDIAL)*, pages 1–8, Tokyo, Japan. Association for Computational Linguistics.

Skantze, G. and Schlangen, D. (2009). Incremental dialogue processing in a micro-domain. In *Proceedings of the Conference of the European Chapter of the Association for Computational Linguistics (EACL)*, pages 745–753, Athens, Greece. Association for Computational Linguistics.

Skuplik, K. (1999). Satzkooperationen. definition und empirische untersuchung. Technical Report 1999/03, Bielefeld University.

Sturt, P. and Crocker, M. (1996). Monotonic syntactic processing: A cross-linguistic study of attachment and reanalysis. *Language and Cognitive Processes*, **11**(5):449–494.

Szczepek Reed, B. (2000). Formal aspects of collaborative productions in English conversation. *Interaction and Linguistic Structures*, **17**.

Thompson, H. (1977). Strategy and tactics: A model for language production. In *Papers from the Regional Meeting of the Chicago Linguistic Society*, volume 13, pages 651–668, Chicago, IL. Chicago Linguistic Society.

van Wijk, C. and Kempen, G. (1987). A dual system for producing self-repairs in spontaneous speech: Evidence from experimentally elicited corrections. *Cognitive Psychology*, **19**(4): 403–440.

Part II

Reference

5 Referability

Kees van Deemter

5.1 Introduction

A key task of almost any natural language generation (NLG) system is to **refer** to entities. Linguists and philosophers have a long tradition of theorizing about reference. In the words of the philosopher John Searle,

Any expression which serves to identify any thing, process, event, action, or any other kind of individual or particular I shall call a referring expression. Referring expressions point to particular things; they answer the questions Who?, What?, Which? (Searle, 1969)

Referring expression generation (REG, sometimes GRE) is the task of producing a (logical or natural language) description of a **referent** that allows the reader to identify it. In producing a **referring expression**, an NLG system can make use of any information that it can safely assume the hearer to possess, based on a model of the world and of the hearer's knowledge about the world (the **knowledge base** (KB)). Given a REG algorithm and a KB, the following questions can be asked:

1. How many entities is the algorithm able to identify? We will call this the **expressive power** of an algorithm. Loosely speaking, the more entities the algorithm is able to single out, the greater its expressive power.
2. How empirically adequate are the referring expressions generated by the algorithm? For example, how **human-like** are they – to what extent do they resemble the human-produced referring expressions in a corpus? How **effective** are they – to what extent do they enable a human recipient to identify the referent easily, quickly, and reliably?

Most work on REG is devoted to algorithms that produce human-like referring expressions (e.g., Passonneau, 1996; Jordan and Walker, 2000; van Deemter *et al.*, 2012; Belz *et al.*, 2010; Viethen and Dale, 2006, 2008; Guhe and Bard, 2008; Goudbeek and Krahmer, 2010), with a far smaller number of studies addressing questions of effectiveness (Paraboni *et al.*, 2007; Gatt and Belz, 2010; Campana *et al.*, 2011; Khan *et al.*, 2012). In this chapter we argue that, although this empirical work is crucial, the question of expressive power deserves the attention of the research community as well, because it can open our eyes to fundamental limitations of REG algorithms, and enable us to extend their expressive power. In examining the expressive power of REG algorithms, we will hit upon another, trickier question as well, one of **logical completeness**:

3. Given a KB, which entities can be referred to (i.e., identified)?

We shall ask how current REG algorithms shape up against the yardstick suggested by this question, and explore novel ways of extending their expressive power. Each extension in expressive power gives rise to new empirical questions because, for every novel type of referring expression, one can ask how and when human speakers use it, and how effective it is for a human recipient. (These empirical considerations, however, will not be the theme of this chapter, except for a brief discussion in Section 5.6.)

We shall restrict ourselves in this chapter to the generation of **first mention** descriptions only (cf. Krahmer and Theune, 2002; Siddharthan and Copestake, 2004; Belz *et al.*, 2010). This is a weighty simplification, particularly where *interactive* NLG systems are concerned (i.e., where the user of the system can respond). Yet, the issues discussed here are as relevant for interactive settings as for non-interactive ones, because the same types of domain knowledge are relevant in both situations, and because any referring expression that can be used in a non-interactive setting can also be used interactively. Everything we say about proper names, sets, and quantifiers, for example, will be relevant for referring expressions generated in interactive contexts.

We follow Dale and Reiter (1995) in focusing on the content of a description (i.e., the problem of **content determination**), assuming that any combination of properties can be expressed by the NLG module responsible for surface realization. Accordingly, when we speak of a referring expression, we will often mean a formal representation, rather than an actual linguistic form. We shall not assume that a description is always expressed by a single noun phrase: if several sentences are needed, then so be it.

In this chapter, we start by summarizing existing work that has sought to extend the expressive power of REG, then present novel work regarding proper names and, particularly, regarding question 3 above. In Section 5.2 we sketch a REG algorithm which, in terms of its expressive power, is representative of the algorithms submitted to recent REG shared tasks (Gatt and Belz, 2010). Strikingly, these algorithms can only generate referring expressions that express conjunctions of atomic one-place predicates (i.e., that refer to individual objects through simple attributes that are true of the referent). However, we shall summarize how these algorithms can be modified to express other boolean connectives. In Section 5.4, we discuss earlier work on relational descriptions and argue that, to do full justice to relational descriptions and implicit knowledge, it is necessary to link REG with serious knowledge representation. This will lead on to a discussion of the curious situation that has now arisen in computational linguistics, where most NLG researchers – who routinely generate text from a KB – choose to ignore modern techniques for representing and manipulating knowledge. In Section 5.3, we make a small detour, addressing a different kind of limitation of REG algorithms by showing, for the first time, how proper names can be incorporated into REG algorithms. In Section 5.5 we address what needs to be done to make REG algorithms logically complete, thereby addressing question 3 (above). In the concluding section we look at how much is actually gained by extending REG algorithms as we propose, arguing that there are substantial benefits to be had here, both practically (in terms of generating intelligible referring expressions) and theoretically (in terms of obtaining insight into the human language competence).

Before we proceed, we introduce some terminology. For our purposes in this chapter, the REG task will be to construct a **description** or **referring expression** (represented as a form in first-order logic) that uniquely identifies a **target referent** (an entity or set of entities) in a **domain** (the set of all entities available for reference). The REG algorithm may use **attributes** of the entities that may consist of **atomic properties** (semantic primitives that are true or false for each entity in the domain) and/or **relational properties** (semantic functions that map sets of entities in the domain to truth values). The entities in the domain other than the target referent are called **distractors**. A description that uniquely identifies a target referent (so that the referent's satellite set contains only the referent itself) is a **distinguishing description**; a referent that has at least one distinguishing description is **distinguishable**. We will measure the **expressive power** of a REG algorithm by looking at the types of referent for which we can construct distinguishing descriptions. Finally, we will say that a REG algorithm is **logically complete** if it permits us to construct distinguishing descriptions for all possible target referents in a domain.

An issue of notation is also worth getting out of the way. One can describe sets both in terms of set theory and in terms of logical symbols. We can talk about a set of objects as either the set-theoretic union $A \cup B$, or in logical jargon as the set of those objects that are members of A or B (or both), that is $\{x \mid x \in A \lor x \in B\}$. In this chapter it will be convenient for us to switch between the two perspectives, sometimes talking about (logical) negation and disjunction, sometimes about (set-theoretic) complementation and union.

5.2   An algorithm for generating boolean referring expressions

It is commonplace to start an article on REG by paraphrasing the Incremental Algorithm of Dale and Reiter (1995), a well-known algorithm aimed at generating human-like referring expressions efficiently.[1] In this chapter, however, our subject is the expressive power of REG algorithms. For this reason, we shall use a different algorithm, first described in van Deemter and Halldórsson (2001), which is based on principles that will be useful to us in the complex situations that we shall study later on.

Suppose we have a domain D containing a target referent r and a number of distractors. Suppose, furthermore, there is a set $\mathbb{P}$ of atomic properties that can be used to identify r. The basic idea of the algorithm is to simply list all properties that are true of the referent r, and intersect their **extensions**:[2]

$$\bigcap_{P \in \mathbb{P}} [[P]]$$

---

[1] For a discussion of these and other classic REG algorithms, see the survey by Krahmer and van Deemter (2012).

[2] The extension $[[P]]$ of a property P denotes the set of entities in the domain for which the property P holds true.

If the resulting intersection equals the singleton $\{r\}$ then the tentative description is a referring expression that identifies r.

To see how this works, consider a KB whose domain D is a set of entities (a, b, c, d, e), over which four properties have been defined:[3]

dog: $\{a, b, c, d, e\}$ (all entities in the domain are dogs)
poodle: $\{a, b\}$
black: $\{a, c\}$
white: $\{b, e\}$

Here is what happens when a and c, respectively, play the role of the target referent. The algorithm finds a distinguishing description for a, but not for c:

Description(a) $= dog \cap poodle \cap black = \{a\}$
Description(c) $= dog \cap black = \{a, c\}$

Is this algorithm logically complete? The answer depends on whether operators other than set intersection (i.e., other than logical conjunction) are taken into account. Classic REG algorithms have tended to disregard set complementation (i.e., negation; van Deemter, 2002). To see that this matters, consider that if it is permitted to describe an animal as *not a poodle* (negating the property of being a poodle), and similarly for all atomic properties, then all five objects in D become distinguishable. Dog c, for example, is identified by the combination $black, \overline{poodle}, \overline{white}$: it is the black dog that is not a poodle and not white.

Is negation important? If all non-poodles were known to be spaniels then the property of not being a poodle would be superfluous. More generally, if for each property P in the KB, the KB also contains a property coextensive with $\overline{P}$ (the complement of P) then negation does not add to the expressive power of the REG algorithm.[4] In other words, if our only target referents are single entities, one might argue that negation is not important for REG.

But what if we want to refer to a *set* of entities? Classic REG algorithms are capable of constructing only descriptions of individual target referents, but references to sets are ubiquitous in natural language. To address this question, it will be useful to look a bit differently at what we did when referring to individual entities such as a. Let us call the tentative description constructed for a target referent r (i.e., the intersection of the extensions of all the properties true of r) the **satellite set** of r. Now let us construct the satellite set for each element of the domain:

Satellites(a) $= dog \cap poodle \cap black = \{a\}$
Satellites(b) $= dog \cap poodle \cap white = \{b\}$
Satellites(c) $= dog \cap black = \{a, c\}$

---

[3] Our assumption – to be made precise later – is that the listing of extensions is complete. For instance, a and b are the only poodles (whatever else they may be), and d is neither black nor white (whatever else it may be).

[4] Descriptions of the form $\overline{X \cup Y}$ do not add expressive power. $\overline{X \cup Y}$ can be written equivalently as $\overline{X} \cap \overline{Y}$: the negation can be "pushed down" until it is only applied to atomic properties.

$Satellites(d) = dog = \{a, b, c, d, e\}$

$Satellites(e) = dog \cap white = \{b, e\}$

Satellite sets are useful constructs, not just because they can be used in algorithms, but because they show us which entities can be referred to at all. The reason is that a satellite set results from conjoining *all* the properties of a given entity. In doing so, we give ourselves the best possible chance of success: if a distractor is not ruled out by this conjunction then it cannot ever be ruled out. Thus, the satellite set of a target referent r is the set of entities from which r cannot be distinguished. If this set contains only r then r can be identified by conjoining the extensions of all r's properties, but if it contains one or more other entities, then it cannot. In our example KB, it is easy to see that only a and b can be distinguished from everything else, namely by conjoining the properties *dog*, *poodle*, and *white*. (The fact that the property *dog* is superfluous is immaterial for present purposes, because we do not aim for "natural" descriptions yet.) In other words, if your goal is to identify a then the satellite set construct gives you a way of doing that. If your goal is to identify c, and the four properties listed above are the only ones available to you, then satellite sets show you that you should give up, because no algorithm will be able to do it for you. If your goal is to identify the set $\{a, c\}$ then satellite sets show you that is possible: for example, the conjunction of *black* and *dog* will do it.

For referring to sets, conjunction and negation of atomic properties alone does not always suffice. This can also be seen from our example KB. it is impossible to refer to the set $\{a, b, d, e\}$ with just these two connectives; we need to add full negation, in which case we can say $black \cap \overline{poodle}$ (which describes the target set as the complement of the set containing just the entity c), or disjunction, using the description $poodle \cup \overline{black}$ (*the poodles and the ones that are not black*).

Let us be a bit more precise and specify an algorithm for referring to individual referents and sets alike, in a manner that allows for all possible boolean combinations. The algorithm uses satellite sets, based on all atomic properties and their negations.

For each $d \in S$ do

$S_d := \{A : A \in \mathbb{P}_{neg} : d \in [[A]]\}$

$Satellites(d) = \bigcap_{A \in S_d} ([[A]])$

First, the algorithm adds to $\mathbb{P}$ the properties whose extensions are the complements of those in $\mathbb{P}$. The resulting set, which contains negative as well as positive properties, is called $\mathbb{P}_{neg}$. For each element d of S, the algorithm finds those properties in $\mathbb{P}_{neg}$ that are true of d, and the intersection of (the extensions of) these properties is formed; this intersection is called $Satellites(d)$. A tentative description is constructed by forming the union of all the $Satellites(d)$ (for each d in S). A description is distinguishing if it evaluates to the target set S; otherwise the algorithm returns Fail. If Fail is returned, no distinguishing boolean description of S is possible.

Description by satellite sets (DBS):

```
Description := ⋃_{d∈S} (Satellites(d))
If Description = S then Return Description else Fail
```

DBS is computationally cheap: it has a worst-case running time of $O(n.p)$, where n is the number of objects in S, and p the number of atomic properties. Rather than searching among many possible unions of sets, a target $S = \{s_1, \ldots, s_n\}$ is described as the union of n satellite sets, the ith one of which equals the intersection of those (at most p) sets in $\mathbb{P}_{neg}$ that contain s_i. Descriptions can make use of the satellite sets computed for earlier descriptions, causing a further reduction of computation time. Satellite sets can even be calculated off-line, for all the elements in the domain, before the need for specific referring expressions arises.[5]

The descriptions produced by DBS are often lengthy. To identify the target set $S = \{c, d, e\}$, for example, the property $\overline{poodle}$ would have sufficed. The algorithm, however, uses three satellite sets:

$S_c = \{dog, black, \overline{poodle}, \overline{white}\}$. $Satellites(c) = \{c\}$
$S_d = \{dog, \overline{white}, \overline{poodle}, \overline{black}\}$. $Satellites(d) = \{d, e\}$
$S_e = \{dog, white, \overline{poodle}, \overline{black}\}$. $Satellites(e) = \{d, e\}$

Consequently, the boolean expression generated is

$(dog \cap black \cap \overline{poodle} \cap \overline{white}) \cup$
$(dog \cap \overline{white} \cap \overline{poodle} \cap \overline{black}) \cup$
$(dog \cap white \cap \overline{poodle} \cap \overline{black})$.

The algorithm's profligacy makes it easy, however, to prove the following completeness theorem:[6]

Theorem 1a. Full Boolean Completeness: For any set S, S is equivalent to a boolean combination of properties in $\mathbb{P}_{neg}$ if and only if $\bigcup_{d \in S}(Satellites(d))$ equals S.
Proof: The implication from right to left is obvious. For the reverse direction, suppose $S \neq \bigcup_{d \in S}(Satellites(d))$. Then for some $e \in S$, $Satellites(e)$ contains an element e' that is not in S. But $e' \in Satellites(e)$ implies that every property in $\mathbb{P}_{neg}$ that holds true of e must also hold true of e'. It follows that S, which contains e but not e', cannot be obtained by a combination of boolean operations on the sets in $\mathbb{P}_{neg}$.

Under reasonable assumptions, the completeness of the DBS algorithm follows directly:

Theorem 1b. Completeness of DBS. Assume there are at most finitely many properties. Then if an individual or a finite set can be individuated by any boolean combination of properties defined on the elements of the domain, then DBS will find such a combination.

The limitation to finite sets is unrelated to the idea of using $\bigcup_{d \in S} \bigcap_{A \in S_d} [[A]]$ but stems from the fact that DBS addresses the elements $d \in S$ one by one. Likewise, the limitation to finite sets of properties stems from an assumption that the set $\{\mathbb{P}_{neg} : d \in [[A]]\}$

[5] Compare Bateman (1999), where a KB is compiled into a format that brings out the commonalities between objects before the content of a referring expression is determined.
[6] The statement and proof of this theorem has been updated from van Deemter and Halldórsson (2001) because of a slight flaw in the original.

needs to be constructed in its entirety. Implementations of DBS that do not make these assumptions are conceivable.

Related algorithms

A logic-based mechanism for optimizing referring expressions, using techniques such as the Quine–McCluskey algorithm (McCluskey, 1986) was proposed in van Deemter (2002). The proposal of van Deemter and Halldórsson (2001) can use the same techniques. Additionally, REG can be streamlined by only adding information to a description that changes its extension. For example, in the computation of S_c above, adding $\overline{white}$ does not change the satellite set S_c.

Gardent (2002) drew attention to situations where earlier proposals produced very lengthy descriptions, and proposed a reformulation of REG as a constraint satisfaction problem. In a similar spirit, Horacek (2004) proposed an algorithm for generating descriptions of sets in disjunctive normal form (DNF; unions of intersections of literals), leading to more naturalistic descriptions; this idea has been taken forward in later work (Gatt, 2007; Khan *et al.*, 2012).

Collective plurals

The proposals discussed above use properties that apply to individual objects. To refer to a set, in this view, is to say things that are true of each member of the set. Such referring expressions may be contrasted with **collective** ones (e.g., *the lines that run parallel to each other*, *the group of 4 people*) whose semantics raises many issues (e.g., Scha and Stallard, 1988; Lønning, 1997). Initial ideas about the generation of collective plurals were proposed in Stone (2000). An algorithm for REG for collective plurals based on Dale and Reiter's incremental algorithm was proposed in van Deemter (2002). The algorithm proposed there can be easily modified to make use of the satellite set of a set (rather than that of an individual, as we presented above).

5.3   Adding proper names to REG

Another type of referring expression that cannot be generated by current REG algorithms is proper names. Some studies have investigated the choice between proper names and other types of referring expression (Henschel *et al.*, 2000; Piwek *et al.*, 2008). Siddharthan *et al.* (2011) studied first-mention referring expressions including proper names, but not in connection with REG as treated in this chapter, where generation starts from a KB rather than a text.

The inclusion of proper names as possible referring expressions for REG algorithms is complicated by semantic and pragmatic issues. First, proper names are sometimes, but not always, known. Suppose, in 1997, you asked me who was the most influential author of the year 1996, and I responded *that must surely be the author of the novel Primary Colors*; this would have been a good description of Joe Klein, the political commentator who had written the book *Primary Colors* anonymously. While this example is extreme, speakers routinely use proper names as part of descriptions, e.g., *the CEO of Honda, the*

Principal of Aberdeen University, particularly in situations where the role of the person is more important than his or her name.

Second, the choice between a proper name and a full description (or between different forms of a proper name) is often pragmatically important. References to people come with social implications to do with personal familiarity, affection, and social hierarchy. Consider a family comprised of a mother and father with a son and a daughter, where the son refers to the daughter. The son can choose between a proper name and a description like *my sister*; the choice would tend to indicate how familiar he believes the hearer to be with the referent. If he wanted to refer to her while addressing his mother, for example, only (a version of) the proper name would be considered normal. Relation-denoting words like *aunt* complicate matters further, by adding a descriptive element to a name. The expression *Aunt Agatha*, for example, would be quite normal to use when referring to someone who is the aunt of the speaker, while addressing another family member (except possibly where that family member is the referent's son or daughter); it would be dispreferred in most other situations. The pragmatic complexities are clearly considerable.

Our approach to treating proper names is simple: we shall assume that each individual in the KB potentially has an attribute `proper name`, with zero or more properties corresponding to its proper name(s). Some individuals may have a proper name, while others may not. A proper name is a property that is true of all individuals that have the name. For example,

– (being named) Joe Klein is a property of all those individuals named Joe Klein
– (being named) Joe is a property of all those individuals named Joe
– (being named) Klein is a property of all those individuals named Klein

To make this work for *parts* of names as well as full proper names, the easiest approach is to associate a *set* of names with each entity. Since longer versions of a person's name will be true of only some of the individuals who have a shorter version of that name, the different values of the NAME attribute will tend to subsume each other. Just like all dogs are canines, and all canines are mammals, so all people who are called Prince Andre Nikolayevich Bolkonsky are also called Prince Andre Bolkonsky, and all of these are called Andre Bolkonsky. Proper names will be assumed to be known to the hearer. We shall not discuss which types of individuals tend to have commonly known proper names (people, cities, and companies come to mind) and which do not (e.g., tables, trees, body parts, atomic particles). Likewise, we will say little about the choice between different versions of a proper name and their pragmatic implications, including the use of titles and honorifics (e.g, the properties of being male and being named Klein might be realized as *Mr. Klein*). We also do not discuss how to use a model of salience to disambiguate several possible referents for a proper name or to allow selection of a proper name as a referring expression even when it is not a distinguishing description; since a mechanism for handling salience can be added to any existing approach to REG, we will consider salience – like other gradable attributes – to be irrelevant to the question of what the expressive power of an REG algorithm is (but see Krahmer and Theune, 2002). In all these areas there is plenty of space for future research.

To sum up our proposal for the treatment of proper names in REG,

- Each individual is associated with an attribute NAMES.
- For a given individual, the set of values of NAMES can be empty (no name is known), singleton (one name), or neither (several names).
- Different individuals can share some or all of their names.
- REG algorithms will treat the NAME attribute in the same way as other attributes.

Even though the focus of this chapter is on questions of expressive power, rather than pragmatic felicity, it is worth mentioning that proper names are often the "canonical" way of referring to an entity. Standard mechanisms could be invoked to favor names at the expense of other properties, including the preference orders of Dale and Reiter (1995), or even (in view of the fact that names are often short) a preference for brevity. However, there can be good reasons for avoiding the use of proper names as well. If you know the name of the Director of Inland Revenue, this does not mean that *Please contact the Director of Inland Revenue* would be better worded as *Please contact Mr X* (where *X* is his name). When reference to sets is considered, the preference for proper names as canonical referring expressions becomes even more doubtful, because listing proper names does not necessarily make for a short description (compare the description *the citizens of China* with a listing of all the elements of this set).

We are now in a position to say more precisely how proper names fit into the story of the expressive power of an REG algorithm. Proper names can evidently add to the store of entities that a noun phrase can individuate: without proper names, one might occasionally struggle to construct a distinguishing description for an entity. In this regard, however, they are no different from other parts of speech: leave out an adjective or a verb from the language and it might become difficult to individuate an entity as well. Since we have shown that proper names can be represented as properties that are treated in the same way as other properties by REG algorithms, the device of using proper names does not add to the expressive power of REG algorithms.

5.4 Knowledge representation

In this section we discuss how recent research has demonstrated the benefits of coupling REG with modern knowledge representation (KR) techniques. We show that these techniques have allowed REG algorithms to produce referring expressions in a much larger range of situations than before; in the terminology of Section 5.1, they have increased the expressive power of REG algorithms.

5.4.1 Relational descriptions

Most REG algorithms are restricted to atomic properties, i.e., they cannot use relations involving two or more arguments. This does not mean that referring expressions like *the man who feeds a dog* cannot be generated by these algorithms: the KB can treat dog-feeding as an atomic property, instead of a relation between two entities. However, it does mean that the algorithms in question are unable to identify one entity via another, as when we say *the man who feeds a dog (who chased a cat, who …)*, and that all instantiations of a relation, e.g., *feeds a dog, feeds a black dog*, and so on, must be treated as logically unrelated properties.

In an early paper, Dale and Haddock (1991) made several important observations about the generation of relational descriptions from a KB with 1-place relations (like *suit*) and 2-place relations (like *to wear*). First, it is possible to uniquely identify an entity through its relations to other entities without uniquely identifying each of the entities separately. Consider a situation involving two cups and two tables, where one cup is on one of the tables. In this situation, neither *the cup* nor *the table* is a distinguishing description, but *the cup on the table* does identify one of the two cups. Secondly, descriptions of this kind can have any level of "depth": for example, one might say *the white cup on the red table in the kitchen*. (The REG algorithm would of course want to avoid descriptive loops, since these do not add information; it would be useless to describe a cup as *the cup to the left of the saucer to the right of the cup to the left of the saucer …*) Dale and Haddock's proposal has been refined and modified in later years, often in relation to the Incremental Algorithm (Horacek, 1996; Krahmer and Theune, 2002; Kelleher *et al.*, 2006; Viethen and Dale, 2008). These extensions were not aimed at increasing the expressive power of the algorithm, but at avoiding linguistically awkward (e.g., lengthy) descriptions.

However, it seems that if a REG algorithm can handle relational descriptions in a general way it *must* have more expressive power than one that cannot. It is nice to be able to generate *the cup on the table* and, analogously, *the table with a cup on it*. But why stop there, instead of pressing on to generate *the table with two cups on it* and *the table that only has a teapot on it*? In short, the algorithms previously proposed for relational descriptions were not *relationally complete*; they could not identify *all* referents that can be identified using relations.

5.4.2   Knowledge representation and REG

Before we consider relational descriptions further, we must discuss some more general limitations of existing REG algorithms.[7] All REG algorithms presented so far in this chapter start from simple, tailor-made KBs, which express only atomic properties. The fact that properties are often grouped into attributes (such as color, size, etc.) does not change the expressive power of the KB, or of the REG algorithms that operate on it. Some algorithms are able to make use of generic information, but this has never gone beyond representing the fact that one property (e.g., being a dog) subsumes another (e.g., being a poodle). To anyone versed in modern knowledge representation (KR), this is surprising. Here are some of the things that these KBs are unable to express:

(a) Kees is Dutch or Belgian.
(b) The relation *part of* is transitive.

---

[7] A more elaborate version of this discussion can be found in Ren *et al.* (2010).

(c) For all x, y, z, if x designed y and z is a part of y then x designed z.
(d) For all x, y, x is to the left of y if and only if y is to the right of x.

(a) is arguably what one can infer after seeing an email message from the author that was written in Dutch; it is a piece of information that cannot be represented if only atomic information is available (unless *being Dutch or Belgian* is treated as a primitive property). Items (b), (c), and (d) might be implicitly present in a KB of atomic facts, but there is substantial mileage to be had from representing rules like these explicitly, because it allows a knowledge engineer to represent information in a much more economical and insightful way. Consider (b), for example, in a situation where a complex machine is described. Suppose part p_1 is part of p_2 which is part of p_3 which is part of p_4. Axiom (b) allows us to leave it at that: four other facts (that p_1 is also part of p_3 and p_4, and that p_2 is part of p_4) do not need to be represented separately but can be deduced from the three facts that are represented. In large domains the savings become enormous. However, existing REG algorithms cannot reason over knowledge of this kind. Moreover, they lack the apparatus for making logical deductions. The latter is hardly surprising given the former: if all knowledge is atomic there is little room for deduction.

Dissatisfaction with existing REG representation formats was perhaps first expressed by Krahmer and colleagues, who proposed the use of labeled directed graphs (Krahmer *et al.*, 2003), a much-studied mathematical formalism that can be used for representing atomic information involving 1- and 2-place relations. While this graph-based framework has since gained in popularity, and avoids certain problems that earlier treatments of relational descriptions had to face,[8] the graph-based framework is not particularly suitable for deduction, or for the expression of logically complex information (van Deemter and Krahmer, 2007).

In the first proposal that aimed explicitly at letting REG use frameworks designed for the representation and manipulation of logically structured information, Croitoru and van Deemter (2007) analyzed REG as a projection problem in Conceptual Graphs. More recently, Areces *et al.* (2008) proposed a REG algorithm that uses Description Logic (DL), a formalism that, like Conceptual Graphs, is designed for representing and reasoning with knowledge. It is this idea (traceable at least to Gardent and Striegnitz, 2007) that we shall use as our starting point here. The idea is to generate a Description Logic concept such as $Dog \cap \exists love.Cat$ (the set of dogs intersected with the set of entities that love at least one cat), and to check how many individuals turn out to be instances of this concept. If the number is 1, the concept is a referring expression referring to this individual, and REG is successful. Figure 5.1 depicts a small KB involving relations between several women, dogs, and cats. Here the above-mentioned formula identifies d_1 as *the dog that loves a cat*, singling out d_1 from the five other entities in the domain.

---

[8] In particular, graphs automatically avoid descriptive loops.

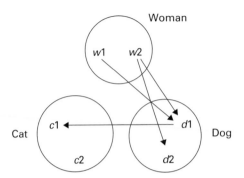

Figure 5.1   An example in which edges from women to dogs denote *feed* relations, while those from dogs to cats denote *love* relations

5.4.3   Description Logic for REG

A DL-based KB describes concepts, relations, and their instances. One of the most prominent DLs is $\mathcal{SROIQ}$ (Horrocks *et al.*, 2006). A $\mathcal{SROIQ}$ ontology consists of a TBox $\mathcal{T}$ and an ABox $\mathcal{A}$. $\mathcal{A}$ contains axioms about specific individuals, e.g.

$a : C$. This means that a is an instance of the concept C, i.e., a has the property C.

$(a, b) : R$. This axiom means that a stands in the R relation to b.

$\mathcal{T}$ contains generic information concerning **concepts** and **relations**. This generic information can either take the form $A \sqsubseteq B$, where A and B are concepts (and certain other conditions are fulfilled), or it can express one of a number of properties of a relation R, such as the fact that R is symmetric, irreflexive, or transitive; it can also say that two relations R and S are disjoint.

The notion of a **concept** is recursively defined. All atomic concepts are concepts. Furthermore, if C and D are concepts, and R is a binary relation (also called a **role**), then so are

$$\top \mid \bot \mid \neg C \mid C \sqcap D \mid C \sqcup D \mid \exists R.C \mid \forall R.C \mid \leq nR.C \mid \geq nR.C \mid \exists R.Self \mid$$
$$\{a_1, \ldots, a_n\},$$

where $\top$ is the top concept (denoting the entire domain), $\bot$ the bottom concept (denoting the empty set), n a non-negative integer, $\exists R.Self$ the self-restriction (the set of objects standing in the relation R to themselves), a_i are individual names, and R is a relation which can be atomic or the inverse of another relation (R^-). We use CN, RN, and IN to denote the set of atomic concept names, relation names, and individual names, respectively.

An *interpretation* $\mathcal{I}$ (also called a model) is a pair $\langle \Delta^{\mathcal{I}}, .^{\mathcal{I}} \rangle$ where $\Delta^{\mathcal{I}}$ is a non-empty set and $.^{\mathcal{I}}$ is a function that maps each atomic concept A to the set $A^{\mathcal{I}} \subseteq \Delta^{\mathcal{I}}$, each atomic role r to $r^{\mathcal{I}} \subseteq \Delta^{\mathcal{I}} \times \Delta^{\mathcal{I}}$, and each individual a to $a^I \in \Delta^{\mathcal{I}}$. The interpretation of complex concepts is defined inductively; for example, $(C \sqcap D)^{\mathcal{I}} = C^{\mathcal{I}} \cap D^{\mathcal{I}}$.

$\mathcal{I}$ is a *model* of the KB Σ, written $\mathcal{I} \models \Sigma$, *iff* all the axioms in Σ are satisfied in $\mathcal{I}$. Notably, a Σ can have multiple models. For example when $\mathcal{T} = \emptyset, \mathcal{A} = \{a : A \sqcup B\}$, there can be a model $\mathcal{I}_1$ such that $\Delta^{\mathcal{I}_1} = \{a\}, a^{\mathcal{I}_1} = a, A^{\mathcal{I}_1} = \{a\}, B^{\mathcal{I}_1} = \emptyset$, and another model $\mathcal{I}_2$ s.t. $\Delta^{\mathcal{I}_2} = \{a\}, a^{\mathcal{I}_2} = a, B^{\mathcal{I}_2} = \{a\}, A^{\mathcal{I}_2} = \emptyset$. For details, see Horrocks *et al.* (2006).

DL is often employed in "open world" situations, where there can exist individuals not mentioned in the KB. However, REG algorithms often assume a closed world. In DL, there are different ways of deviating from the open world perspective. One solution is to close an ontology partly or wholly with a DBox, $\mathcal{D}$ (Seylan *et al.*, 2009). A DBox is syntactically similar to the ABox, except that $\mathcal{D}$ contains only atomic formulas. Every concept or relation appearing in $\mathcal{D}$ is closed. Their extension is exactly defined by the contents of $\mathcal{D}$, i.e., if $D \not\models a : A$ then $a : \neg A$. Concepts and relations not appearing in $\mathcal{D}$ remain open.

However, closing ontologies by means of the DBox restricts the use of implicit knowledge (from $\mathcal{T}$). The interpretations of the concepts and relations appearing in $\mathcal{D}$ are fixed, so no implicit knowledge concerning them can be inferred. We introduce the notion of an NBox (with N short for Negation as Failure). The effect of an NBox on a KB is as follows:

Suppose $\emptyset$ is a KB $(\mathcal{T}, \mathcal{A}, \mathcal{N})$, where $\mathcal{T}$ is a TBox, $\mathcal{A}$ an ABox, and $\mathcal{N}$ an NBox that is a subset of $CN \cup RN$. Then, firstly, $\forall x \in IN, \forall A \in \mathcal{N} \cap CN$, if $\emptyset \not\models a : A$ then $\emptyset \models a : \neg A$. Secondly, $\forall x, y \in IN, \forall r \in \mathcal{N} \cap RN$, if $\emptyset \not\models (x, y) : r$, then $O \models (x, y) : \neg r$.

Like the DBox approach, the NBox $\mathcal{N}$ defines conditions in which "unknown" should be treated as "failure." But, instead of hard-coding such conditions, it specifies a vocabulary to which this treatment should be applied. Unlike the DBox approach, inferences on this Negation as Failure (NAF) vocabulary are still possible.

Applying DL to familiar REG problems.

DL forces one to be explicit about a number of assumptions that are usually left implicit in REG. In discussing these, it will be useful to consider our old example KB again:

TYPE: dog $\{a, b, c, d, e\}$, poodle $\{a, b\}$
COLOR: black $\{a, c\}$, white $\{b, e\}$

The Closed World Assumption (CWA): REG tacitly assumes a closed world. For example, if there are poodles other than those listed in the KB, then these may include c, and if this is the case then *black poodle* no longer identifies a uniquely. Our REG proposal does not depend on a specific definition of CWA. In what follows, we use the NBox to illustrate our idea. The domain will be considered to be finite and consisting of only individuals appearing in $\mathcal{A}$.[9]

---

[9] These assumptions cause no problems for the treatment of proper names of Section 4. If the proper name *Shona Doe* is in the NBox, for example, then only those individuals that can be deduced to have this name have it. A proper name that is not listed may or may not be a name of an individual in the KB, just like any other property that is not listed.

The Unique Name Assumption (UNA): REG tacitly assumes that different names denote different individuals. This matters when there are entities that share all their properties. For example, if a and b were the same individual (and CWA applied) then the property *poodle* alone would individuate this individual, and reference would become trivial, which is not the case if $a \neq b$.[10]

In our discussion here we will adopt both the CWA and the UNA. We assume that the KB includes $\mathcal{A}$, $\mathcal{T}$, and $\mathcal{N}$. In the usual syntax of $\mathcal{SROIQ}$, negation of relations is not allowed in concept expressions. So, strictly speaking, one cannot compose a concept $\exists \neg feed . Dog$. However, if $feed \in \mathcal{N}$, then we can interpret $(\neg feed)^{\mathcal{I}} = \Delta^{\mathcal{I}} \times \Delta^{\mathcal{I}} \setminus feed^{\mathcal{I}}$. For this reason, we will here use negated relations as part of concept expressions, while remaining aware that their handling is computationally less straightforward than that of other constructs.

Given a KB, every DL concept denotes a set. If this set is not empty (i.e., it is either a singleton set or a set with at least two elements) then the concept can be seen as referring to this set (Areces *et al.*, 2008). For example, the domain in Figure 5.1 can be formalized as follows:

$$\mathcal{T}_1 = \emptyset$$
$$\mathcal{A}_1 = \{w1 : Woman, \ w2 : Woman, \ d1 : Dog, \ d2 : Dog, \ c1 : Cat,$$
$$c2 : Cat, \ (w1, d1) : feed, \ (w2, d1) : feed, \ (w2, d2) : feed, \ (d1, c1) : love\}$$

The algorithm proposed by Areces *et al.* (2008) computes all the **similarity sets** associated with a given KB. Similarity sets generalize our notion of satellite sets by taking 2-place relations into account, and by being explicitly parameterized to a given logic.[11] For consistency with our earlier discussion, we will refer to the similarity sets of Areces *et al.* (2008) as satellite sets. Thus, if L is a DL and M a model then

j is an *L-satellite* of i relative to M iff,
for every concept φ in L such that, given M, $i \epsilon |\varphi|$ implies $j \epsilon |\varphi|$

The *L-satellite set* of an individual i relative to M is defined as
$\{j : j$ is an L-satellite of i relative to $M\}$.

Theorem 2: L contains a referring expression for i relative to the domain M iff the L-satellite set of i relative to M equals $\{i\}$.
Proof: Trivial.

The algorithm of Areces *et al.* (2008) first identifies which (sets of) objects are describable through increasingly large conjunctions of (possibly negated) atomic concepts, then

---

[10]   UNA concerns the names (i.e., individual constants) of DL, not the proper names of any natural language (see Section 5.3).

[11]   The parameterization to a logic is important, because which referring expressions are expressible, and which sets of objects are referable, depends on the logic that provides the referring expressions. We return to this issue in Section 5.5.

tries to extend these conjunctions with complex concepts of the form $(\neg)\exists R1.Concept$, then with concepts of the form $(\neg)\exists R2.(Concept \sqcap (\neg)\exists R1.Concept)$, and so on. At each stage, only those concepts that have been found through earlier stages are used. For example, a partial listing of the concepts generated at each step of this algorithm for the KB above is:

1. $Dog = \{d1, d2\}$, $Woman = \{w1, w2\}$, $Cat = \{c1, c2\}$.
2. $Dog \sqcap \exists love.Cat = \{d1\}$, $Dog \sqcap \neg\exists love.Cat = \{d2\}$.
3. $Woman \sqcap \exists feed.(Dog \sqcap \neg\exists love.Cat) = \{w2\}$,
   $Woman \sqcap \neg\exists feed.(Dog \sqcap \neg\exists love.Cat) = \{w1\}$.

We will take this algorithm as the starting point for our discussion of the proposal of Ren *et al.* (2010) in the next section.

Applying DL to new REG problems. Since Areces *et al.* consider only the ABox, the KB always has a single fixed model. Consequently, their algorithm essentially uses model-checking, rather than full reasoning. However, when implicit information is involved, reasoning has to be taken into account. To illustrate this, we extend the KB in Figure 5.1 with background knowledge saying that *one should always feed any animal loved by an animal whom one is feeding*, while also adding an edge to the *love* relation (Figure 5.2) between $d2$ and $c2$.

We close the domain, using an NBox, as follows:

$T_2 = \{feed \circ love \sqsubseteq feed\}$
$A_2 = A_1 \cup \{(d2, c2) : love\}$
$N_2 = \{Dog, Woman, feed, love\}$

The TBox axiom enables the inference of implicit information over this KB: $(w1, c2)$: *feed*, $(w2, c1)$: *feed*, and $(w2, c2)$: *feed* can be inferred using DL reasoning under the above NBox N_2. However, the REG algorithm we currently have fails to identify *any* individual in the KB of Figure 5.2, because none of the relevant satellite sets are singletons. For example, the satellite set of $w1$ (and $w2$) is $\{w1, w2\}$, a set with two elements. In this section, we extend our REG algorithm to allow quantifiers other than

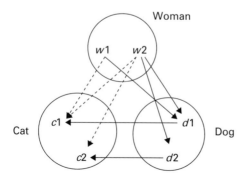

Figure 5.2 An extension of the KB in Figure 5.1. Dashed edges denote implicit relations, inferred using the TBox

the existential quantifier, and to allow inverse relations (Ren *et al.*, 2010). We show that with these extensions all the individuals in the KB of Figure 5.2 become referable.

But first, we need to ask what the appropriate level of expressivity is for REG tasks. Here we can benefit from the theory of Generalized Quantifiers (GQ), where quantifiers other than *all* and *some* are studied (e.g., Mostowski, 1957; van Benthem, 1986; Peters and Westerståhl, 2006). GQs are these researchers' answer to the question of what quantifiers are possible (in logic, and in natural language). GQs can occur in many different contexts, for example in *All (some, ten, most, etc.) cats slept*. The most general format for referring expressions that involve a relation R is The N1 who R Q N2's, where N1 and N2 denote sets, R denotes a relation, and Q a generalized quantifier, as in *the women who feed* some *dogs*. A referring expression of this form refers to an individual entity if it denotes a singleton set, i.e., if the following set has a cardinality of 1:

$$\{y \in N1 : Qx \in N2(Ryx)\}$$

It will be convenient to write this in the standard GQ format where quantifiers are cast as relations between sets of domain objects A and B. Using the universal quantifier as an example, instead of writing $\forall x \in A(x \in B)$, we write $\forall(AB)$. Thus, the formula above is written:

$$\{y \in N1 : Q(N2 \{z : Ryz\})\}$$

Instantiating this with $Q = \exists$, $A = Dog$, and $B = \{z : Feed \ yz\}$ for some y, we get $\{y \in Woman : \exists(Dog \{z : Feed \ yz\})\}$, or *women who feed a dog*. If Q is the quantifier *at least two*, we get *women who feed at least two dogs*.

To show which quantifiers are expressible in the logic that we are using, let us think of quantifiers in terms of simple quantitative constraints on the sizes of the sets $A \cap B$, $A - B$, and $B - A$, asking what types of constraints can be expressed in referring expressions based on $\mathcal{SROIQ}$. The findings are summarized in Table 5.1, varying on the example of people feeding dogs. OWL2, a widely used ontology language for the semantic web that is based on $\mathcal{SROIQ}$, can express any of the following types of descriptions.

For example, when $n = 1$, constraint type 1 becomes $\exists Feed.Dog$, i.e., the *existential* quantifier. When $n = 0$, constraint type 7 becomes $\forall Feed.Dog$, i.e., the quantifier *only*, and constraint type 6 becomes $\forall\neg Feed.\neg Dog$, i.e., the quantifier *all*. (In constraint types 2, 4, 6, and 8, negation of a relation is used. This is not directly supported in $\mathcal{SROIQ}$ but, as we indicated earlier, given $Feed \in \mathcal{N}$, $\neg Feed$ can be used in concepts.)

Together, these constructs allow the expression of a description such as *women who feed at least 1 but at most 7 dogs*, by conjoining constraint type 1 (with $n = 1$) with constraint type 5 (with $n = 7$). Using negation, we can say *women who do not feed all dogs and who feed at least one non-dog* ($Woman \sqcap \neg\forall\neg Feed.\neg Dog \sqcap \exists Feed.\neg Dog$). In addition to Table 5.1, $\mathcal{SROIQ}$ can even represent relations to self, such as *the dog who loves itself* ($Dog \sqcap \exists Love.Self$). Using conjunctions, one can express exact numbers, as in $\geq 1Feed.Cat \sqcap \leq 1Feed.Cat$ (*Men who feed exactly one cat*), or intervals, as in $Man \sqcap \geq 10Sell.Car \sqcap \leq 20Sell.Car$ (*Men who sell between 10 and 20 cars*), and so on. Disjunctions extend expressivity even further.

Table 5.1. Expressing quantified referring expressions in DL

	QAB	DL
1	$\geq n\,Dog\,\{z|Feed(y,z)\}$	$y:\geq nFeed.Dog$
2	$\geq n\,Dog\,\neg\{z|Feed(y,z)\}$	$y:\geq n\neg Feed.Dog$
3	$\geq n\neg\,Dog\,\{z|Feed(y,z)\}$	$y:\geq nFeed.\neg Dog$
4	$\geq n\neg\,Dog\,\neg\{z|Feed(y,z)\}$	$y:\geq n\neg Feed.\neg Dog$
5	$\leq n\,Dog\,\{z|F(y,z)\}$	$y:\leq nFeed.Dog$
6	$\leq n\,Dog\,\neg\{z|Feed(y,z)\}$	$y:\leq n\neg Feed.Dog$
7	$\leq n\neg\,Dog\,\{z|Feed(y,z)\}$	$y:\leq nFeed.\neg Dog$
8	$\leq n\neg\,Dog\,\neg\{z|Feed(y,z)\}$	$y:\leq n\neg Feed.\neg Dog$

Comparing the quantifiers that become expressible through OWL2 with classes of quantifiers studied in the theory of GQ, it is clear that OWL2 is highly expressive. It is true that some quantifiers routinely expressed in natural language referring expressions are not expressible in OWL2; examples include *most* and *many*, and *infinitely many* (van Deemter, 1984). It appears, however, that few of these matter in terms of the power of referring expressions to identify their referent, at least in domains of finite size. To see why, consider this simple example: c_1 is the only cat that is loved by most women, but it is also the only cat that is not loved by two women. In this way, the problematic quantifier *most* is replaced by one that is easily expressible in OWL2, namely the quantifier *two*.

Generating $\mathcal{SROIQ}$-enabled referring expressions. Now we show how the descriptions of the previous section can be generated in REG. We will present the algorithm of Ren *et al.* (2010), which extends that of Areces *et al.*, (2008) by using DL reasoning to test whether a description denotes the referent, allowing it to make use of TBox information. Moreover, it uses a large variety of quantifiers, instead of only the existential quantifier. Because it applies a generate-and-test strategy that composes increasingly complicated descriptions from the vocabulary, this algorithm gives preference to syntactically simple descriptions over complex ones.

Different descriptions can have the same denotation, e.g. $\neg\forall R.A \equiv \exists R.\neg A$. We are uninterested in the question which of these should be preferred, so we generate descriptions in their unique *negation normal form* (NNF). An NNF has $\neg$ in front of only atomic concepts (including $\top$ and $\bot$), or self restrictions. The NNF of $\neg C$ ($\neg R$) is denoted $\sim C$ ($\sim R$).

The algorithm takes a KB $\pm$ as its input and outputs a queue D of descriptions. It uses the following sets:

1. The set RN of relation names. For any relation name R, RN contains R, $\tilde{}R$, R^-, and $\tilde{}R^-$. RN contains nothing else.
2. The set CN of concept names. CN contains $\top$ and all expressions of the form $\exists R.Self$ where $R \in RN$. For all of these, and for any concept name A, CN also contains the NNF ($\tilde{}A$). CN contains nothing else.

3. The natural numbers set N containing $1, 2, \ldots, n$ where n is the number of individuals in $\pm$.
4. The construct set S containing all the connectives supported by $\mathcal{SROIQ}$, namely $\{\neg, \sqcap, \sqcup, \exists, \forall, \leq, \geq, =\}$.

We are now ready to sketch the algorithm, which is called GROWL.

$Construct - Description(\pm, CN, RN, N, S)$
INPUT: $\pm, CN, RN, N, S$
OUTPUT: Description Queue D

1: $D := \emptyset$
2: **for** each $e \in CN$ **do**
3:      $D := Add(D, e)$
4: **end for**
5: **for** each $d = fetch(D)$ **do**
6:      **for** each $s \in S$ **do**
7:          **if** $s = \sqcap$ or $s = \sqcup$ **then**
8:              **for** each $d' \in D$ **do**
9:                  $D := Add(D, d\ s\ d')$
10:             **end for**
11:         **end if**
12:         **if** $s = \exists$ or $s = \forall$ **then**
13:             **for** each $r \in RN$ **do**
14:                 $D := Add(D, s\ r.d)$
15:             **end for**
16:         **end if**
17:         **if** $s = \leq$ or $s = \geq$ or $s\ is\ =$ **then**
18:             **for** each $r \in RN$, each $k \in N$ **do**
19:                 $D := Add(D, s\ k\ r.d)$
20:             **end for**
21:         **end if**
22:     **end for**
23: **end for**
24: **return** D

In step 1, D is initialized to $\emptyset$. Steps 2 to 3 use the procedure Add to add suitable elements of CN to D. The details of Add depend on the requirements of the application; the version presented in Ren *et al.* (2010) uses a simple heuristic, whereby more complex descriptions are only added if they have smaller extensions than all elements of the existing D. Crucially, all these procedures take implicit knowledge (i.e., TBox axioms) into account.

From step 4 to step 14, elements of D are processed recursively, one by one. $fetch(D)$ retrieves the first unprocessed element of D. New elements are added to the end of D, making D a first-come-first-served queue. Processed elements are not removed from D. For each element d of D, Steps 6 to 14 generate a more complex description and add it at the end of D. Steps 7 and 8 conjoin or disjoin d with each element of D. Steps 10 and

11 extend d with all relations of RN. Steps 13 and 14 extend d with all relations in RN and all numbers in N. During the extension of D, we do not consider $s = \neg$ because, in NNF form, negation $\neg$ appears only in front of atomic concept names.

D, RN, N, S are all finite, hence steps 5 to 14 terminate for a particular $d \in D$. Steps 5 to 14 generate descriptions of increasing complexity.

GROWL generates certain types of referring expressions for the first time, for example ones with universal and numerical quantifiers. Although this may be of linguistic interest, it does not necessarily mean that we have achieved greater expressive power. It is easy to see, however, that GROWL also makes some entities **referable** for the first time. For example, if we apply GROWL to the KB in Figure 5.2, the following descriptions emerge:

1. $\{w1\} = Woman \sqcap \exists \neg feed . Cat$, the woman that does not feed all cats.
2. $\{w2\} = \leq 0 \neg feed . Cat$, the woman that feeds all cats.
3. $\{d1\} = Dog \sqcap \leq 0 \neg feed^- . Woman$, the dog that is fed by all women.
4. $\{d2\} = Dog \sqcap \exists \neg feed^- . Woman$, the dog that is not fed by all women.
5. $\{c1\} = Cat \sqcap \leq 0 \neg feed^- . Woman$, the cat that is fed by all women.
6. $\{c2\} = Cat \sqcap \exists \neg feed^- . Woman$, the cat that is not fed by all women.

We will discuss the importance of this added expressive power in Section 5.6. Like the other algorithms discussed in this chapter, GROWL focuses on creating referring expressions, leaving aside the question of which of several alternative referring expressions for an object is "best." We leave this empirical issue for future research.

The examples in this section show only the simplest kind of use that can be made of implicit information. Consider an example from Croitoru and van Deemter (2007). Suppose a KB in Description Logic expressed the TBox axiom that every table has exactly one cup on it:

$$Table \sqsubseteq (\leq 1 HasOn . Cup \sqcap \geq 1 HasOn . Cup)$$

If there is only one wooden table in the domain then this axiom justifies the referring expression *the cup on the wooden table*, even if the KB does not say which table this is. Examples of this kind show how radical a departure from standard REG we are proposing, because it enables the generation of references to individuals whose existence is inferred rather than explicitly stated. We shall return briefly to this issue at the end of this chapter.

5.5   Referability

In the introduction to this chapter, we listed two questions that research into REG has sought to answer, and we added a third one, which had not been tackled explicitly before:

3. Given a KB, which entities can be referred to (i.e., identified)?

We have seen that the early REG algorithms were unable to refer to certain entities in a KB. These limitations arose from an inability to represent relations, negation, disjunction, and most quantifiers. We extended a standard REG algorithm to handle all of these phenomena. So where are we now? For instance, is the GROWL algorithm logically complete?

The question of logical completeness cannot be addressed without reference to the types of information that may be contained in a KB and the types of information that may be expressed by referring expressions. A good way to think about this is in terms of the model within which the target referent "lives." It would be unfair to complain that GROWL is unable to exploit epistemic modalities, for example, since these are not in the KB. In the same way, we cannot expect a REG algorithm to use 3-place relations, as in *the person who gave a present to a child* (which involve a giver, a present, and a beneficiary), until such relations are in the KB. In what follows, we shall show that there does exist an absolute sense in which a REG algorithm can be logically complete, provided some assumptions are made concerning the information expressed in the KB.

In Section 5.4, where we were using Description Logic, the notion of an *interpretation* was defined as a pair $\langle \Delta^{\mathcal{I}}, \cdot^{\mathcal{I}} \rangle$ where $\Delta^{\mathcal{I}}$ is a non-empty set and $\cdot^{\mathcal{I}}$ is a function that maps atomic concepts A to $A^{\mathcal{I}} \subseteq \Delta^{\mathcal{I}}$, atomic roles r to $r^{\mathcal{I}} \subseteq \Delta^{\mathcal{I}} \times \Delta^{\mathcal{I}}$ and individuals a to $a^I \in \Delta^{\mathcal{I}}$. Implicit in this definition is the idea that a model contains individuals, which belong to concepts (i.e., properties) and are connected to each other by roles (i.e., 2-place relations). Let us assume our present models to be exactly like that. In this way, we are simplifying matters, because we are disregarding functions, and relations of higher arity (though we believe these particular simplifications not to be crucial, see below). The resulting models are graphically depictable as in Figure 5.1.[12] In connection with REG it is, moreover, reasonable to assume that models have only finitely many entities. Under these assumptions, what would it mean for a REG algorithm to be logically complete in an absolute sense, that is, regardless of the logical language employed?

Is it ever possible to prove that one element of $\Delta^{\mathcal{I}}$ cannot be distinguished from another one? More generally, is there such a thing as "the" satellite set of an entity, regardless of the logical language under consideration? We can simplify this question by making it symmetrical, asking under what conditions one can prove that two entities, x and y, cannot be distinguished *from each other*. It is obvious that some models contain elements that are indistinguishable in an absolute sense. Consider a model that contains two men and two women, happily arranged as follows:

$m_1 : Man,\ w_1 : Woman,\ (m_1, w_1) : Love$
$m_2 : Man,\ w_2 : Woman,\ (m_2, w_2) : Love$

There is nothing to separate the two men, or the two women for that matter; the model can be seen as consisting of two "halves" that are completely isomorphic to each other (i.e., there exists a 1–1 mapping between the two halves that respects all 1- and 2-place

---

[12] Equivalently, one could use the graphs of Krahmer *et al.* (2003), where properties are represented as "looping" arcs, which connect a node with itself.

relations). The same is true in a variant of the model that has only one woman, loved by both men:

m_1 : *Man*, w_1 : *Woman*, (m_1, w_1) : *Love*
m_2 : *Man*, w_1 : *Woman*, (m_2, w_1) : *Love*

This example shows overlapping halves (because w_1 belongs to each half), which are nonetheless isomorphic. What these observations suggest is (put informally) that two objects can only be told apart if (unlike, for example, m_1 and m_2 in the two examples above) they take up different parts in the "topology" of the model. This idea will underlie the following account.

Given an entity d in M, we will define the model *generated* by d (abbreviated as $M(d)$) as the directed part of M that is reachable from d. More precisely, $M(d)$ is the result of restricting the model M (with its set of individuals Δ) to the set $Reach(\Delta, d)$ ("the subset of Δ reachable from d") consisting of all those objects in Δ to which one can "travel" from d following the arcs regardless of their direction, also including the starting point d itself:

1. $d \in Reach(\Delta, d)$.
2. For all objects x and y and for every role r,
   if $x \in Reach(\Delta, d)$ and $(x, y) \in r^I$ then $y \in Reach(\Delta, d)$.
3. For all objects x and y and for every role r,
   if $x \in Reach(\Delta, d)$ and $(y, x) \in r^I$ then $y \in Reach(\Delta, d)$.
4. Nothing else is in $Reach(\Delta, d)$.

Using this perspective, we shall prove that an object is referable if and only if it can be referred to by means of a formula of *First-Order Predicate Logic With Identity* (FOPL). The key is the following theorem, which states that two elements, a and b, are indistinguishable using FOPL if and only if the models $M(a)$ and $M(b)$ generated by a and b are isomorphic, with a and b taking up analogous positions in their respective submodels. We write $\varphi(x)^M$ to denote the set of objects in M (i.e., in Δ) that satisfy the FOPL formula $\varphi(x)$.

The proof will make use of the idea that, given finite model M, a FOPL formula describing M completely up to isomorphism can be "read off" M.

Reading off a description of r from a submodel model M' of a model M:

1. Logically conjoin all atomic p such that $M' \models p$.
2. Close off all properties and relations by adding to the conjunction, for each constant a and property C in M such that $M' \not\models C(a)$, the new proposition that $\neg C(a)$. Similarly, for all constants a and b in M and for each relation R in M, such that $M' \not\models aRb$, add the new proposition $\neg aRb$.
3. Add inequalities $a \neq b$ for all pairs of constants a, b occurring in the conjunction.
4. In the resulting conjunction, replace all constants, except r, by (different) variables.
5. Quantify existentially over all the variables in the conjunction by adding $\exists$ at the start of the formula.

Step 2 takes into account all constants and relations that occur in M, regardless of whether they occur in M'. To see that this is necessary suppose M is as follows, and our referent is m_1:

$m_1 : Man, w_1 : Woman, (m_1, w_1) : Love$

$m_2 : Man, w_2 : Woman, w_3 : Woman$

$(m_2, w_2) : Love, (m_2, w_3) : Love$

Then m_1 cannot be identified without noting that he does *not* love w_2 and w_3, individuals living outside $M(m_1)$. Step 2 accomplishes his by adding $\neg Love(m_1, c_2)$ and $\neg Love(m_1, c_3)$. After step 5, this comes out as the information that the referent m_1 (unlike the distractors m_2, w_1, w_2, and w_3) loves exactly one woman. The complexity of step 2 is linear in the size of Δ; the complexity of the entire "Reading off" process is quadratic in the size of Δ (because of step 3).

Theorem 3. Referability Theorem: Consider a finite model M and arbitrary elements a and b of this model. Now a and b are FOPL-indistinguishable $\Leftrightarrow$ there exists a bijection f from $M(a)$ to $M(b)$, such that f respects all properties and 2-place relations and such that $f(a) = b$.

Proof: We prove the two directions of the theorem separately.

$\Leftarrow$ Let f be a bijection from $M(a)$ to $M(b)$ as specified. Then one can prove using induction following the structure of formulas that $a \in \varphi(x)^M$ iff $b \in \varphi(x)^M$, for any FOPL formula $\varphi(x)$ (i.e., a and b are FOPL-indistinguishable). Note that this equivalence holds across all of M (i.e., we do not just prove that $a \in \varphi(x)^{M(a)}$ iff $b \in \varphi(x)^{M(b)}$). The key factor is that objects in M that lie outside $M(a)$ are irrelevant to the description of a, and analogously for $M(b)$.

$\Rightarrow$ Suppose there does not exist a function f as specified. Then there exists a FOPL formula $\varphi(x)$ such that $\varphi(a)$ is true and $\varphi(b)$ is false (hence a and b are not FOPL indistinguishable) and conversely. A suitable $\varphi(x)$ can be *read off $M(a)$*.

Equality ($=$) is crucial here because, without it, one cannot distinguish between, on the one hand, an object that stands in a given relation to one object and, on the other hand, an object that stands in that relation to two different objects. To be referable, an object has to be distinguishable from all other objects in the domain. It therefore follows from the Referability Theorem that an element a of a graph M is referable using FOPL *unless* there exists some other element b of M such that there exists a bijection f from $M(a)$ to $M(b)$ with $f(a) = b$. Thus, FOPL can characterize precisely those domain elements that take up a unique place in the model. In other words, given the assumptions that we made, then if FOPL cannot refer to a domain element, this element cannot be referred to at all. In other words, FOPL is a good yardstick to measure REG algorithms against.

The finiteness assumption is crucial too, because for infinite models the second half ($\Rightarrow$) of the Referability Theorem does not always hold. To see this, consider a large model M where there are (for simplicity) no properties, where R is the only 2-place relation, and where a and b take up different but similar positions in the graph of the model, with $\forall x(aRx \leftrightarrow x \in \{a_i : i \in \mathbb{R}\})$ and $\forall x(bRx \leftrightarrow x \in \{b_i : i \in \mathbb{N}\})$, causing b to be related to *countably* many objects, while *a* is related to *uncountably* many objects (such as $a_1, a_2, \ldots, a_\pi, \ldots$). Now the generated models $M(a)$ and $M(b)$ are clearly not isomorphic, yet no FOPL formula φ exists such that $\varphi(a) \wedge \neg\varphi(b)$ (or the other way round). When only finite models are taken into account, the theorem does hold; moreover, the proof of its second half ($\Rightarrow$) is *constructive*. This suggests the possibility of new REG algorithms, which is an issue we shall turn to soon.

If proper names are treated as properties (rather than individual constants) in the fashion of Section 5.3, then proper names do not change this story. Consider a model just like the first model of this section, but with proper names added:

$m_1 : Man, w_1 : Woman, (m_1, w_1) : Love$
$m_2 : Man, w_2 : Woman, (m_2, w_2) : Love$
$w_1 : Shona, w_1 : ShonaBlogs$
$w_2 : Shona, w_2 : ShonaDoe.$

If w_1 and w_2 are only known by the name of Shona then, once again, the two women cannot be told apart, and neither can the two men. But if the two women are known by their different family names as well (as in the KB above) then all four inhabitants of the model can be distinguished from each other, and the submodel generated by m_1 is not isomorphic to the one generated by m_2.

Our statement of the Referability Theorem was limited to properties and 2-place relations, but it appears that some generalizations can be proven straightforwardly. If models can include relations of higher arity, for example, then the concept $Reach(\Delta, d)$ can be redefined to include all objects that can be reached through any kind of relation. For instance, to cover 3-place relations, if $x \in Reach(\Delta, d)$ and $(x, y, z) \in r^I$ then $y \in Reach(\Delta, d)$ and $z \in Reach(\Delta, d)$. The Referability Theorem itself should be reformulated analogously, after which it appears that its proof will hold in the same way as before.

In recent years, REG algorithms have made great strides forward, in terms of their empirical adequacy but also in terms of their expressive power. To show that even the GROWL algorithm is not complete, however, let us consider a type of situation mentioned by Gardent and Striegnitz (2007) in the context of a different DL-based algorithm. Consider a model M containing two men and three children. One man adores and criticizes the same child. The other adores one child and criticizes the other:

$m_1 : Man, c_1 : Child, (m_1, c_1) : Adore,$
$(m_1, c_1) : Criticize$
$m_2 : Man, c_2 : Child, c_3 : Child,$
$(m_2, c_2) : Adore, (m_2, c_3) : Criticize$

$M(m_1)$ is not isomorphic to $M(m_2)$, so m_1 and m_2 are distinguishable from each other in FOPL. A suitable formula φ true for m_1 but false for m_2 can be read off $M(m_1)$. Part of this (lengthy) formula says that $\exists x(Adore(m_1, x) \wedge Criticize(m_1, x))$. This information about m_1 would be false of m_2, hence FOPL with equality can separate the two men.

So what about GROWL? The DL-style formula *Man* $\sqcap$ $\exists$*adore* $\sqcap$ *criticize.Child* (*the man who adores and criticizes one and the same child*) comes to mind, but conjunctions of relations (such as *adore* $\sqcap$ *criticize*) are not supported by OWL2, so this is not an option for GROWL. OWL2 is expressive enough to produce (*Man* $\sqcap$ $\exists$*adore.Child*) $\sqcap$ *Man* $\sqcap$ $\exists$*criticize.Child* but, in the model at hand, this denotes $\{m_1, m_2\}$, without distinguishing between the two. We state here without proof that OWL2 (and hence also GROWL) is unable to distinguish m_1 and m_2 from each other.[13] This is not so much the fault of the GROWL algorithm, but of OWL2 (i.e., basically the logic $\mathcal{SROIQ}$), but from the point of view of the present section, which focuses on the question of which entities can be individuated in *any* logic, it is a limitation. In practical terms, it means that the search for more expressive yet still computationally efficient DLs must go on.

It is worth noting that the procedure of "reading off" a formula from a submodel implements the idea of a satellite set in a setting where full FOPL (with identity) is available. Because the formula read off the submodel generated by an entity m_1 spells out all that is known about m_1, this formula represents a logically exhaustive attempt at distinguishing this entity from one particular distractor (m_2) *and* from every other entity in the model from which it can be distinguished at all. If the formula happens to be true for any other entity, m, in the model, then m cannot be distinguished from m_1 at all, hence m is a member of the satellite set of m_1.

Scope for new algorithms.   Although the "reading off" process employed by the Referability Theorem was not intended to produce a REG algorithm, and although FOPL does not have the efficient reasoning support that exists for DL fragments such as $\mathcal{SROIQ}$, the fact that reading off can be done constructively suggests at least the *possibility* of new REG algorithms. One straightforward algorithm would proceed in three steps. To generate a referring expression of an entity d given domain M:

1. Compute the submodel $M(d)$ generated by d.
2. Read a FOPL formula φ off $M(d)$.[14] Finally,
3. If d is the only object for which φ is true then optimize φ and produce a referring expression, else no referring expression exists.

Step 2 will often produce unnecessarily lengthy descriptions, so optimization would be necessary. A promising direction for solutions is local optimization (Reiter, 1990), by removing conjuncts that are not necessary for producing a distinguishing description. To find descriptions that meet the empirical requirements associated with question 2 of the Introduction – which is not the aim of this chapter – additional optimization would likely have to be applied. As we shall see in the next section, it may be difficult to select

---

[13] No similar restriction holds regarding the disjunction of relations (i.e., roles). For even though role disjunction, as in *Man* $\sqcap$ $\exists$*Adore* $\sqcup$ *Criticize.Child* (*the men who adore or criticize a child*), is inexpressible, an equivalent concept can be expressed through disjunctions of concepts, as in (*Man* $\sqcap$ $\exists$*adore.Child*) $\sqcup$ (*Man* $\sqcap$ $\exists$*criticize.Child*).

[14] "Reading off" proceeds as specified earlier in this section. Note that φ can contain a wide range of formulas, including negated ones such as $\neg\exists xA(c, x)$. Hence, to test whether d is the only object for which φ is true (step 3 above), it does not suffice to proceed analogously to Krahmer *et al.* (2003), testing whether $M(d)$ stands in the relation of a subgraph isomorphism to any other parts of M.

the empirically most appropriate referring expression without considering its realization in actual words.

This section has made explicit, for the first time, what it means for a REG algorithm to be logically complete in an "absolute," FOPL-related, sense, and observed that against this yardstick, existing REG algorithms still fall short, because certain types of referring expressions are still beyond our grasp, causing some entities to be un-referable even though a FOPL-based formula can identify them uniquely. The outlines of a new FOPL-based algorithm have been sketched, which is logically complete in an absolute sense, but whose details are yet to be worked out.

5.6 Why study highly expressive REG algorithms?

We have seen how the expressive power of REG algorithms has grown substantially in recent years, allowing these algorithms to generate novel kinds of referring expressions and allowing them to identify referring expressions that no previous algorithm was able to identify. The mechanisms investigated include proper names, boolean connectives, and a range of generalized quantifiers. How useful are these additions? Could it be that our additions amount to a complex mechanism whose practical pay-offs are relatively modest? In what follows, we offer a number of reasons why this is not the case, and why the potential pay-offs from working on these issues are substantial.

5.6.1 Sometimes the referent could not be identified before

Sometimes, the new algorithms allow us to refer (i.e., individuate) where this was previously impossible. This is, of course, the argument that we have used time and again in this chapter: referents that were not referable in the simple algorithm we first described have become referable because of various later extensions. We will not elaborate on these arguments here.

5.6.2 Sometimes they generate simpler referring expressions

Some of the referring expressions that recent REG extensions can generate for the first time are simple, and occur frequently. This almost goes without saying for proper names, and likewise for quantifiers like *all* (*The woman who programmed all these functions*), *two* (*The man with two lovers*), and *only* (*The dog that only chases cats*). Referring expressions of these forms seem fairly common, and easy to understand.

It may well be that speakers and hearers have difficulties with some of the constructs that play a role in the new algorithms: our cognitive difficulties with negation are well attested, for instance. Still, a description that contains a negation may be *less* complex than any other description that manages to single out the referent. Consider a car park full of vehicles, of many different brands. Given the choice between *The cars that are not Hondas need to be removed* and *The Fords, Lexuses, Toyotas, Audis, (...), Seats need to be removed*, the former is shorter and probably preferable. The claim here is not that negation should be used at every opportunity but that, in some situations, negated referring expressions are preferable to ones that contain only conjunctions and disjunctions. This argument is not limited to negation, but applies to quantified referring expressions as well. Expressions like *The man who adores* `all` *cars* or *The woman with* `two` *suitors* may be somewhat complex, but in a particular situation, they may well be simpler than the simplest unquantified distinguishing description.

5.6.3   Simplicity is not everything

So far, we have tacitly assumed that a complex description is worse than a simple one, suggesting that a REG algorithm should always prefer a simpler one over a complex one. But simplicity is not everything. Suppose, for example, we consider

– *The man who loves all cars* (when a car exhibition is discussed)
– *John's uncle* (when John is salient, or after discussing someone else's uncle)
– *The cars that are not Hondas* (in front of a Honda factory).

In all these situations, the referring expression contains a source of complexity (the quantifier *all*, a relation, or a negation) for which it should be rewarded rather than penalized, because it makes the referring expression more contextually relevant than a simpler referring expression would have been.

5.6.4   Complex content does not always require a complex form

As explained in Section 5.1, we have been assuming that the generator performs content determination before choosing the syntactic and lexical form of the referring expression. We therefore should not assume that the natural language expression coming out at the end of the generation pipeline was constructed by mapping each logical connective in the logical description to a separate natural language expression. A complex formula can sometimes be expressed through simple words. For example,

– $Man \sqcap \leq 0\neg Love. Car$ (*The man who loves all cars*): the car lover
– $Woman \sqcap 2Givebirth. Baby$ (*The woman who gave birth to two babies*): the woman with twins

Furthermore, there is no reason to assume that all the information in the referring expression must be realized within the same utterance. This point is particularly relevant in interactive settings, where reference can result from a sequence of turns in the dialogue. For example, suppose the generator produces a description $Woman \sqcap \exists Feed. (Dog \sqcap \forall Chase. Cat)$. Parts of this referring expression may end up scattered over various dialogue turns, as in

(5.1)   A: *This woman*
         B: *uh-huh*
         A: *she feeds a dog*
         B: *and?*
         A: *The dog chases only cats*

By breaking complex information down into bite-size chunks, the referring expression as a whole may be easier to understand; it also becomes easier for the speaker to verify the hearer's understanding. Further research is needed to spell out the logical constraints on the dispersal of referential information, which have to do with the monotonicity properties of the different types of quantifier. To see the problem, consider the information in a slightly different referring expression, *Woman* $\sqcap \leq 0 \neg Feed.(Dog \sqcap \forall Chase.Cat)$. An unthinkingly direct realization might be *the woman such that there are no dogs that chase only cats whom she does not feed*. After some simple logical manipulation, this becomes *the woman who feeds all dogs that chase only cats*. The information in this referring expression cannot, however, be dispersed over three separate utterances, as in

(5.2)   *This woman*
        she feeds all dogs
        the dogs chase only cats

because the information in the second utterance may well be false.

5.6.5   Characterizing linguistic competence

Finally, exploring the space of possible referring expressions is of theoretical interest. Let us assume, against our better judgment, that empirical research will show beyond reasonable doubt that most or all of the types of referring expressions our most advanced REG algorithm can generate are useless: no human speaker would ever generate them, and no human hearer would ever benefit from them. I would contend that our understanding of human language and language use would be considerably advanced by this insight, because it would show us which logically possible referential strategies people actually use.

Two analogies come to mind. First, consider center-embedding, for example of relative clauses. This syntactic phenomenon can give rise to arbitrarily deep nestings, which are easily covered by recursive grammar rules yet extremely difficult to understand (e.g., *a man that a woman that a child knows loves*; Karlsson, 2007); moreover, there has to exist a finite upper bound on the depth of any noun phrase ever encountered. Does this mean there is no need for recursive rules? A standard response is that recursive grammar rules model the human linguistic **competence** (i.e., what can be done in principle), whereas human linguistic **performance** is best understood by additional constraints (e.g., to reflect limitations on human memory and calculation). I believe that the semantically complex expressions of the previous sections may be viewed in the same light. The second analogy transports us back to our discussion of quantifiers. The theory of Generalized Quantifiers does not stop at a logical characterization of the class of all possible quantifiers. Instead, the theory focuses on the question of what class of quantifiers can be expressed through syntactically simple means (e.g., in one word, or within one noun phrase; Peters and Westerståhl, 2006). It is this theory that gave rise to the apparatus that enabled us to extend REG algorithms as proposed in Section 5.4.3. Once we know what referring expressions are possible, it might be useful to ask – broadly in the spirit of the theory of Generalized Quantifiers – which of these are actually used.

5.7   Whither REG?

In their most radical – and logically speaking most natural – form, the proposals of this chapter amount to a radical extension of REG, in which atomic facts are no longer the substance of the KB, but merely its seeds. The change is particularly stark if we drop the assumption that all individuals in the domain must be mentioned by name in the KB (see the example at the end of Section 5.4). If this assumption is dropped, it becomes possible to refer to individuals whose existence is not stated but merely inferred. Referring expressions that do this are extremely frequent, for example when a newspaper writes about *President Obama's foreign policy*, *The first day of the year*, *The mother of the newly born*, or *Those responsible for (a crime, a policy, etc.)*. All such referring expressions were out of reach of REG algorithms until recently, but appear on the horizon now. It remains to be seen how the axioms on which these REG algorithms are based should be obtained, and how the aim of empirical adequacy (i.e., question 2 in the introduction to this chapter) can be achieved as well.

Acknowledgments

This chapter has benefitted from discussions in Aberdeen's Natural Language Generation group and from comments by Helmut Horacek, Emiel Krahmer, Chris Mellish, Graeme Ritchie, and Martin Caminada.

References

Areces, C., Koller, A., and Striegnitz, K. (2008). Referring expressions as formulas of description logic. In *Proceedings of the International Conference on Natural Language Generation (INLG)*, pages 42–49, Salt Fork, OH. Association for Computational Linguistics.

Bateman, J. (1999). Using aggregation for selecting content when generating referring expressions. In *Proceedings of the Annual Meeting of the Association for Computational Linguistics (ACL)*, pages 127–134, College Park, MD. Association for Computational Linguistics.

Belz, A., Kow, E., Viethen, J., and Gatt, A. (2010). Generating referring expressions in context: The GREC task evaluation challenges. In Krahmer, E. and Theune, M., editors, *Empirical Methods in Natural Language Generation*, pages 294–327. Springer, Berlin, Germany.

Campana, E., Tanenhaus, M. K., Allen, J. F., and Remington, R. (2011). Natural discourse reference generation reduces cognitive load in spoken systems. *Natural Language Engineering*, **17**(3):311–329.

Croitoru, M. and van Deemter, K. (2007). A conceptual graph approach to the generation of referring expressions. In *Proceedings of the International Joint Conference on Artificial Intelligence (IJCAI)*, pages 2456–2461, Hyderabad, India. International Joint Conference on Artificial Intelligence.

Dale, R. and Haddock, N. (1991). Content determination in the generation of referring expressions. *Computational Intelligence*, **7**(4):252–265.

Dale, R. and Reiter, E. (1995). Computational interpretation of the Gricean maxims in the generation of referring expressions. *Cognitive Science*, **19**(2):233–263.

Gardent, C. (2002). Generating minimal definite descriptions. In *Proceedings of the Annual Meeting of the Association for Computational Linguistics (ACL)*, pages 96–103, Philadelphia, PA. Association for Computational Linguistics.

Gardent, C. and Striegnitz, K. (2007). Generating bridging definite descriptions. In Bunt, H. and Muskens, R., editors, *Computing Meaning*, volume 3, pages 369–396. Springer, Dordrecht, The Netherlands.

Gatt, A. (2007). *Generating Coherent References to Multiple Entities*. PhD thesis, Department of Computing Science, University of Aberdeen.

Gatt, A. and Belz, A. (2010). Introducing shared tasks to NLG: The TUNA shared task evaluation challenges. In Krahmer, E. and Theune, M., editors, *Empirical Methods in Natural Language Generation*, pages 264–293. Springer, Berlin, Heidelberg.

Goudbeek, M. and Krahmer, E. (2010). Preferences versus adaptation during referring expression generation. In *Proceedings of the Annual Meeting of the Association for Computational Linguistics (ACL)*, pages 55–59, Uppsala, Sweden. Association for Computational Linguistics.

Guhe, M. and Bard, E. G. (2008). Adapting referring expressions to the task environment. In *Proceedings of the Annual Conference of the Cognitive Science Society (CogSci)*, pages 2404–2409, Washington, DC. Cognitive Science Society.

Henschel, R., Cheng, H., and Poesio, M. (2000). Pronominalization revisited. In *Proceedings of the International Conference on Computational Linguistics (COLING)*, pages 306–312, Saarbrücken, Germany. International Committee on Computational Linguistics.

Horacek, H. (1996). A new algorithm for generating referential descriptions. In *Proceedings of the European Conference on Artificial Intelligence (ECAI)*, pages 577–581, Budapest, Hungary. John von Neumann Computer Society.

Horacek, H. (2004). On referring to sets of objects naturally. In *Proceedings of the International Conference on Natural Language Generation (INLG)*, pages 70–79, Brockenhurst, UK. Springer.

Horrocks, I., Kutz, O., and Sattler, U. (2006). The even more irresistible SROIQ. In *Proceedings of the International Conference on Principles of Knowledge Representation and Reasoning*, pages 57–67, Lake District, UK. Principles of Knowledge Representation and Reasoning, Inc.

Jordan, P. W. and Walker, M. A. (2000). Learning attribute selections for non-pronominal expressions. In *Proceedings of the Annual Meeting of the Association for Computational Linguistics (ACL)*, pages 181–190, Hong Kong. Association for Computational Linguistics.

Karlsson, F. (2007). Constraints on multiple center-embedding of clauses. *Journal of Linguistics*, **43**(2):365–392.

Kelleher, J. D., Kruijff, G.-J. M., and Costello, F. J. (2006). Incremental generation of spatial referring expressions in situated dialog. In *Proceedings of the International Conference on Computational Linguistics and the Annual Meeting of the Association for Computational Linguistics (COLING-ACL)*, pages 745–752, Sydney, Australia. Association for Computational Linguistics.

Khan, I. H., van Deemter, K., and Ritchie, G. (2012). Managing ambiguity in reference generation: The role of surface structure. *Topics in Cognitive Science*, **4**(2):211–231.

Krahmer, E. and Theune, M. (2002). Efficient context-sensitive generation of referring expressions. In van Deemter, K. and Kibble, R., editors, *Information Sharing: Reference and Presupposition in Language Generation and Interpretation*, pages 223–264. CSLI Publications, Stanford, CA.

Krahmer, E. and van Deemter, K. (2012). Computational generation of referring expressions: A survey. *Computational Linguistics*, **38**(1):173–218.

Krahmer, E., van Erk, S., and Verleg, A. (2003). Graph-based generation of referring expressions. *Computational Linguistics*, **29**(1):53–72.

Lønning, J. T. (1997). Plurals and collectivity. In van Benthem, J. and ter Meulen, A., editors, *Handbook of Logic and Language*, pages 1009–1053. Elsevier, Amsterdam, The Netherlands.

McCluskey, E. J. (1986). *Logic Design Principles with Emphasis on Testable Semicustom Circuits*. Prentice-Hall, Upper Saddle River, NJ.

Mostowski, A. (1957). On a generalization of quantifiers. *Fundamenta Mathematicae*, **44**:12–36.

Paraboni, I., van Deemter, K., and Masthoff, J. (2007). Generating referring expressions: Making referents easy to identify. *Computational Linguistics*, **33**(2):229–254.

Passonneau, R. (1996). Using centering to relax Gricean informational constraints on discourse anaphoric noun phrases. *Language and Speech*, **39**(2–3):229–264.

Peters, S. and Westerståhl, D. (2006). *Quantifiers in Language and Logic*. Oxford University Press, Oxford, UK.

Piwek, P., Beun, R.-J., and Cremers, A. (2008). 'Proximal' and 'distal' in language and cognition: Evidence from deictic demonstratives in Dutch. *Journal of Pragmatics*, **40**(4):694–718.

Reiter, E. (1990). The computational complexity of avoiding conversational implicatures. In *Proceedings of the Annual Meeting of the Association for Computational Linguistics (ACL)*, pages 97–104, Pittsburgh, PA. Association for Computational Linguistics.

Ren, Y., van Deemter, K., and Pan, J. Z. (2010). Charting the potential of description logic for the generation of referring expressions. In *Proceedings of the International Conference on Natural Language Generation (INLG)*, pages 115–124, Trim, Ireland. Association for Computational Linguistics.

Scha, R. and Stallard, D. (1988). Multi-level plurals and distributivity. In *Proceedings of the Annual Meeting of the Association for Computational Linguistics (ACL)*, pages 17–24, Buffalo, NY. Association for Computational Linguistics.

Searle, J. (1969). *Speech Acts: An Essay in the Philosophy of Language*. Cambridge University Press, Cambridge, UK.

Seylan, I., Franconi, E., and De Bruijn, J. (2009). Effective query rewriting with ontologies over DBoxes. In *Proceedings of the International Joint Conference on Artificial Intelligence (IJCAI)*, pages 923–925, Pasadena, CA. International Joint Conference on Artificial Intelligence.

Siddharthan, A. and Copestake, A. (2004). Generating referring expressions in open domains. In *Proceedings of the Annual Meeting of the Association for Computational Linguistics (ACL)*, pages 407–414, Barcelona, Spain. Association for Computational Linguistics.

Siddharthan, A., Nenkova, A., and McKeown, K. (2011). Information status distinctions and referring expressions: An empirical study of references to people in news summaries. *Computational Linguistics*, **37**(4):811–842.

Stone, M. (2000). Towards a computational account of knowledge, action and inference in instructions. *Journal of Language and Computation*, **1**(2):231–246.

van Benthem, J. (1986). *Essays in Logical Semantics*. Reidel, Dordrecht, The Netherlands.

van Deemter, K. (1984). Generalized quantifiers: Finite versus infinite. In van Benthem, J. and ter Meulen, A., editors, *Generalized Quantifiers in Natural Language*, pages 145–159. Foris Publications, Dordrecht, The Netherlands.

van Deemter, K. (2002). Generating referring expressions: Boolean extensions of the incremental algorithm. *Computational Linguistics*, **28**(1):37–52.

van Deemter, K., Gatt, A., van der Sluis, I., and Power, R. (2012). Generation of referring expressions: Assessing the Incremental Algorithm. *Cognitive Science*, **36**(5):799–836.

van Deemter, K. and Halldórsson, M. M. (2001). Logical form equivalence: The case of referring expressions generation. In *Proceedings of the European Workshop on Natural Language Generation (EWNLG)*, pages 21–28, Toulouse, France. Association for Computational Linguistics.

van Deemter, K. and Krahmer, E. (2007). Graphs and booleans: On the generation of referring expressions. In Bunt, H. and Muskens, R., editors, *Computing Meaning*, volume 3, pages 397–422. Springer, Dordrecht, The Netherlands.

Viethen, J. and Dale, R. (2006). Algorithms for generating referring expressions: Do they do what people do? In *Proceedings of the International Conference on Natural Language Generation (INLG)*, pages 63–72, Sydney, Australia. Association for Computational Linguistics.

Viethen, J. and Dale, R. (2008). The use of spatial relations in referring expression generation. In *Proceedings of the International Conference on Natural Language Generation (INLG)*, pages 59–67, Salt Fork, OH. Association for Computational Linguistics, Association for Computational Linguistics.

6 Referring expression generation in interaction: A graph-based perspective

Emiel Krahmer, Martijn Goudbeek, and Mariët Theune

6.1 Introduction

Buying new chairs can be complicated. Many constraints have to be kept in mind, including your financial situation, the style and color of the furniture you already own and possibly also the taste of your partner. But once you have made a tentative choice (say, the chair in Figure 6.1 on the left), there is one final hurdle: you have to inform the seller of your desire to buy it. Furniture stores tend to contain many chairs, so somehow you need to refer to your chair of choice, for example *I'd like to buy the wooden chair with the thin legs and solid seat, the red one with the open back*. It is hardly surprising that many people in this situation resort to pointing (*that* one). Of course, it would be helpful to know that salespeople might refer to this chair as *the red Keystone chair*, because that would allow you to adapt to their way of referring.

This problem illustrates the importance of **reference** in everyday interactions: people can only exchange information about objects when they agree on how to refer to those objects. How this agreement may arise, and how we can model this in natural language generation, is the topic of this chapter. We argue that two possibly competing forces play a role. On the one hand, speakers may have inherent preferences for certain properties when referring to objects in a given domain. On the other, they may also have a tendency to adapt to the references produced by their dialogue partner. We describe how preferences can be determined, and how they interact with adaptation. We model this tradeoff using a graph-based referring expression generation algorithm (Krahmer *et al.*, 2003).

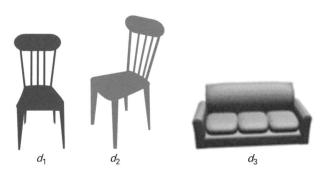

d_1 d_2 d_3

Figure 6.1   Three pieces of furniture. In order from left to right, the pieces are red, blue, and green

6.1.1   Referring expression generation

Given the centrality of reference in interaction, it is hardly surprising that one of the first things that children learn when acquiring language is how to refer to the objects surrounding them (Matthews *et al.*, 2012). Similarly, when researchers develop a natural language generation (NLG) application, they typically also require a module that generates referring expressions (Reiter and Dale, 2000; Mellish *et al.*, 2006). Such a referring expression generation (REG) module is typically dedicated to identifying **target** objects with respect to a set of **distractor** objects using natural language, and to do so the module needs to make a series of related choices. First, it needs to select the form of reference, for example, deciding whether to refer to the chair using a deictic pronoun (*that one*) or a full description (*the chair with the armrests*). If the REG module decides to generate a description, two additional choices need to be made: which **properties** of the target should be included in the description, and how the selected property set can be expressed as a natural language description. These two processes are often referred to as **attribute selection** and **surface realization**, respectively. Of these, attribute selection has received by far the most scholarly attention, perhaps because researchers tend to assume that a standard surface realizer for a given language can be used to express a set of selected properties. Attribute selection is a complex balancing act (Reiter and Dale, 2000): we need to include enough properties that an addressee will be able to determine the target, but including all known properties of the target may be awkward or misleading. Hence a selection of properties needs to be made, and this selection should take as little time as possible. This is especially crucial in NLG for interactive settings, where a system needs to respond to a user in near real time.

6.1.2   Preferences versus adaptation in reference

Typically, a target can be distinguished using many different properties; for example, a chair can be referred to as wooden, having armrests, being red, or facing right. Many REG algorithms, including Dale and Reiter's well-known Incremental Algorithm (Dale and Reiter, 1995), assume that some properties or attributes[1] are preferred, and will be selected first by the content determiner. This heuristic allows REG algorithms to ignore some potential property sets during attribute selection. It may also lead to **overspecified** referring expressions, i.e., those that contain more properties than are necessary to uniquely identify the target, but we have considerable evidence that humans also overspecify (e.g., Dale and Reiter, 1995). So how can we determine a preference ordering over a set of properties or attributes? Dale and Reiter stress that constructing a preference ordering is essentially an empirical question, which will differ from one domain to another, but they do point to psycholinguistic research (especially Pechmann, 1989) suggesting that, in general, absolute attributes (such as color) are preferred over relative

---

[1] In this chapter, we use the term **attribute** to refer to concepts, such as size and color, that can be used in referring expressions. When referring to an attribute–value pair, e.g., color = blue, we use the term **property**. As we will see, the distinction is important, because some REG algorithms operate over attributes while others, in particular the graph-based algorithm, rank properties.

ones (such as size). After all, to determine the color of an object we only need to look at the object itself, while to determine whether it is large or small all domain objects need to be inspected.

Even though the Incremental Algorithm is probably unique in assuming a complete preference ordering of attributes, many other REG algorithms rely on preferences as well. This became apparent, for example, during the REG Challenges (see Gatt and Belz (2010) for an overview); virtually all participating systems relied on training data to determine preferences of one form or another. However, relevant training data is hard to find. It has been argued that determining which properties to include in a referring expression requires a "semantically transparent" corpus (van Deemter *et al.*, 2006): a corpus that contains the actual properties of all domain objects as well as the properties that were selected for each referring expression. Obviously, text corpora hardly ever meet this requirement. The few existing semantically transparent corpora were collected by the time-consuming exercise of asking human participants to produce referring expressions in a particular language (typically, English) for targets in controlled visual scenes for a particular domain (see, e.g., Gorniak and Roy, 2004; Viethen and Dale, 2006; Gatt *et al.*, 2007; Guhe and Bard, 2008). An important question therefore is how many human-produced references are needed to achieve a certain level of accuracy in preference ordering. One way to answer this question is by training a REG algorithm on subsets of a (semantically transparent) corpus of various sizes, and measuring the performance differences. This is precisely what we do in this chapter, in Section 6.3.

Another question is how stable preference orderings are in interactive settings, e.g., for applications such as spoken dialogue systems or interactive virtual characters. In these cases, it seems likely that referring expressions produced earlier in the interaction are also important. We know, for instance, that if one dialogue partner refers to a couch as a *sofa*, the other is more likely to use the word *sofa* as well (Branigan *et al.*, 2010). This kind of micro-planning or **lexical entrainment** (Brennan and Clark, 1996) can be seen as a specific form of **alignment** (Pickering and Garrod, 2004) in interaction. But what if dialogue partners' preference orderings differ? Do they adapt to the other's preference ordering, or stick to their own? And what if one dialogue partner's preference ordering leads to an overspecified referring expression – will the other partner reproduce this overspecified form due to alignment? These questions are also addressed in this chapter, in Section 6.4, where we report on two experimental studies using an interactive reference production paradigm and discuss how a REG algorithm could model our findings.

As our REG model of choice we use the graph-based algorithm, originally proposed by Krahmer, van Erk, and Verleg (2003), and described in Section 6.2. This algorithm models domain information about potential target objects in a graph structure and treats REG as a graph-search problem, where a cost function is used to prefer some solutions over others. The graph-based algorithm is a state-of-the-art REG algorithm; it was among the best-scoring algorithms on attribute selection in the 2008 REG Challenge (Gatt *et al.*, 2008), and emerged as the best-performing algorithm in the most recent REG Generation Challenge (Gatt *et al.*, 2009). In this chapter we argue that the use of cost functions makes the algorithm well suited to deal with the trade-off between preference orderings and alignment.

6.2   Graph-based referring expression generation

6.2.1   Scene graphs

Figure 6.1 depicts an example **domain** with three potential referents or **objects**, ($\mathcal{D} = \{d_1, d_2, d_3\}$), a set of **properties** ($Prop = \{$chair, blue, facing-left, $\ldots\}$), and a set of **relations** ($Rel = \{$left-of, right-of$\}$). This domain can be modeled as the labeled directed **scene graph** shown in Figure 6.2. Properties are modeled as loops, i.e., edges that start and end in the same node, whereas relations are modeled as edges between nodes.

Formally, scene graphs are defined as follows. Let $\mathcal{D}$ be the **domain**, and $L = Prop \cup Rel$ the set of **labels**. Then, the scene graph $G = \langle V_G, E_G \rangle$ is a labeled directed graph, where $V_G \subseteq \mathcal{D}$ is the set of nodes or vertices (the objects) and $E_G \subseteq V_G \times L \times V_G$ is the set of labeled directed edges (in this chapter, subscripts are omitted whenever this can be done without creating confusion).

6.2.2   Referring graphs

Now imagine that given our example domain we want to generate a **distinguishing description**, i.e., a referring expression that uniquely identifies d_1. We need to select a set of properties and relations that single out the **target** d_1 from the other two domain objects (the **distractors**). In the graph-based REG approach, this is done by constructing **referring graphs**. Each referring graph includes at a minimum a vertex representing the target. Referring graphs are defined in exactly the same way as scene graphs, which allows us to view REG as a graph construction exercise. Informally, a target node v

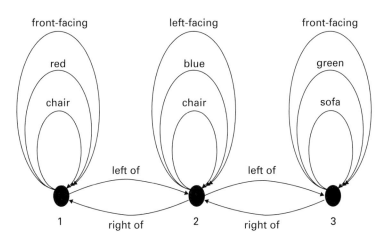

Figure 6.2   A simple scene graph

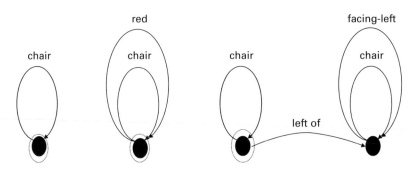

Figure 6.3   Three referring graphs

in a referring graph refers to a node w in a scene graph if the referring graph can be "placed over" the scene graph in such a way that v can be placed over w, and each edge from the referring graph labeled with some property or relation can be placed over a corresponding edge in the scene graph with the same label. If there is only one way in which a referring graph can be placed over a scene graph, we have found a distinguishing description.

Figure 6.3 shows three potential referring graphs for d_1, with the target circled. The first, which could be realized as *the chair*, can be placed over node d_1, but also over d_2, and hence is not distinguishing. The other two, which could be realized as *the red chair* and *the chair to the left of the chair facing left*, respectively, can only be placed over the scene graph in one way, and hence represent possible distinguishing descriptions for target d_1. Clearly, the second would be a more natural description for d_1 than the third; below we will discuss how cost functions can be used to rank different descriptions such as these.

6.2.3   Formalizing reference in terms of subgraph isomorphism

Let us make the "placed over" notion a bit more precise. The informal notion of one graph being placed over another corresponds to a well-known construct in graph theory, namely subgraph isomorphism.

A graph G' is a **subgraph** of G if and only if $V_{G'} \subseteq V_G$ and $E_{G'} \subseteq E_G$. A **subgraph isomorphism** between graphs H and G exists if there is a subgraph G' of G such that H is isomorphic to G'. H is **isomorphic** to G' if and only if there exists a bijection $\pi : V_H \rightarrow V_{G'}$, such that for all vertices $v, w \in V_H$ and all labels $l \in L$:

$$(v, l, w) \in E_H \Leftrightarrow (\pi . v, l, \pi . w) \in E_{G'}.$$

Given a graph H and a vertex v in H, and a graph G with a vertex w in G, we will say that the pair (v, H) **refers to** the pair (w, G) if and only if (a) H is a connected graph (that is: each vertex in H has at least one edge that links it to another vertex in H), and (b) H is mapped to a subgraph of G by an isomorphism π with $\pi . v = w$. A vertex–graph pair (v, H) **uniquely refers to** (w, G) if and only if (v, H) refers to (w, G) and there is no other vertex w' in G such that (v, H) refers to (w', G).

6.2.4   Cost functions

As with most REG algorithms, the graph-based algorithm requires a mechanism to give some solutions preference over others. It does so by using **cost functions**, which assign costs to the edges and nodes of a referring graph, and sums these:

$$\text{cost}(G) = \Sigma_{v \in G_V} \text{cost}(v) + \Sigma_{e \in G_E} \text{cost}(e)$$

The only a priori assumption that we make is that the cost function should be **monotonic**: extending a graph G with an edge e (notation: $G + e$) should never result in a graph that is cheaper than G. Formally,

$$\forall H \subseteq G, \forall e \in E_G : \text{cost}(H) \leq \text{cost}(H + e)$$

As we will see below, cost functions can be defined in various ways, and this is one of the attractive properties of the graph-based REG algorithm.

6.2.5   Algorithm

Figure 6.4 contains the sketch of a basic graph-based REG algorithm, called **makeReferringExpression**. It takes as input a target v in a scene graph G. The algorithm constructs a referring graph H, which is initialized as the graph consisting of only

```
makeReferringExpression(v, G) {
    bestGraph := ⊥
    H := ⟨{v}, Ø⟩
    return findGraph(v, bestGraph, H, G)
}


findGraph(v, bestGraph, H, G) {
    if [bestGraph ≠ ⊥ and cost(bestGraph) ≤ cost(H)]
    then return bestGraph
    C := {v' | v' ∈ V_G & (v, H)refers to(v', G)}
    if C = {v} then return H
    for each adjacent edge e do
        I := findGraph(v, bestGraph, H + e, G)
        if [bestGraph = ⊥ or cost(I) ≤ cost(bestGraph)]
        then bestGraph := I
    rof
    return bestGraph
}
```

Figure 6.4   Sketch of the main function (**makeReferringExpression**) and the subgraph construction function (**findGraph**), based on Krahmer *et al.* (2003)

one node: the target v. In addition, a variable *bestGraph* is introduced, for the best solution found so far. Since none has been found at this stage, *bestGraph* is initialized as the empty graph $\perp$. In the **findGraph** function the algorithm systematically tries expanding H by adding adjacent edges (i.e, edges from v, or possibly from any of the other vertices added to the referring graph H under construction). For each H the algorithm finds the set of nodes $C \subseteq G$ to which H could refer. A successful distinguishing description is found if and only if H can only refer to the target (i.e., $C = \{v\}$). The first distinguishing description that is found is stored in *bestGraph* (best solution found so far). At that point the algorithm only looks for referring graphs that are cheaper than the best (cheapest) solution found so far, performing a complete, depth-first search.[2] It follows from the monotonicity requirement on cost functions that the algorithm outputs the cheapest distinguishing description graph, if one exists. Otherwise it returns the empty graph.

6.2.6   Discussion

The graph-based REG algorithm has a number of attractive properties. For example, graphs are a well-understood mathematical formalism, and there are many efficient algorithms for dealing with graph structures (see, for instance, Gibbons, 1985; Chartrand and Oellermann, 1993). In addition, because relational properties are handled in the same way as other properties (namely as edges in a graph), the treatment of relations between objects does not suffer from some of the problems of earlier REG approaches; there is, for instance, no need to make any *ad hoc* stipulations (e.g., that a property can only be attributed to a given object once per referring expression; Dale and Haddock, 1991). Relational properties cause testing for a subgraph isomorphism to have exponential complexity (Garey and Johnson, 1979), but special cases are known in which the problem has lower complexity, for example when considering graphs that are **planar** (that is, drawable without crossing edges). Krahmer *et al.* (2003) sketch a greedy algorithm that can simplify any graph into a planar equivalent.

Variations and extensions

Van Deemter and Krahmer (2007) discuss how the basic graph-based algorithm can be extended with, for example, plurals (e.g., *the chairs*) and boolean expressions (e.g., *the blue chairs and the coach that is not red*), while van der Sluis and Krahmer (2007) present an extension of the algorithm that is capable of generating multimodal referring expressions (e.g., *that chair* accompanied by a pointing gesture). Their account allows for pointing degrees of various precisions, where less precise pointing gestures are accompanied by more extensive verbal descriptions (see de Ruiter *et al.* (2012) for a discussion).

   The approach to graphs outlined here offers an attractive account of binary relations, but not of more complex relations (e.g., *the chair given by John to Mary*, *the fan in*

---

[2] Naturally, graph-based generation is compatible with different search strategies as well.

between the chair and the couch). Croitoru and van Deemter (2007) offer an alternative way of constructing graphs, using insights from conceptual graph theory, which can account for relations of arbitrary complexity. Their approach also allows for a logic-based interpretation of reference graphs, which enables more complex knowledge representation and reasoning.

Computing costs

The use of cost functions is an important ingredient of the graph-based algorithm. Various alternative ways of computing costs have been considered. Perhaps the most straightforward option is to assign a cost of 1 to each edge and vertex. It is easily seen that in this way the cheapest distinguishing graph will also be the smallest one; this would make the graph-based algorithm equivalent (in terms of its input–output behavior) to the well-known Full Brevity algorithm (Dale, 1989). The Full Brevity algorithm has been criticized, however, for lacking "humanlikeness," since human speakers frequently produce overspecified referring expressions (Olson, 1970; Sonnenschein, 1984; Arts, 2004; Engelhardt *et al.*, 2006; Pechmann, 1989). The graph-based algorithm can model overspecification by allowing some properties to be included for free (Viethen *et al.*, 2008).[3] However, if a graph contains zero-cost edges, the order in which the graph-based algorithm tries to add properties to a referring expression must be explicitly controlled, to ensure that "free" distinguishing properties are included (Viethen *et al.*, 2008).

Which properties should be available for free inclusion? One option is to link cost to frequency; properties that speakers use often should be cheap, while those that are less frequent should be more expensive. In other words, if we assume that $p(e)$ is the probability that an edge (property) will be used in a referring expression, then we could define $cost(e) = -\log(p(e))$. Properties that are very cheap (costs below a certain threshold) can manually be set to 0. Explicit control of property inclusion order can also be tied to frequency; edges can be tried in order of their corpus frequency.

The probabilities of properties can be estimated by counting frequencies of occurrence in a semantically transparent corpus such as the TUNA corpus (Gatt *et al.*, 2007; van Deemter *et al.*, 2012). The TUNA corpus consists of two domains: one containing pictures of people (all famous mathematicians), the other containing furniture items in different colors depicted from different orientations, such as those in Figure 6.1.[4] Viethen *et al.* (2008) compare a cost function where costs are directly derived from frequencies in the TUNA corpus (in terms of log probabilities) with a "Free Naïve" one that just assigns three costs based on the frequencies (with 0 = free, 1 = cheap, 2 = expensive), and found that the latter results in more human-like referring expressions. The graph-based algorithm with the Free Naïve cost function, combined with a dedicated linguistic realizer, was the best performing REG algorithm in the 2009 REG

---

[3] Notice that this does not violate the monotonicity assumption.
[4] The pictures of furniture items were taken from the Object Databank, courtesy of Michael J. Tarr, Center for the Neural Basis of Cognition and Department of Psychology, Carnegie Mellon University, and distributed under the Creative Commons License at `http://www.tarrlab.org`.

Challenge (Krahmer *et al.*, 2008; Gatt *et al.*, 2009): the referring expressions this algorithm produced were overall most similar to the referring expressions in the test set, were subjectively judged to be most adequate and most fluent, and resulted in the highest identification accuracies (ability of human readers to identify the target referent given the generated referring expression). Although it is still a matter of some debate how REG algorithms should best be evaluated (see Chapter 13), and the test data used in the 2009 REG Challenge contains some idiosyncratic referring expressions (Gatt and Belz, 2010), it is clear that the graph-based REG algorithm produces results that are state-of-the-art, and that the cost function, giving preference to some properties over others, plays an important part in this. In the next section we will discuss in more detail how these cost functions can be determined automatically, and how much data is required to obtain accurate cost functions.

6.3   Determining preferences and computing costs

In the work of Krahmer *et al.* (2008) and Viethen *et al.* (2008), cost functions for properties for the two TUNA domains were determined in two ways: (a) by directly using frequencies from the TUNA corpus of about 450 human-produced referring expressions; and (b) by manually clustering properties into cost clusters based on the corpus frequencies. This raises two related questions: how can we achieve an optimal cost clustering, and how much training data is necessary for accurate cost function estimation? In this section we address these two questions.

Mapping frequencies to cost clusters

One way to identify an optimal clustering of frequencies is to systematically compare the performance of cost functions derived from various clusterings on a held-out test set. We use k-means clustering (Hartigan and Wong, 1979), which partitions n data points into k clusters (S_1 to S_k), with $k \leq n$, by assigning each point to the cluster with the nearest mean. The total intra-cluster variance V is minimized using the function

$$V = \sum_{i=1}^{k} \sum_{x_j \in S_i} (x_j - \mu_i)^2$$

where μ_i is the centroid of all the points $x_j \in S_i$. In our case, the points n are properties, and μ_i is the average frequency of the properties in S_i. The cluster-based costs were defined as follows:

$$\forall x_j \in S_i, \operatorname{cost}(x_j) = i - 1$$

where S_1 is the cluster with the most frequent properties, S_2 is the cluster with the next most frequent properties, and so on. Using this approach, properties from cluster S_1 get cost 0 and thus can be added for free to a referring expression.

The training and test data on which we performed our experiment were taken from the TUNA corpus (Gatt *et al.*, 2007; van Deemter *et al.*, 2012). For training, we

Table 6.1. Cost clustering results for 2-means costs and the Free Naïve costs of Krahmer *et al.* 2008

Costs	Furniture		People	
	Dice	Accuracy	Dice	Accuracy
k-means	0.810	0.50	0.733	0.29
Free Naïve	0.829	0.55	0.733	0.29

used the -LOC data from the REG Challenge 2009 training data (Gatt *et al.*, 2009): 165 Furniture referring expressions and 136 People referring expressions.[5] For testing, we used the -LOC data from the TUNA 2009 development set: 38 Furniture referring expressions and 38 People referring expressions.

We clustered the training data repeatedly using $k = 2$ to $k = 6$. Then we evaluated the performance of the graph-based algorithm with the resulting cost functions on the TUNA 2009 development data. We used two metrics, Dice (overlap between sets of properties) and Accuracy (perfect match between sets of properties), as evaluation metrics. For comparison, we also ran the graph-based algorithm on this data set using the Free Naïve cost function of Viethen *et al.* (2008). In all our tests, we used decreasing frequency for explicit control of property inclusion to ensure that free properties would be considered, i.e., the algorithm always examined more frequent properties first.

The best results were achieved with $k = 2$, for both TUNA domains. Interestingly, this is the coarsest possible k-means function: with only two costs (0 and 1) it is even less fine-grained than the Free Naïve cost functions. The results for the k-means costs with $k = 2$ and the Free Naïve costs of Krahmer *et al.* (2008) are shown in Table 6.1. A repeated measures analysis of variance (ANOVA) on the Dice and Accuracy scores, using cost function as a within-subjects variable (with levels Free Naïve and 2-means) revealed no statistically significant differences between the two cost functions. This suggests that k-means clustering offers a good and systematic alternative to manual clustering of frequency-based costs.

Varying training set size

To find out how much training data is required to achieve an acceptable property selection performance, we derived cost functions and preference orderings from different sized training sets, and evaluated them on our test data.

For training data, we used randomly selected subsets of the -LOC data from the REG Challenge 2009 training data (Gatt *et al.*, 2009), with set sizes of 1, 5, 10, 20, and 30 items. Because the accidental composition of a training set may strongly influence the results, we created five different sets of each size. The training sets were built up in a cumulative fashion: we started with five sets of size 1, then added 4 items to each of

---

[5] The -LOC data was collected by explicitly instructing participants not to use location information (e.g., *in the top left corner*) when referring to targets in the grid.

Table 6.2. Mean results for different training set sizes

Set size	Furniture		People	
	Dice	Accuracy	Dice	Accuracy
1	0.693	0.25	0.560	0.13
5	0.756	0.34	0.620	0.15
10	0.777	0.40	0.686	0.20
20	0.788	0.41	0.719	0.25
30	0.782	0.41	0.718	0.27
Entire set	0.810	0.50	0.733	0.29

them to create five sets of size 5, and so on. This resulted in five series of increasingly sized training sets. As before, for testing we used the -LOC data from the TUNA 2009 development set and the Dice and Accuracy metrics.

We derived cost functions (using k-means clustering with $k = 2$) and preference orderings for each of the training sets, following the method outlined earlier in this chapter. In doing so, we had to deal with missing data: not all properties were present in all data sets. This problem mostly affected the smaller training sets. By set size 10, only a few properties were missing, while by set size 20, all properties were present in all sets. For the cost functions, we simply assigned the highest cost (1) to the missing properties. For the sake of comparison, the algorithm selected properties with the same frequency (0 for missing properties) always in alphabetical order.

To determine significance, we calculated the means of the scores of the five training sets for each set size, so that we could compare them with the scores of the entire set. We applied repeated measures analyses of variance (ANOVA) to the Dice and Accuracy scores, using set size (1, 5, 10, 20, 30, entire set) as a within-subjects variable. The mean results for the different training set sizes are shown in Table 6.2. The general pattern is that the scores increase with the size of the training set, but the increase gets smaller as the set sizes become larger.

In the Furniture domain, we found a main effect of set size (Dice: $F_{(5,185)} = 7.209$, $p < 0.001$; Accuracy: $F_{(5,185)} = 6.140$, $p < 0.001$). To see which set sizes performed differently as compared to the entire set, we conducted Tukey's HSD post hoc comparisons. For Dice, the scores of set size 10 ($p = 0.141$), set size 20 ($p = 0.353$), and set size 30 ($p = 0.197$) did not differ significantly from the scores of the entire set of 165 items. The Accuracy scores show a slightly different pattern: the scores of the entire training set were still significantly higher than those of set size 30 ($p < 0.05$).

In the People domain we also found a main effect of set size (Dice: $F_{(5,185)} = 21.359$, $p < 0.001$; Accuracy: $F_{(5,185)} = 8.074$, $p < 0.001$). Post hoc pairwise comparisons showed that the scores of set size 20 (Dice: $p = 0.416$; Accuracy: $p = 0.146$) and set size 30 (Dice: $p = 0.238$; Accuracy: $p = 0.324$) did not significantly differ from the performance of the full set of 136 items.

The results suggest that small data sets can be sufficient for training the graph-based REG algorithm. However, domain differences play a role as well in how much training data is needed: using Dice as an evaluation metric, training sets of 10 sufficed in the relatively simple Furniture domain, while in the People domain it took a set size of 20 to achieve similar results as when using the full training set. Using the full training sets did give numerically higher scores, but the differences were not statistically significant. Furthermore, the accidental composition of the training sets may strongly influence attribute selection performance. In the Furniture domain, there were clear differences between the results of specific training sets, with "bad sets" pulling the overall performance down. This affected Accuracy but not Dice, perhaps because the latter is a less strict metric.

Of course, Dice and Accuracy are not the only evaluation metrics. It would be particularly interesting to see how the use of small training sets affects effectiveness and efficiency of target identification by human subjects; as shown by Belz and Gatt (2008), task-performance measures do not necessarily correlate with similarity measures such as Dice (although the graph-based algorithm scored well on both dimensions).

It is interesting to see which preferences were learned using the graph-based algorithm with corpus-based cost functions. If we focus on attributes, we find that generally color is preferred and orientation and size less so (in the Furniture domain), while having glasses is highly preferred, and for example, wearing a tie or a suit is not (in the People domain). Color and glasses are good examples of attributes that can be added for free. It also interesting to observe that orientation is far less dispreferred than wearing a tie; in fact, hasTie $= 1$ is *never* used in the TUNA data. Many of these distinctions can already be observed in set sizes as small as 5.

This is the situation when we look at the level of attributes. The graph-based REG algorithm, however, operates with preferences on the level of properties (i.e, attribute–value combinations). The potential advantage of this is that it becomes possible to prefer some colors (e.g., red) and disprefer others (mauve, taupe); the intuition is that it may be simpler to describe a mauve chair in terms of its size (certainly when assuming that the addressee may not know what color mauve is). Indeed, if we look at the preferences that were learned from the data, we see that for instance having glasses (hasGlasses $= 1$) is strongly preferred (costs 0), while not having glasses (hasGlasses $= 0$) is not (costs 1). Of course, it can be conjectured that learning preferred attributes will require less data (fewer referring expressions) than learning preferred properties.

So far, we have been working on the assumption that some properties are preferred over others, and we have just shown that a limited set of referring expressions may be enough to determine these preferences. However, is this basic assumption tenable when we consider the production of referring expressions in interaction? Unfortunately, data for this has been lacking. We now describe two experiments looking at the relation between adaptation and interaction.

6.4   Adaptation and interaction

In this section we report on two experiments that study the tradeoff between domain-dependent preferences and adaptation to referring expressions that have been used earlier in an interaction. Experiment I studies what speakers do when referring to a target that can be distinguished in a preferred (e.g., *the red chair*) or a dispreferred way (e.g., *the left-facing chair*), when earlier in the interaction either the preferred or the dispreferred variant was **primed**, or used by a dialogue partner. Experiment II studies overspecification, where participants were again asked to refer to a target, which can be distinguished using a minimal referring expression containing only a preferred attribute (e.g., *the red chair*), while earlier overspecified references (e.g., *the red front-facing chair*) were primed. Both studies use a novel interactive reference production paradigm, applied to the Furniture and People domains of the TUNA corpus, to see to what extent adaptation may be domain dependent.

6.4.1   Experiment I: adaptation and attribute selection

This experiment studies whether and how adaptation influences attribute selection in REG in interactive settings.

Method
Participants.   Participants were 26 native speakers of Dutch (two males, mean age 20 years, 11 months) who participated in the experiment in exchange for partial course credit.

Materials.   The stimulus pictures were taken from the TUNA corpus (Gatt, 2007). We relied on a Dutch version of the TUNA corpus (Koolen *et al.*, 2009) to determine which properties our participants would prefer and disprefer (i.e., which properties they use frequently and infrequently). It was found that Dutch speakers, like English ones, have a preference for color in the Furniture domain and wearing glasses in the People domain, and disprefer orientation of a furniture piece and wearing a tie, respectively. These properties were used as primes.

Procedure.   Each experimental trial consisted of four turns in an interactive reference understanding and production experiment: a prime, two fillers, and the experimental referring expression (see Figure 6.5 for an example of an experimental trial). In each trial, the prime and final turns were from one domain (Furniture or People), while the filler turns were from the other domain. The two filler turns were intended to prevent too direct a connection between the prime and the target. In the prime, the participant listened to a referring expression pre-recorded by a female voice and had to select a referent from three possibilities in the trial domain. In this turn, referring expressions used either a preferred or a dispreferred property; each property alone would be enough to uniquely identify the referent. In the two filler turns, the participant him/herself first produced a referring expression for a target given three objects in the other domain, and then had to select, from three possibilities in the other domain, the referent for a spoken referring expression. Finally, the participant produced a referring expression for a target object in the trial domain, which could always be distinguished from its two distractors using a preferred (e.g., *the blue fan*) or a dispreferred (e.g., *the left-facing fan*) property.

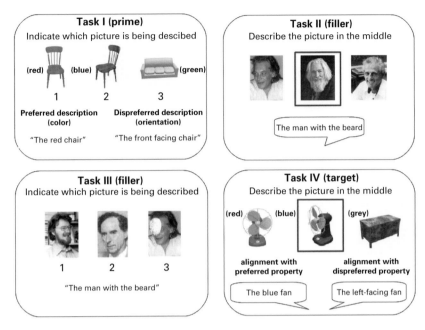

Figure 6.5   The four turns that constitute a trial in experiment I. This figure shows a Furniture trial; People trials have an identical structure. Colors are indicated with labels for print viewing; participants saw color images for Furniture

Note that *attributes* were primed, not properties; a participant may have heard *front-facing* in the prime turn, while the target referent had a different value for the orientation attribute in the experimental turn (as in Figure 6.5). In addition, in the Furniture domain but not in the People domain, the type values could differ; for example, when primed with a (preferred or dispreferred) referring expression for a chair, participants did not necessarily have to describe a chair in the experimental turn.

For each domain, there were 20 preferred and 20 dispreferred trials, resulting in $2 \times (20 + 20) = 80$ critical trials. These were presented in counterbalanced blocks, and within blocks each participant received a different random order. In addition, there were 80 filler trials (each following the same structure as outlined in Figure 6.5); filler trials never involved the attributes of interest. During debriefing, none of the participants indicated they had been aware of the experiment's true purpose.

Results and discussion

The proportions of preferred and dispreferred attributes used by participants as a function of prime and domain are shown in Figure 6.6. The black bars indicate use of the preferred attribute and the white bars indicate use of the dispreferred attribute. In both domains, the preferred attribute is used more frequently than the dispreferred attribute with the preferred prime, which serves as a manipulation check (our participants indeed overall preferred the preferred attributes to the dispreferred ones). The results show a clear effect of prime for the Furniture domain: participants used the

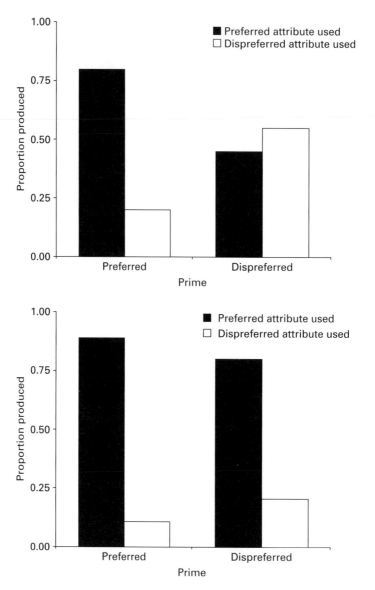

Figure 6.6   Proportions of preferred and dispreferred attributes in the Furniture (top) and People (bottom) domains

preferred attribute (color, as in *the red fan*) more when they were primed with it, and the dispreferred attribute (orientation, as in *the fan seen from the front*) more when it was in the prime. The results for the People domain reveal a similar picture (when participants were primed with the dispreferred attribute, they used it more often), but much less pronounced.

For our statistical analysis, we use the proportion of attribute **alignment** as the dependent measure. Alignment occurs when a participant uses the same attribute in

Table 6.3. Alignment means (and standard deviations) as a function of domain (Furniture and People) and prime (preferred and dispreferred)

Domain	Prime	Alignment mean (SD)
Furniture	Preferred	0.89 (0.32)
	Dispreferred	0.60 (0.49)
People	Preferred	0.97 (0.16)
	Dispreferred	0.25 (0.43)

the target as occurred in the prime. Table 6.3 displays the alignment mean and standard deviation per prime (preferred versus dispreferred) for the Furniture and People domains. We conducted a 2×2 repeated measures analysis of variance (ANOVA) with alignment as the dependent variable, and domain (Furniture versus People) and prime (preferred versus dispreferred) as independent variables. A statistically significant main effect was found for prime ($F(1, 25) = 6.43, p < 0.05$), showing that the prime influenced the selection of attributes in the experimental turn: when primed with dispreferred attributes, our participants used the dispreferred attributes more often than when they were primed with preferred attributes. A statistically significant main effect was found for domain ($F(1, 25) = 10.88, p < 0.01$), confirming that there is significantly more alignment in the Furniture domain. Finally, a statistically significant interaction was found ($F(1, 25) = 5.74, p < 0.05$), confirming our observation that the effect of the prime was less pronounced in the People domain. Interestingly, this is very much in line with the observations made in the previous section, where we saw that orientation is less dispreferred than wearing a tie.

6.4.2   Experiment II: adaptation and overspecification

This experiment looks at overspecification: participants were primed with overspecified referring expressions that included both preferred and dispreferred attributes, and were then asked to produce a referring expression for a target that could be distinguished using a minimal referring expression including only a preferred attribute.

Method
Participants.   Particpants were 28 native speakers of Dutch (8 males, mean age 20 years, 6 months) who participated in exchange for partial course credit. None had participated in experiment I.

Materials and procedure.   The materials and procedure were identical to those in experiment I, with the exception of the referring expressions in the prime turn (see Figure 6.5). In experiment II, the referring expressions in the prime turn were always overspecified. Thus, in the Furniture domain participants heard referring expressions such as *the red chair seen from the front* and in the People domain they heard referring expressions such as *the man with the glasses and the tie*. All these referring expressions were overspecified in that they use two attributes (in addition to the type attribute), including a preferred and a dispreferred one, while either attribute would be sufficient to uniquely identify the referent. All referring expressions in the prime turns were produced by the same speaker as in experiment I.

Results and discussion

Figure 6.7 and Table 6.4 show the proportions of overspecified references in experiment I ("single prime") and experiment II ("dual prime") for both domains. A referring expression was considered overspecified when both the preferred and the dispreferred attribute were used. The results show that when participants were primed with both the preferred and the dispreferred attribute, 52% of the Furniture trials and 57% of the People trials were produced with both attributes, even though the preferred attribute would be sufficient to distinguish the target. By contrast, in experiment I speakers produced overspecified referring expressions in only 11% to 15% of the experimental turns.

To analyze these results, we combined the data for experiment I (single prime) and experiment II (dual prime) and conducted a mixed-effects ANOVA with proportion of overspecification as the dependent variable, domain (Furniture versus People) as a within-subjects variable, and prime (single prime versus dual prime) as a between-subjects variable. The results show a statistically significant effect for prime ($F(1, 52) = 32.50, p < 0.001$); the dual primes result in more overspecified referring expressions (and thus a more frequent use of the dispreferred property) than the single primes. There was no statistically significant effect for domain, and no statistically significant interaction between domain and prime.

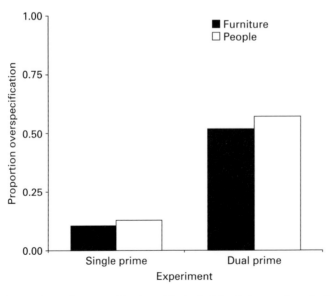

Figure 6.7   Proportions of overspecification with single (*the chair seen from the front / the man with the tie*) and dual primes (*the blue chair seen from the front / the man with the glasses and the tie*) in the People and Furniture domains

Table 6.4. Overspecification means (and standard deviations) for experiment I (selection) and experiment II (overspecification) per domain (Furniture and People) and prime (preferred and dispreferred).

	Experiment I		Experiment II
	Preferred	Dispreferred	
Furniture	0.13 (0.34)	0.11 (0.31)	0.52 (0.37)
People	0.15 (0.36)	0.13 (0.33)	0.57 (0.34)

6.5 General discussion

In this chapter we showed that REG in interactive settings is a balancing act between relatively stable domain-dependent preferences and relatively dynamic interactive factors. We first asked how much data is required to determine the preference ordering for a domain. Our experiment in Section 6.3 showed that with 20 or fewer training instances, acceptable attribute selection results can be achieved; that is, results that do not differ significantly from those obtained using many more training instances. This is good news, because collecting such small amounts of training data should not take too much time and effort, making it relatively easy to apply the graph-based REG algorithm for new domains and languages. Next, we examined the relation between preferences and adaptation, describing two experiments in Section 6.4. Experiment I (looking at attribute selection) showed that speakers were more likely to include a (preferred or dispreferred) attribute in a referring expression when this attribute was primed. Experiment II (looking at referential overspecification) revealed that alignment and overspecification are closely related. While some participants were reluctant to select a dispreferred attribute in experiment I, participants in experiment II aligned frequently with an overspecified referring expression that contained *both* a preferred and a dispreferred attribute, even though including only the preferred one would have been sufficient to produce a distinguishing description.

It could be argued that the interactive nature of our experimental paradigm in Section 6.4 is limited, in that participants did not truly interact with the speaker of the referring expressions they had to comprehend. Rather, participants interacted with an imaginary dialogue partner, which allowed us to guarantee that all participants were primed in exactly the same way. Using an imaginary audience is a standard experimental procedure to study interactive communication, and recent studies have shown that the differences between a real audience and an imagined audience are small (Ferreira *et al.*, 2005; Van Der Wege, 2009).

It is also worth emphasizing that our experimental results in Section 6.4 cannot readily be explained in terms of well-understood phenomena such as lexical or syntactic alignment. In experiment I, what are primed are not lexical items, but attributes. A prime in the Furniture domain may be *the front-facing chair*, where *front-facing* is the relevant value of the orientation attribute, while in the experimental turn participants should produce a referent for, say, a fan whose orientation is to the left. Arguably, what is being primed is a way to look at an object, thereby making certain attributes of the object more salient.

The graph-based algorithm, as described in Section 6.2, does not yet capture the alignment effects we found in experiments I and II (and the same applies to other state-of-the-art REG algorithms such as the Incremental Algorithm: Dale and Reiter, 1995). The graph-based algorithm, as it stands, predicts that a dispreferred property would never be used if a preferred property would be sufficient to uniquely characterize a target. And while the algorithm can account for some amount of overspecification (by allowing some properties to be included for free), it would never redundantly use a dispreferred (expensive) property, even though our participants did this in over half of the cases in experiment II.

What is missing in the algorithm is a sensitivity to the references produced during an earlier interaction. In fact, the use of cost functions offers an elegant way to account for this. In interactive settings, costs could be thought of as a composition of relatively stable domain-dependent preferences (as formalized above) combined with relatively dynamic costs, modeling the activation of properties in the preceding interaction. The latter costs can be derived from Buschmeier *et al.* (2010), who study alignment in micro-planning. Inspired by Buschmeier and colleagues, we can make the costs of an attribute a much cheaper when it is mentioned repeatedly, after which the costs gradually increase as the activation of a decays. The net result is that dispreferred properties become relatively cheap when they have been used in the previous interaction, and hence are more likely to be selected by the graph-based algorithm.

Gatt *et al.* (2011) go one step further than we have here, proposing a new model for alignment in referring expression production that integrates alignment and preference-based attribute selection. This model consists of two parallel processes: a preference-based search process, and an alignment-based process. These two processes run concurrently and compete to contribute attributes to a limited-capacity working memory buffer that will produce the referring expression. This model was tested against the data of experiment II and showed a similar amount of overspecification as the human participants produced.

Use of the graph-based algorithm in an interactive setting has a number of other theoretical and practical advantages. First of all, alignment may reduce the search space for the algorithm; not all alternatives need to be explored, because the search process can be driven by the edges that were used previously in the interaction. In addition, as we have seen, preference orders need to be empirically determined for each new domain. But what to do when your REG algorithm is applied in a domain for which the preference order is unknown? Our experiments suggest that a good strategy might be to simply model alignment.

Various studies have demonstrated both the existence of and the benefits of alignment in human–computer interaction. For a recent survey, see Branigan *et al.* (2010), who argue that a lot is to be gained from computers that align: "Speakers should also feel more positive affect when interacting with a computer that aligns with them than with one that does not" (p. 2359).

The approach outlined in this chapter is limited to generation of distinguishing descriptions, which have identification of a target as their main function. Even though this kind of referring expression has received the most attention in the REG literature, they are certainly not the only kind of referring expression that occur. Various studies (Passonneau, 1996; Di Eugenio et al., 2000; Gupta and Stent, 2005; Jordan and Walker, 2005) confirm that references in interactive settings may serve other functions besides identification. The Coconut corpus (Di Eugenio et al., 2000), for example, is a set of task-oriented dialogues in which participants negotiate the furniture items they want to buy on a fixed, shared budget. Referring expressions in this corpus (e.g., *a yellow rug for 150 dollars*) not only identify a particular piece of furniture, but also include properties that directly pertain to the task (e.g., the amount of money that is still available and the state of agreement between the negotiators). More recently, other researchers have started exploring the generation of referring expressions in interactive settings as well. Stoia et al. (2006), for example, presented a system that generates referring expressions in situated dialogue, taking into account both the dialogue history and the visual context (defined in terms of which distractors are in the current field of vision of the speakers and how distant they are from the target). Janarthanam and Lemon (2010) present a REG policy that automatically adapts to the expertise level of the intended addressee (for example, using *the router* when communicating with an expert user, and *the black block with the lights* when communicating with a novice). These lines of research fit in well with another, more general, strand of research concentrating on choice optimization during NLG based on user data (Walker et al., 2007; White et al., 2009).

6.6 Conclusion

When speakers want to identify a target, such as a chair in a furniture shop, using a distinguishing description, they tend to prefer certain properties over others. We have shown that only a limited number of (semantically transparent) example descriptions is required to be able to determine these preferences, although this also depends on the size and complexity of the domain. In interactive settings, however, the generation of distinguishing descriptions depends not only on preferences but also on the descriptions that were produced earlier in the interaction, as we have shown in two experiments, one dedicated to attribute selection and the other to overspecification. We argue that the graph-based REG algorithm is a suitable candidate to model this balancing act, since its use of cost functions enables us to weigh the different factors in a dynamic way.

Acknowledgments

Emiel Krahmer and Martijn Goudbeek thank The Netherlands Organisation for Scientific Research (NWO) for the VICI grant "Bridging the Gap between Computational Linguistics and Psycholinguistics: The Case of Referring Expressions" (277-70-007). The research described in Section 6.3 is based on Theune et al. (2011); for more information on the research presented in Section 6.4 we refer the reader to Goudbeek and Krahmer (2010). We would like to thank Ruud Koolen and Sander Wubben for their help with the preferences experiment (Section 6.3) and Albert Gatt for discussions on REG and adaptation.

References

Arts, A. (2004). *Overspecification in Instructive Texts*. PhD thesis, Tilburg University.

Belz, A. and Gatt, A. (2008). Intrinsic vs. extrinsic evaluation measures for referring expression generation. In *Proceedings of the Annual Meeting of the Association for Computational Linguistics: Human Language Technologies (ACL-HLT)*, pages 197–200, Columbus, OH. Association for Computational Linguistics.

Branigan, H. P., Pickering, M. J., Pearson, J., and McLean, J. F. (2010). Linguistic alignment between people and computers. *Journal of Pragmatics*, **42**(9):2355–2368.

Brennan, S. E. and Clark, H. H. (1996). Conceptual pacts and lexical choice in conversation. *Journal of Experimental Psychology*, **22**(6):1482–1493.

Buschmeier, H., Bergmann, K., and Kopp, S. (2010). Modelling and evaluation of lexical and syntactic alignment with a priming-based microplanner. In Krahmer, E. and Theune, M., editors, *Empirical Methods in Natural Language Generation*, pages 85–104. Springer, Berlin.

Chartrand, G. and Oellermann, O. R. (1993). *Applied and Algorithmic Graph Theory*. McGraw-Hill, New York, NY.

Croitoru, M. and van Deemter, K. (2007). A conceptual graph approach to the generation of referring expressions. In *Proceedings of the International Joint Conference on Artificial Intelligence (IJCAI)*, pages 2456–2461, Hyderabad, India. International Joint Conference on Artificial Intelligence.

Dale, R. (1989). Cooking up referring expressions. In *Proceedings of the Annual Meeting of the Association for Computational Linguistics (ACL)*, pages 68–75, Vancouver, Canada. Association for Computational Linguistics.

Dale, R. and Haddock, N. (1991). Content determination in the generation of referring expressions. *Computational Intelligence*, **7**(4):252–265.

Dale, R. and Reiter, E. (1995). Computational interpretation of the Gricean maxims in the generation of referring expressions. *Cognitive Science*, **19**(2):233–263.

de Ruiter, J. P., Bangerter, A., and Dings, P. (2012). The interplay between gesture and speech in the production of referring expressions: Investigating the tradeoff hypothesis. *Topics in Cognitive Science*, **4**(2):232–248.

Di Eugenio, B., Jordan, P. W., Thomason, R. H., and Moore, J. D. (2000). The agreement process: An empirical investigation of human–human computer-mediated collaborative dialogue. *International Journal of Human-Computer Studies*, **53**(6):1017–1076.

Engelhardt, P. E., Bailey, K. G. D., and Ferreira, F. (2006). Do speakers and listeners observe the Gricean maxim of quantity? *Journal of Memory and Language*, **54**(4):554–573.

Ferreira, V. S., Slevc, L. R., and Rogers, E. S. (2005). How do speakers avoid ambiguous linguistic expressions? *Cognition*, **96**(3):263–284.

Garey, M. R. and Johnson, D. S. (1979). *Computers and Intractability: A Guide to the Theory of NP-Completeness*. W. H. Freeman, New York, NY.

Gatt, A. (2007). *Generating Coherent References to Multiple Entities*. PhD thesis, Department of Computing Science, University of Aberdeen.

Gatt, A. and Belz, A. (2010). Introducing shared tasks to NLG: The TUNA shared task evaluation challenges. In Krahmer, E. and Theune, M., editors, *Empirical Methods in Natural Language Generation*, pages 264–293. Springer, Berlin, Heidelberg.

Gatt, A., Belz, A., and Kow, E. (2008). The TUNA challenge 2008: Overview and evaluation results. In *Proceedings of the International Workshop on Natural Language Generation (INLG)*, pages 198–207, Salt Fork, OH. Association for Computational Linguistics.

Gatt, A., Belz, A., and Kow, E. (2009). The TUNA-REG challenge 2009: Overview and evaluation results. In *Proceedings of the European Workshop on Natural Language Generation (ENLG)*, pages 174–182, Athens, Greece. Association for Computational Linguistics.

Gatt, A., Goudbeek, M., and Krahmer, E. (2011). Attribute preference and priming in reference production: Experimental evidence and computational modelling. In *Proceedings of the Annual Conference of the Cognitive Science Society (CogSci)*, pages 2627–2632, Boston, MA. Cognitive Science Society.

Gatt, A., van der Sluis, I., and van Deemter, K. (2007). Evaluating algorithms for the generation of referring expressions using a balanced corpus. In *Proceedings of the European Workshop on Natural Language Generation (ENLG)*, pages 49–56, Saarbrücken, Germany. Association for Computational Linguistics.

Gibbons, A. M. (1985). *Algorithmic Graph Theory*. Cambridge University Press, Cambridge, UK.

Gorniak, P. J. and Roy, D. (2004). Grounded semantic composition for visual scenes. *Journal of Artificial Intelligence Research*, **21**:429–470.

Goudbeek, M. and Krahmer, E. (2010). Preferences versus adaptation during referring expression generation. In *Proceedings of the Annual Meeting of the Association for Computational Linguistics (ACL)*, pages 55–59, Uppsala, Sweden. Association for Computational Linguistics.

Guhe, M. and Bard, E. G. (2008). Adapting referring expressions to the task environment. In *Proceedings of the Annual Conference of the Cognitive Science Society (CogSci)*, pages 2404–2409, Washington, DC. Cognitive Science Society.

Gupta, S. and Stent, A. (2005). Automatic evaluation of referring expression generation using corpora. In *Proceedings of the Workshop on Using Corpora for Natural Language Generation (UCNLG)*, Birmingham, UK. ITRI, University of Brighton.

Hartigan, J. and Wong, M. (1979). Algorithm AS 136: A k-means clustering algorithm. *Journal of the Royal Statistical Society. Series C (Applied Statistics)*, **28**(1):100–108.

Janarthanam, S. and Lemon, O. (2010). Learning adaptive referring expression generation policies for spoken dialogue systems. In Krahmer, E. and Theune, M., editors, *Empirical Methods in Natural Language Generation*, pages 67–84. Springer, Berlin.

Jordan, P. W. and Walker, M. A. (2005). Learning content selection rules for generating object descriptions in dialogue. *Journal of Artificial Intelligence Research*, **24**(1):157–194.

Koolen, R., Gatt, A., Goudbeek, M., and Krahmer, E. (2009). Need I say more? On factors causing referential overspecification. In *Proceedings of the CogSci workshop on the Production of Referring Expressions*, Amsterdam, The Netherlands. Available from http://pre2009.uvt.nl/. Accessed on 11/24/2013.

Krahmer, E., Theune, M., Viethen, J., and Hendrickx, I. (2008). GRAPH: The costs of redundancy in referring expressions. In *Proceedings of the International Conference on Natural Language Generation (INLG)*, pages 227–229, Salt Fork, OH. Association for Computational Linguistics.

Krahmer, E., van Erk, S., and Verleg, A. (2003). Graph-based generation of referring expressions. *Computational Linguistics*, **29**(1):53–72.

Matthews, D., Butcher, J., Lieven, E., and Tomasello, M. (2012). Two- and four-year-olds learn to adapt referring expressions to context: Effects of distracters and feedback on referential communication. *Topics in Cognitive Science*, **4**(2):184–210.

Mellish, C., Scott, D., Cahill, L., Paiva, D., Evans, R., and Reape, M. (2006). A reference architecture for natural language generation systems. *Natural Language Engineering*, **12**(1):1–34.

Olson, D. R. (1970). Language and thought: Aspects of a cognitive theory of semantics. *Psychological Review*, **77**(4):257–273.

Passonneau, R. (1996). Using centering to relax Gricean informational constraints on discourse anaphoric noun phrases. *Language and Speech*, **39**(2–3):229–264.

Pechmann, T. (1989). Incremental speech production and referential overspecification. *Linguistics*, **27**(1):89–110.

Pickering, M. and Garrod, S. (2004). Toward a mechanistic psychology of dialogue. *Behavioral and Brain Sciences*, **27**(2):169–226.

Reiter, E. and Dale, R. (2000). *Building Natural Language Generation Systems*. Cambridge University Press, Cambridge, UK.

Sonnenschein, S. (1984). The effect of redundant communication on listeners: Why different types may have different effects. *Journal of Psycholinguistic Research*, **13**(2):147–166.

Stoia, L., Shockley, D. M., Byron, D., and Fosler-Lussier, E. (2006). Noun phrase generation for situated dialogs. In *Proceedings of the International Conference on Natural Language Generation (INLG)*, pages 81–88, Sydney, Australia. Association for Computational Linguistics.

Theune, M., Koolen, R., Krahmer, E., and Wubben, S. (2011). Does size matter: How much data is required to train a REG algorithm? In *Proceedings of the Annual Meeting of the Association for Computational Linguistics: Human Language Technologies (ACL-HLT)*, pages 660–664, Portland, OR. Association for Computational Linguistics.

van Deemter, K., Gatt, A., van der Sluis, I., and Power, R. (2012). Generation of referring expressions: Assessing the Incremental Algorithm. *Cognitive Science*, **36**(5):799–836.

van Deemter, K. and Krahmer, E. (2007). Graphs and booleans: On the generation of referring expressions. In Bunt, H. and Muskens, R., editors, *Computing Meaning*, volume 3, pages 397–422. Springer, Dordrecht, The Netherlands.

van Deemter, K., van der Sluis, I., and Gatt, A. (2006). Building a semantically transparent corpus for the generation of referring expressions. In *Proceedings of the International Conference on Natural Language Generation (INLG)*, pages 130–132, Sydney, Australia. Association for Computational Linguistics.

van der Sluis, I. and Krahmer, E. (2007). Generating multimodal referring expressions. *Discourse Processes*, **44**(3):145–174.

Van Der Wege, M. (2009). Lexical entrainment and lexical differentiation in reference phrase choice. *Journal of Memory and Language*, **60**(4):448–463.

Viethen, J. and Dale, R. (2006). Algorithms for generating referring expressions: Do they do what people do? In *Proceedings of the International Conference on Natural Language Generation (INLG)*, pages 63–72, Sydney, Australia. Association for Computational Linguistics.

Viethen, J., Dale, R., Krahmer, E., Theune, M., and Touset, P. (2008). Controlling redundancy in referring expressions. In *Proceedings of the International Conference on Language Resources and Evaluation (LREC)*, Marrakech, Morocco. European Language Resources Association.

Walker, M. A., Stent, A., Mairesse, F., and Prasad, R. (2007). Individual and domain adaptation in sentence planning for dialogue. *Journal of Artificial Intelligence Research*, **30**(1):413–456.

White, M., Rajkumar, R., Ito, K., and Speer, S. R. (2009). Eye tracking for the online evaluation of prosody in speech synthesis: Not so fast! In *Proceedings of the International Conference on Spoken Language Processing (INTERSPEECH)*, pages 2523–2526, Brighton, UK. International Speech Communication Association.

Part III

Handling uncertainty

7 Reinforcement learning approaches to natural language generation in interactive systems

Oliver Lemon, Srinivasan Janarthanam, and Verena Rieser

In this chapter we will describe a new approach to generating natural language in interactive systems – one that shares many features with more traditional planning approaches but that uses statistical machine learning models to develop *adaptive* natural language generation (NLG) components for interactive applications. We employ statistical models of users, of generation contexts, and of natural language itself. This approach has several potential advantages: the ability to train models on real data, the availability of precise mathematical methods for optimization, and the capacity to adapt robustly to previously unseen situations. Rather than emulating human behavior in generation (which can be sub-optimal), these methods can find strategies for NLG that improve on human performance. Recently, some very encouraging test results have been obtained with real users of systems developed using these methods.

In this chapter we will explain the motivations behind this approach, and will present several case studies, with reference to recent empirical results in the areas of information presentation and referring expression generation, including new work on the generation of temporal referring expressions. Finally, we provide a critical outlook for future work on statistical approaches to adaptive NLG.

7.1 Motivation

A key problem in Natural Language Generation (NLG) is choosing "how to say" something once "what to say" has been determined. For example, if the specific domain of an interactive system is restaurant search, the dialogue act inform(price=cheap) might be realized as *This restaurant is economically priced* or *That's a cheap place to eat*, and the act present_results(item3,item4,item7) could be realized as *There are 3 places you might like. The first is El Bar on Quixote Street, the second is Pancho Villa on Elm Row, and the third is Hot Tamale on Quality Street*, or else perhaps by contrasting the differing features of the three restaurants. The central issue for NLG in interactive systems is that the context (linguistic or extra-linguistic) within which generation takes place is dynamic. This means that different expressions might need to be generated at different moments, even to convey the same concept or refer to the same object for the same user. That is, *good NLG systems must appropriately adapt their output to the changing context*. This requires modeling the interaction context, and determining policies, plans, or strategies that are sensitive to transitions in this context.

In principle, NLG in an interactive system such as a spoken dialogue system (SDS) comprises a wide variety of decisions, from content structuring, choice of referring expressions, use of ellipsis, aggregation, and choice of syntactic structure, to the choice of intonation markers for synthesized speech. In existing computational dialogue systems, "what to say" is usually determined by a dialogue manager (DM) component, via planning, hand-coded rules, finite-state machines, or learned strategies, and "how to say it" is then very often determined by simple templates or hand-coded rules that define appropriate word strings to be sent to a speech synthesizer or screen.

Reinforcement learning-based approaches to NLG for interactive systems could in theory have several key advantages over template-based and rule-based approaches (we discuss other approaches to "trainable" NLG in Section 7.1.2):

- The ability to adapt to fine-grained changes in dialogue context
- A data-driven development cycle
- Provably optimal action policies with a precise mathematical model for action selection
- The ability to generalize to unseen dialogue states.

In the remainder of this chapter we will present the background concepts for these approaches (Section 7.1.1), previous work on adaptive NLG (Section 7.1.2), and then detailed case studies in the areas of information presentation (Section 7.2) and referring expression generation (Section 7.3), including recent work on adaptive generation of temporal referring expressions (Section 7.4). We maintain a particular focus on results of recent evaluations with real users. We also provide an outlook for this research area in Section 7.5.

7.1.1   Background: Reinforcement learning approaches to NLG

The basic model for each of the approaches that we will discuss below is the Markov Decision Process (MDP). Here a stochastic system interacting with its environment (in our case, the user of the interactive system) through its actions is described by a number of states $\{s_i\}$ in which a given number of actions $\{a_j\}$ can be performed. In an interactive system, the states represent the possible contexts (e.g., how much information we have so far obtained from the user or what has previously happened in the interaction, etc.; Frampton and Lemon, 2006; Henderson *et al.*, 2008[1]), and the actions are system interaction actions (e.g., to display a graphic or speak an utterance). See Rieser and Lemon (2011) for a full presentation of such methods in the context of dialogue management and NLG.

Each state–action pair is associated with a transition probability $P_a(s, s') = P(s_{t+1} = s'|s_t = s, a_t = a)$: the probability of moving from state s at time t to state s' at time $t+1$ after having performed action a when in state s. These probabilities are acquired by observing how users respond to the system's actions. The transitions are also associated

---

[1] Note that a common misunderstanding is that the Markov property constrains models of state to exclude the history. However, we can employ variables in the current state that explicitly represent features of the history.

with a reinforcement signal (or "reward") $r_a(s, s')$ describing how good the result of action a was when performed in state s. In interactive systems these reward signals are most often associated with task completion and the length or duration of the interaction, and they have been defined by the system designer (e.g., 100 points for successful task completion, -1 per turn) to optimize some *a priori* objective of the system. However, as we shall discuss, researchers have also developed data-driven methods for defining reward functions (Walker *et al.*, 2000; Rieser and Lemon, 2008, 2011).

To control a system described in this way, one needs a strategy or policy π mapping all states to actions. In this framework, a reinforcement learning agent is a system aiming at optimally mapping states to actions, i.e., finding the best strategy π' so as to maximize an overall reward R that is a function (most often a weighted sum) of all the immediate rewards. In interaction the reward for an action is often not immediate but is delayed until successful completion of a task. Of course, actions affect not only the immediate reward but also the next state and thereby all subsequent rewards.

In general, then, we try to find an action policy π that maximizes the value $Q^{\pi}(s, a)$ of choosing action a in state s, which is given by the Bellman equation:

$$Q^{\pi}(s, a) = \sum_{s'} P_a(s, s')[r_a(s, s') + \gamma V^{\pi}(s')] \tag{7.1}$$

(Here γ is a discount factor between 0 and 1, and $V^{\pi}(s')$ is the value of state s' according to π; see Sutton and Barto, 1998).

If the transition probabilities are known, an analytic solution can be computed by dynamic programming. Otherwise the system has to learn the optimal strategy by a trial-and-error process, for example using reinforcement learning methods (Sutton and Barto, 1998) as we discuss below. Trial-and-error search and delayed rewards are the two main features of reinforcement learning.

With these concepts in place, we can discuss recent advances made in the application of such models to NLG problems. The overall framework of NLG for interaction as planning under uncertainty was first set out in Lemon (2008) and developed further in Rieser and Lemon (2009) and Lemon (2011). In this framework, each NLG action is a sequential decision point, based on the current interaction context and the expected long-term utility or "reward" of the action. This approach has recently been adopted by other researchers (Dethlefs and Cuayáhuitl, 2011b; Dethlefs *et al.*, 2011). Other recent approaches describe NLG for interaction as planning (e.g., Koller and Stone, 2007; Koller and Petrick, 2011), or as contextual decision making according to a cost function (van Deemter, 2009), but not as a statistical planning problem, where uncertainty in the stochastic environment is explicitly modeled.

7.1.2   Previous work in adaptive NLG

Most approaches to generating natural language have focused on static, non-interactive settings, such as summarizing numerical time-series data (Reiter *et al.*, 2005), and the majority have generated text meant for human consumption via reading, rather than for listening to via synthesized speech in an interactive dialogue system.

Numerous NLG systems have the capability to adapt the language used, e.g., based on the discourse context (Stoyanchev and Stent, 2009), on the user's needs or preferences (Stent et al., 2002; Reiter et al., 2003; Moore et al., 2004; Walker et al., 2004), or on the speaker style targeted for the system (Belz, 2007; Mairesse, 2008; Mairesse and Walker, 2010). However, very few systems can adapt both the content and the language generated by the system in a combined fashion.

One of the most sophisticated adaptive NLG approaches in recent years generated different types of spoken information presentations for different types of users (in the flight information domain; Moore et al., 2004). This system determined theme/rheme information structure (including contrasts for different user preferences) and employed word-level emphasis for spoken output. It used a rule-based planning system for content structuring based on predefined user models, and a Combinatory Categorial Grammar (CCG) realizer component (White et al., 2007). However, this system, like other user-adaptive systems (Dale, 1989; Paris, 1988; Isard et al., 2003; Walker et al., 2004), depends on a static user model, which the system is given before performing generation. A new challenge is generation based on uncertain and dynamic models of the user and the interaction context. These models need to be estimated during interaction with the system, and cannot be preloaded (Maloor and Chai, 2000; Janarthanam and Lemon, 2010a).

In addition, the great majority of currently deployed NLG modules within interactive systems use only template-based generation (fixed patterns of output where some phrases are variables filled at runtime). Here, the key decisions of "what to say and how to say it" are still being made by non-adaptive template-based systems, which always say the same things in the same way, no matter who the user is, or where they are in space, time, or social context. Recent advances made in the CLASSiC[2] project have shown that statistical planning methods can outperform traditional template-based and rule-based adaptive approaches (Rieser and Lemon, 2009; Janarthanam and Lemon, 2010a; Rieser et al., 2010).

Rule-based and supervised learning approaches have both been proposed for learning and adapting dynamically to users during interaction. Such systems learn from a user at the start of an interaction and then later adapt to the domain knowledge of the user. However, they either require expensive expert knowledge resources to hand-code the inference rules for adaptation (Cawsey, 1993), or else a large corpus of expert–layperson interactions from which adaptive strategies can be learned and modeled, using methods such as Bayesian networks (Akiba and Tanaka, 1994). In contrast, our approach learns in the absence of these expensive resources. It is also not clear how supervised approaches choose between when to seek more information from the user and when to adapt to them. We show how to learn this decision automatically, in Section 7.3.

We now examine case studies that illustrate some of the advantages of this new adaptive approach to NLG in interactive systems: information presentation (in Section 7.2) and referring expression generation (in Section 7.3), including temporal referring expression generation (in Section 7.4).

---

[2] http://www.classic-project.org.

7.2   Adaptive information presentation

In information-seeking dialogues the information presentation phase is the primary contributor to dialogue duration (Walker *et al.*, 2001), and as such, is a central aspect of the design of such a system. During this phase the system presents a set of options, retrieved from a database, that match the user's current search constraints. The information presentation strategy should support the user in choosing between all the available options and ultimately in selecting the most suitable option. A central problem in this task is the tradeoff between presenting "enough" information to the user (for example, helping them to feel confident that they have a good overview of the search results) and keeping the utterances short and understandable, especially when using spoken output.

In Rieser *et al.* (2010) we showed that information presentation for spoken dialogue systems can be treated as a data-driven joint optimization problem. The information presentation problem contains many decisions available for exploration; for instance, which presentation strategy to apply (*NLG strategy selection*), how many attributes of each item to present (*attribute selection*), how to rank the items and attributes according to different models of user preferences (*attribute ordering*), how many (specific) items to tell the user about (*coverage*), how many sentences to use when doing so (*syntactic planning*), which words to use (*lexical choice*), and so on. All these parameters (and potentially many more) can be varied, and ideally, jointly optimized based on user judgments.

Broadly speaking, previous research on information presentation (IP) for spoken dialogue systems has been concerned with two major problems: (1) IP strategy selection and (2) content or attribute selection. Researchers have presented a variety of **IP strategies** for structuring information (see examples in Table 7.1). For example, the SUMMARY strategy is used to guide the user's "focus of attention" to relevant attributes by grouping the current results from the database into clusters (e.g., Demberg and Moore, 2006; Polifroni and Walker, 2008). Other studies investigate a COMPARE strategy (e.g., Walker *et al.*, 2007; Nakatsu, 2008), while most work in SDS uses a RECOMMEND strategy, e.g., Young *et al.* (2007).

Prior work on **content or attribute selection** has used a "Summarize and Refine" approach (Chung, 2004; Polifroni and Walker, 2006, 2008). This approach employs utility-based attribute selection with respect to how each attribute of a set of items (e.g., price or food type in restaurant search) helps to narrow down the user's goal to a single item. Related work explores a user modeling approach, where attributes are ranked according to user preferences (Demberg and Moore, 2006; Winterboer *et al.*, 2007). The data collection and training environment we present in this chapter incorporate these approaches.

Previous work on information presentation for interaction only focused on individual aspects of the problem (e.g., how many attributes to generate, or when to use a SUMMARY), using a pipeline model with features from the dialogue manager as input, and where NLG has no knowledge of lower-level features (e.g., the expected behavior of the surface realizer).

Rieser *et al.* (2010) were the first to apply a data-driven method to this whole decision space (i.e., combinations of information presentation strategies and attribute selection),

Table 7.1. Example realizations, generated when the user provided `cuisine=Indian`, and where the wizard has also selected the additional attribute `price` for presentation to the user

Strategy	Example utterance
SUMMARY no UM	*I found 26 restaurants, which have Indian cuisine. 11 of the restaurants are in the expensive price range. Furthermore, 10 of the restaurants are in the cheap price range and 5 of the restaurants are in the moderate price range.*
SUMMARY UM	*26 restaurants meet your query. There are 10 restaurants which serve Indian food and are in the cheap price range. There are also 16 others which are more expensive.*
COMPARE by item	*The restaurant called Kebab Mahal is an Indian restaurant. It is in the cheap price range. And the restaurant called Saffrani, which is also an Indian restaurant, is in the moderate price range.*
COMPARE by attribute	*The restaurant called Kebab Mahal and the restaurant called Saffrani are both Indian restaurants. However, Kebab Mahal is in the cheap price range while Saffrani is moderately priced.*
RECOMMEND	*The restaurant called Kebab Mahal has the best overall quality amongst the matching restaurants. It is an Indian restaurant, and it is in the cheap price range.*

and to show the utility of both lower-level features (e.g., from the surface realizer) and higher-level features (e.g., from the dialogue manager) for this problem. We used reinforcement learning (RL; Sutton and Barto, 1998) as a statistical planning framework to explore contextual features for making NLG decisions, and proposed a new joint optimization method for information presentation strategies combining content structuring and attribute selection (Rieser *et al.*, 2010). Here, the task of the spoken dialogue system was to present a set of search results (e.g., restaurants) to users. In particular, the NLG module had seven possible policies for structuring the content (see Figure 7.1): recommending one single item, comparing two items, summarizing all of them, or ordered combinations of those actions, e.g., first summarize all the retrieved items and then recommend one of them. It had to decide which action to take next, how many attributes to mention, and when to stop generating. We used a sentence generator based on the stochastic sentence planner SPARKY (Stent *et al.*, 2004) for surface realization.

In particular, we implemented the following information presentation actions (see examples in Table 7.1):

- SUMMARY of all matching restaurants with or without a user model (UM), following Polifroni and Walker (2008). With a UM, the system only tells the user about items matching the user's preferences (e.g., cheap), whereas with no UM, the system lists all matching items.
- COMPARE the top two restaurants *by item* (i.e., listing all the attributes for the first item and then for the other) or *by attribute* (i.e., directly comparing the different attribute values).
- RECOMMEND the top-ranking restaurant for this user (based on the UM).

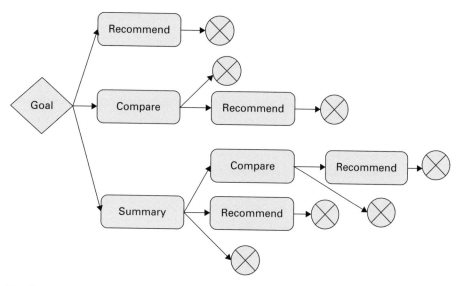

Figure 7.1   Possible information presentation action paths (X = stop generation)

Recall that for a Markov Decision Process (MDP), we need to define the state space, the actions that can be taken from each state, and a reward function. The MDP can be trained using data to learn the optimal action to take in each state. For the information presentation problem, our actions are as defined above and depicted in Figure 7.1. Each state is defined by a set of dialogue context features including the user's most recent action, features from the dialogue manager, and predicted features from the surface realizer. To avoid having to take the time of real users, we train a *user simulation* (described in Section 7.2.2) from a small amount of dialogue data (summarized in Section 7.2.1). We also use the collected data to train a reward function (presented in Section 7.2.3).

7.2.1   Corpus

We conducted a Wizard of Oz data collection, in which trained humans ("wizards") made information presentations about restaurants to users who believed they were talking with a spoken dialogue system. We collected 213 dialogues with 18 participants and 2 wizards. We used that data to train the models described below.

7.2.2   User simulations for training NLG

User simulations are commonly used to train strategies for dialogue management (e.g., Georgila *et al.*, 2005; Schatzmann *et al.*, 2006, 2007; Ai and Litman, 2007). A user simulation for NLG is very similar, in that it is a predictive model of the most likely next user action. However, this NLG-predicted user act does not actually change the overall dialogue state (e.g., by filling slots); it only changes the generator state. In other words, the NLG user simulation tells us what the user would most likely do next, *if we were to stop generating now*.

In information-seeking systems we are most interested in these user actions:

1. `select`: The user chooses one of the presented items, e.g., *Yes, I'll take that one*. This reply type indicates that the information presentation was sufficient for the user to make a choice.
2. `addInfo`: The user provides more attributes, e.g., *I want something cheap*. This reply type indicates that the user has a more specific request, which she or he wants to specify after being presented with the current information.
3. `requestMoreInfo`: The user asks for more information, e.g., *Can you recommend me one?*, *What is the price range of the last item?*. This reply type indicates that the system failed to present the information for which the user was looking.
4. `askRepeat`: The user asks the system to repeat the same message again, e.g., *Can you repeat?*. This reply type indicates that the information presentation was either too long or confusing for the user to remember, or the speech synthesis quality was not good enough, or both.
5. `silence`: The user does not say anything. In this case it is up to the system to take the initiative.
6. `hangup`: The user closes the interaction.

For the experiments presented below, we built user simulations from the Wizard of Oz data using n-gram models of system (s) and user (u) acts, as first done by Eckert *et al.* (1997). In order to account for data sparsity, we applied different discounting ("smoothing") techniques including back-off, using the CMU Statistical Language Modeling toolkit (Clarkson and Rosenfeld, 1997). We constructed a bigram model for predicting user reactions to the system's information presentation strategy decisions: $P(a_{u,t}|IP_{s,t})$; and a trigram (i.e., IP strategy + attribute selection) model for predicting user reactions to the system's combined information presentation strategy and attribute selection decisions: $P(a_{u,t}|IP_{s,t}, attributes_{s,t})$.[3]

We evaluated the performance of these models by measuring dialogue similarity to the original data, based on the Kullback–Leibler (KL) divergence, as also used by, for example, Cuayáhuitl *et al.* (2005), Jung *et al.* (2009), and Janarthanam and Lemon (2009). We compared the raw probabilities as observed in the data with the probabilities generated by our n-gram models using different discounting techniques for each context (Rieser *et al.*, 2010). All the models diverged only slightly from the original data (especially the bigram model), suggesting that they are reasonable simulations for training and testing NLG strategies. We used the most similar user models for system training, and the most dissimilar user models for testing NLG strategies, in order to test whether the learned strategies are robust and can adapt to unseen dialogue contexts.

---

[3] $a_{u,t}$ is the predicted next user action at time t; $IP_{s,t}$ is the system's information presentation action at time t; and $attributes_{s,t}$ are the attributes selected by the system at time t.

7.2.3   Data-driven reward function

The reward function was constructed from the Wizard of Oz data, using a step-wise linear regression, following the PARADISE framework (Walker *et al.*, 2000). This model selects the features that significantly correlate with users' ratings for the NLG strategy in the Wizard of Oz data. We also assign a value to the user's reactions (*valueUserReaction*), similar to optimizing task success for dialogue management (Young *et al.*, 2007). This reflects the fact that good information presentation strategies should help the user select an item (*valueUserReaction* = +100) or provide more constraints (*valueUserReaction* = ±0), but should not necessitate any other type of user response (*valueUserReaction* = −100).

The regression in Equation (7.2) ($R^2 = 0.26$) indicates that users' ratings are influenced by higher-level and lower-level features: users prefer to be focused on a small set of database hits (where *#DBhits* ranges over [1–100]), which will enable them to choose an item (*valueUserReaction*), while keeping the information presentation utterances short (where *#sentence* is in the range [2–18]). Database hits approximate the user's cognitive load; see Rieser *et al.* (2010).

$$Reward = (-1.2) * \#DBhits + (0.121) * valueUserReaction$$
$$- (1.43) * \#sentence \tag{7.2}$$

Note that the worst possible reward for an NLG move is therefore $(-1.20 * 100) - (0.121 * 100) - (18 * 1.43) = -157.84$. This is achieved by presenting 100 items to the user in 18 sentences,[4] in such a way that the user ends the conversation unsuccessfully. The top possible reward is achieved in the rare cases where the system can immediately present one item to the user using just two sentences, and the user then selects that item, i.e., $-(1.20 * 1) + (0.121 * 100) - (2 * 1.43) = 8.06$.

7.2.4   Reinforcement learning experiments

We have successfully formulated the information presentation problem for the restaurant-search domain as a Markov Decision Process (MDP), where states are NLG dialogue contexts and actions are NLG decisions. Each state–action pair is associated with a transition probability, which is the probability of moving from state s at time t to state s' at time $t + 1$ after having performed action a. This transition probability is computed by the environment model (i.e., the user simulation and realizer), and explicitly captures the uncertainty in the generation environment. This is a major difference compared to other non-statistical planning approaches to NLG. Each transition is also associated with a reinforcement signal (or "reward") r_{t+1} describing how good the result of action a was when performed in state s. The aim of the MDP is to maximize the

---

[4] Note that the maximum possible number of sentences generated by the realizer is 18 for the full information presentation sequence SUMMARY+COMPARE+RECOMMEND using all the attributes.

long-term expected reward of its decisions, resulting in a policy that maps each possible state to an appropriate action in that state.

We treat information presentation as a hierarchical joint optimization problem, where first one of the information presentation actions (SUMMARY, COMPARE, or RECOM-MEND) is chosen and then the number of attributes is decided. At each generation step, the MDP can choose from one to five attributes (e.g., cuisine, price range, location, food quality, and/or service quality). Generation stops as soon as the user is predicted to select an item, i.e., the information presentation task is successful. (Note that the same constraint was operational for the Wizard of Oz data collection.)

States are represented as sets of context features. The state space comprises lower-level features about realizer behavior (two discrete features representing the number of attributes and sentences generated so far) and three binary features representing the user's predicted next action, as well as higher-level features provided by the dialogue manager. In particular, in some experiments we use a "focus-of-attention" model comprising additional "higher-level" features, e.g., current database hits covering the user's query and current number of attributes specified by the user (*#DBhits* and *#attributes*; Rieser *et al.*, 2010). We trained the strategy using the SHARSHA algorithm (Shapiro and Langley, 2002) with linear function approximation (Sutton and Barto, 1998), and the simulation environment described in Section 7.2.1. The strategy was trained for 60 000 iterations.

We compare the learned information presentation strategies against a Wizard of Oz baseline, constructed using supervised learning over the Wizard of Oz data. This baseline selects only higher-level features (Rieser *et al.*, 2010).

For attribute selection we choose a majority baseline (randomly choosing between three or four attributes) since the attribute selection models learned by supervised learning on the Wizard of Oz data didn't show significant improvements over this baseline.

For training, we used the user simulation model most similar to the Wizard of Oz data (see Section 7.2.1). For testing, we used the user simulation model most *different* from the Wizard of Oz data.

We first investigated how well information presentation (IP) strategies (without attribute choice) can be learned in increasingly complex generation **scenarios**. A generation scenario is a combination of a particular kind of surface realizer (template vs. stochastic) along with different levels of variation introduced by certain features of the dialogue context (e.g., *#DBhits* and *#attributes*). We therefore investigated the following information presentation **policies**:

1.1 **IP strategy choice, template realizer**: Predicted next user action varies according to the bigram user simulator: $P(a_{u,t}|IP_{s,t})$. Number of sentences and attributes per information presentation strategy is set by defaults, reflecting the use of a template-based realizer.

1.2 **IP strategy choice, stochastic realizer**: The number of attributes per NLG turn is given at the beginning of each episode (e.g., set by the dialogue manager). Sentences are generated using the SPARKY stochastic realizer.

We then investigated different scenarios for *jointly* optimizing information presentation strategy (IPS) and attribute selection (Attr) decisions. We investigated the following policies:

2.1 **IPS+Attr choice, template realizer**: Predicted next user action varies according to the trigram model: $P(a_{u,t}|IP_{s,t}, attributes_{s,t})$. Number of sentences per information presentation strategy set by default.

2.2 **IPS+Attr choice, template realizer+focus model**: Trigram user simulation with the template-based realizer, and focus of attention model with respect to *#DBhits* and *#attributes*.

2.3 **IPS+Attr choice, stochastic realizer**: Trigram user simulation with sentence/attribute relationship according to the stochastic realizer.

2.4 **IPS+Attr choice, stochastic realizer+focus model**: Predicted next user action varies according to the trigram model, the focus of attention model, *and* the sentence/attribute relationship according to the stochastic realizer.

7.2.5   Results: Simulated users

We compared the average final reward (see Equation 7.2) gained by the baseline strategy against the reinforcement learning-based policies in the different scenarios for each of 1000 test runs, using a paired sample t-test. The results are shown in Table 7.2. In five out of six scenarios the reinforcement learning-based policy significantly ($p < 0.001$) outperformed the supervised learning ("wizard") baseline. We also report the percentage of the top possible reward gained by the individual policies, and the raw percentage improvement over the baseline strategy. (The best possible [100%] reward can only be achieved in rare cases.)

The reinforcement learning-based policies show that lower-level features are important in gaining significant improvement over the baseline. The more complex the scenario, the harder it is to gain higher rewards for the policies in general (as more variation is introduced), but the relative improvement in rewards also increases with complexity: the baseline does not adapt well to the variations in lower-level features, whereas reinforcement learning can adapt to the more challenging scenarios.[5]

For example, the reinforcement learning-based policy for scenario 1.1 starts with a SUMMARY if the initial number of items returned from the database is high (> 30). It will then stop generating if the user is predicted to select an item. Otherwise, it continues with a RECOMMEND. If the number of database items is low, it will start with a COMPARE and then continue with a RECOMMEND, unless the user selects an item.

The reinforcement learning-based policy for scenario 1.2 learns to adapt to a more complex scenario: the number of attributes requested by the dialogue manager and produced by the stochastic sentence realizer. It learns to generate the whole sequence (SUMMARY+COMPARE+RECOMMEND) if *#attributes* is low (<3), because the overall

---

[5] The baseline does reasonably well in scenarios with variation introduced by only higher-level features (e.g., scenario 2.2).

Table 7.2. Experimental results for 1000 simulated dialogues, where *** denotes that the reinforcement learning (RL)-based strategy is significantly ($p < 0.001$) better than the baseline strategy

Scenario	Wizard baseline average reward	RL average reward	RL % − baseline % = % improvement
1.1	$-15.82(\pm15.53)$	$-9.90^{***}(\pm15.38)$	**89.2%** − 85.6% = 3.6%
1.2	$-19.83(\pm17.59)$	$-12.83^{***}(\pm16.88)$	**87.4%** − 83.2% = 4.2%
2.1	$-12.53(\pm16.31)$	$-6.03^{***}(\pm11.89)$	**91.5%** − 87.6% = 3.9%
2.2	$-14.15(\pm16.60)$	$-14.18(\pm18.04)$	**86.6%** − 86.6% = 0.0%
2.3	$-17.43(\pm15.87)$	$-9.66^{***}(\pm14.44)$	**89.3%** − 84.6% = 4.7%
2.4	$-19.59(\pm17.75)$	$-12.78^{***}(\pm15.83)$	87.4% − 83.3% = 4.1%

generated information presentation (final *#sentences*) is still relatively short. Otherwise it is similar to the policy for scenario 1.1.

The reinforcement learning-based policies for jointly optimizing information presentation strategy and attribute selection learn to select the number of attributes according to the generation scenarios 2.1–2.4. For example, the reinforcement learning-based policy for scenario 2.1 generates a RECOMMEND with 5 attributes if the number of database hits is low (<13). Otherwise, it will start with a SUMMARY using 2 attributes. If the user is predicted to narrow down his/her focus after the summary, the strategy continues with a COMPARE using one attribute only, otherwise it helps the user by presenting 4 attributes. It then continues with RECOMMEND(5), and stops as soon as the user is predicted to select one item.

The Wizard of Oz baseline strategy generates all the possible information presentation actions (with three or four attributes) but is restricted to use only higher-level features. Nevertheless, this wizard strategy achieves up to 87.6% of the possible reward on this task, and so can be considered a serious baseline against which to measure performance. By beating this baseline we show the importance of the lower-level features for optimizing NLG for interactive systems. The only case (scenario 2.2) where reinforcement learning does not improve significantly over the baseline is where lower-level features do not play an important role for learning good strategies: scenario 2.2 is only sensitive to higher-level features (*#DBhits*).

7.2.6   Results: Real users

A reinforcement learning-based information presentation strategy based on the results of the previous experiment was deployed in an extensive online user study, involving 131 users and more than 800 test dialogues, which explores the contribution of adaptive information presentation to overall "global" task success in a spoken dialogue system.

The information presentation strategy was integrated into the CAMINFO system, a spoken dialogue system developed in the CLASSIC project[6] and providing tourist information for Cambridge (Young *et al.*, 2010). This baseline system has been made

---

[6] http://www.classic-project.org.

accessible by phone using VoIP technology, enabling out-of-lab evaluation with large numbers of users. The speech recognizer (ASR), semantic parser (SLU), and dialogue manager (DM) were all developed at Cambridge University. For speech synthesis (TTS), the Baratinoo synthesizer, developed at France Telecom, was used.

The DM uses a POMDP (Partially Observable Markov Decision Process) framework, allowing it to process n-best lists of ASR hypotheses and keep track of multiple dialogue state hypotheses. The DM strategy is trained to select system dialogue acts given a probability distribution over possible dialogue states. It has been shown that such dialogue managers can exploit the information in the n-best lists (as opposed to only using the top ASR hypothesis) and are therefore particularly effective in noisy conditions (Young *et al.*, 2010).

The NLG component of this baseline system is a standard rule-based surface realizer covering the full range of dialogue acts that the dialogue manager can produce. It has only one information presentation strategy, i.e., the *recommend* strategy, see Table 7.1. The attributes of the venue to be presented are selected heuristically. In the extended version of the system, the information presentation strategy is replaced by our trained NLG component, which is optimized to decide between different information presentation actions as described above.

The participants in this study were directed to a webpage with detailed instructions and, for each task, a phone number to call and a scenario to follow. The scenario describes a place to eat in town, with some constraints, for example: *You want to find a moderately priced restaurant and it should be in the Riverside area. You want to know the address, phone number, and type of food.* After interacting with the system to complete the scenario, the participants were asked to fill in a short questionnaire.

For objective evaluation we focused on measuring task completion rates. We found that the reinforcement learning-based information presentation strategy significantly improves task completion rates for real users, with up to a 9.7% increase (30% relative) compared to the baseline, hand-coded, dialogue system. This result establishes the benefits of this overall methodology for building NLG systems. A full presentation of this study is given in Rieser *et al.* (2011).

7.3   Adapting to unknown users in referring expression generation

Now that we have illustrated the reinforcement learning-based approach to NLG for interactive systems using the information presentation problem, we present another case study, for a reinforcement learning approach to user-adaptive referring expression generation (REG) (Janarthanam and Lemon, 2010b). Our task here is to give the system a method for choosing appropriate expressions to refer to domain entities in a dialogue setting, where users' levels of knowledge and expertise may vary. For instance, in a technical support conversation, an interactive system could choose to use more technical terms with an expert user, or to use more descriptive and general expressions with novice users, and a mix of the two with intermediate users of various sorts (see examples in Table 7.3).

Table 7.3. Referring expression examples for two entities (from a corpus of Wizard of Oz dialogues about setting up a broadband internet connection)

Jargon: *Please plug one end of the* `broadband cable` *into the* `broadband filter`.

Descriptive: *Please plug one end of the* `thin white cable with grey ends` *into the* `small white box`.

In natural human–human conversations, dialogue partners learn about each other and adapt their language accordingly (Clark and Murphy, 1982; Bell, 1984; Isaacs and Clark, 1987). A spoken dialogue system should be capable of observing the user's dialogue behavior, modeling his/her domain knowledge, and adapting accordingly, just like a human interlocutor. We present a corpus-driven framework for learning a user-adaptive referring expression generation strategy from a small corpus of non-adaptive human–machine interactions. We show that the learned strategy performs better than a simple hand-coded adaptive strategy in terms of accuracy of adaptation, dialogue time, and task completion rate when evaluated with real users.

In Section 7.3.1 we summarize the data we used for this set of experiments. In Sections 7.3.2 and 7.3.3 we describe the dialogue system framework and the user simulation model we implemented. In Section 7.3.4, we describe how we trained the reinforcement learning-based referring expression generation strategy, and in Section 7.3.5, we present an evaluation of the referring expression generation strategy with real users.

7.3.1   Corpus

We used as training data a small corpus of 12 non-adaptive dialogues between real users and a dialogue system, in which the system instructed the user in setting up a broadband internet connection. There were six dialogues in which users interacted with a system that used only jargon expressions and six with a system that used descriptive expressions. For more details about the corpus, see Janarthanam and Lemon (2010b).

Domain entities whose jargon expressions raised clarification requests from users were listed; those that had more than the mean number of clarification requests were classified as `difficult`, and the others as `easy` entities (for example, *power adaptor* is easy because all users understood this expression, while *broadband filter* is `difficult`).

7.3.2   Dialogue manager and generation modules

In this section, we describe the different modules of the overall dialogue system we used for our adaptive referring expression generation experiments. The dialogue system presents the user with instructions to set up a broadband connection at home. In our machine learning setup, the system and the user simulation interact at the level of abstract dialogue actions and abstract referring expressions. Our objective is to learn to choose the appropriate referring expressions to refer to the domain entities in the instructions.

Dialogue manager

The dialogue manager (DM) identifies the next dialogue act ($a_{s,t}$ where t denotes turn and s denotes system) to give to the user based on the dialogue management strategy π_{DM}. Here, the dialogue management strategy is fixed, and is coded in the form of a finite-state machine. In this dialogue task, the system provides instructions to either observe or manipulate the environment. When users ask for clarification on referring expressions, the system clarifies by giving information to enable the user to associate the expression with the intended referent. When the user responds in any other way, the system's instruction is simply repeated. The dialogue manager is also responsible for updating and managing the system state $S_{s,t}$. At each system turn, the system passes to the user simulation both the system action $a_{s,t}$ and the referring expressions $REC_{s,t}$.

The dialogue state and dynamic user model

The dialogue state $S_{s,t}$ represents the current state of the conversation at system turn t. In addition to maintaining an overall dialogue state, the system maintains a user model $UM_{s,t}$ which records the initial domain knowledge of the user. This is a dynamic model that starts with a state where the system does not have any information about the user. Since the model is updated according to the user's behavior, it may be inaccurate if the user's behavior is itself uncertain. Hence, the user model is not always an accurate model of the user's knowledge and reflects a level of uncertainty about the user.

Each jargon referring expression x is represented by a three-valued variable in the dialogue state: user_knows_x, which can take a value from {yes, no, unknown}. The variables are updated using a simple update algorithm after each user turn. Initially each variable is set to unknown. If the user responds to an instruction containing the referring expression x with a clarification request, then user_knows_x is set to no. Similarly, if the user responds with appropriate information to the system's instruction, the dialogue manager sets user_knows_x to yes. Only the user's initial knowledge is recorded. This is based on the assumption that an estimate of the user's initial knowledge helps to predict the user's knowledge of the other referring expressions in the domain.

7.3.3   Referring expression generation module

The referring expression generation module identifies the list of domain entities to be referred to and chooses the appropriate referring expression for each of the domain entities for each given dialogue act. It chooses between two types of referring expressions – jargon and descriptive. For example, the domain entity broadband_filter can be referred to using the jargon expression *broadband filter* or using the descriptive expression *small white box*. Although adaptation is the primary goal, it should be noted that in order to get information for the user model, the system needs to seek information by using jargon expressions.

The referring expression generation module operates in two modes – **learning** and **evaluation**. In **learning mode**, it learns to associate dialogue states with optimal referring expressions. That is, it learns a referring expression generation strategy $\pi_{reg} : UM_{s,t} \rightarrow REC_{s,t}$, which maps the state of the user model to optimal referring expressions. Each referring expression is represented as a pair identifying a referent r and a type of expression rt. For instance, the pair (broadband filter, desc) represents the descriptive expression *small white box*. The set of referring expressions used in the current system turn is $REC_{s,t} = (r_1, rt_1), \ldots, (r_n, rt_n)$. In **evaluation mode**, a trained referring expression generation system interacts with unknown users. It consults the learned strategy π_{reg} to choose referring expressions based on the dialogue state and current user model.

7.3.4   User simulations

In this section, we present several user simulation models that simulate the dialogue behavior of a human user and are sensitive to a system's choices of referring expressions.

Corpus-driven action selection model

The user simulation (US) receives the system action $a_{s,t}$ and its referring expression choices $REC_{s,t}$ at each turn. The US responds with a user action $a_{u,t}$ (u denoting user). This can either be a clarification request (cr) or an instruction response (ir). The US produces a clarification request cr based on the class of the referent $C(r_i)$, type of the referring expression rt_i, and the current domain knowledge of the user for the referring expression, $DK_{u,t}(r_i, rt_i)$. Clarification requests are produced using a model that aims to maximize the following:

$$P(a_{u,t} = cr(r_i, rt_i)|C(r_i), rt_i, DK_{u,t}(r_i, rt_i))$$

where $(r_i, rt_i) \in REC_{s,t}$. The user identification of the entity is signified when there is no clarification request produced (i.e., $a_{u,t} = none$). When no clarification request is produced, the environment action $EA_{u,t}$ is generated using a model that aims to maximize

$$P(EA_{u,t}|a_{s,t}) \text{ if } a_{u,t}! = cr(r_i, t_i)$$

Finally, the user action is an instruction response that is determined by the system action $a_{s,t}$. Instruction responses can be either provide_info, acknowledgement, or other based on the system's instruction. The instruction response is chosen to maximize

$$P(a_{u,t} = ir|EA_{u,t}, a_{s,t})$$

All the above models were trained on our corpus using maximum likelihood estimation and smoothed using a variant of Witten–Bell discounting. The corpus contained dialogues between a non-adaptive dialogue system and real users. According to the data, clarification requests are much more likely when jargon expressions are used to refer to referents that belong to the difficult class and that the user does not know about. When the system uses expressions that the user knows, the user generally responds to the instruction given by the system.

User domain knowledge

The user domain knowledge is set to one of several models at the start of every conversation. The models, which range from novices to experts, were identified from the corpus using k-means clustering. A novice user knows only descriptive expressions like *power adaptor*; an intermediate user knows some jargon expressions; an expert knows all the jargon expressions. We assume that all users can interpret the descriptive expressions and resolve their references. Therefore, they are not explicitly represented. We only code the user's knowledge of jargon expressions (see Section 7.3.2).

7.3.5   Training the referring expression generation module

The referring expression generation module was trained (operated in learning mode) using the above simulations to learn reinforcement learning-based policies that select referring expressions based on the user's domain knowledge. In this section, we discuss how to code the reward for learning. We then discuss how the policy of the referring expression generation module is learned.

Reward function

We designed a reward function for the goal of adapting to each user's domain knowledge. We call this function *adaptation accuracy* (AA); it calculates how accurately the agent chose the appropriate expressions for each referent r, with respect to the user's knowledge. So, when the user knows the jargon expression for r, the appropriate expression to use is jargon, and if she or he does not know the jargon, a descriptive expression is appropriate. Although the user's domain knowledge is dynamically changing due to learning, we base appropriateness on the user's initial state, because our objective is to adapt to the initial state of the user, $DK_{u,0}$. However, in reality, designers might want their system to account for users' changing knowledge as well. We calculate accuracy per referent, RA_r, and then calculate the overall mean adaptation accuracy (AA) over all referents as shown below:

$$RA_r = \frac{\#(appropriate\ expressions(r))}{\#(instances(r))}$$

$$AA = \frac{1}{\#r} \sum_r RA_r$$

Learning

The referring expression generation module was trained in learning mode using the above reward function and the SARSA reinforcement learning algorithm (with linear function approximation; Sutton and Barto, 1998; Shapiro and Langley, 2002). The user simulation was calibrated to produce three types of users – novice, intermediate, and expert – randomly but with equal probability. The training produced approximately 5000 dialogues.

Initially, during training, the referring expression generation strategy chooses randomly between the referring expression types for each domain entity in the system utterance, irrespective of the user model state. Once the referring expressions are chosen, the system presents the user simulation with both the dialogue act and the referring expression choices. The choice of referring expression affects the user's dialogue behavior. For instance, choosing a jargon expression could evoke a clarification request from the user, based on which the dialogue manager updates the internal user model ($UM_{s,t}$) with the new information that the user is ignorant of that particular expression. (Recall that using a jargon expression is an information-seeking move that enables the referring expression generation module to estimate the user's knowledge level.) The same process is repeated for every dialogue instruction. At the end of the dialogue, the system is rewarded based on its referring expression choices. If the system chooses jargon expressions for novice users or descriptive expressions for expert users, penalties are incurred; conversely, if the system chooses referring expressions appropriately, the reward is high. On the one hand, those actions that fetch more reward are reinforced, and on the other hand, the agent tries out new state–action combinations to explore the possibility of greater rewards. Over time, it stops exploring new state–action combinations and exploits those actions that contribute to higher reward. Figure 7.2 shows how the referring expression generation module learns during training. It can be

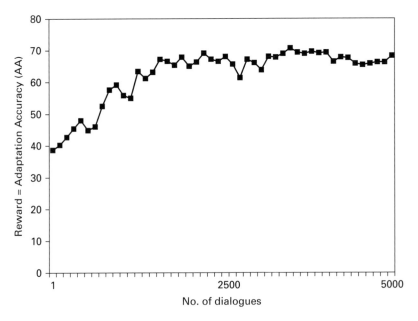

Figure 7.2   Learning curve for adaptive referring expression generation

seen in the figure that towards the end the curve plateaus, signifying that learning has converged.

7.3.6   Evaluation with real users

In order to compare the performance of the learned strategy with a baseline, a simple rule-based strategy was built. This baseline was chosen because it performed better in simulation than a variety of other baselines (Janarthanam and Lemon, 2010b). It uses jargon for all referents by default and provides clarifications when requested. It exploits the user model in subsequent references after the user's knowledge of the expression has been set to either yes or no. Therefore, although it is a simple strategy, it does adapt to a certain extent. It should be noted that this strategy was built in the absence of expert domain knowledge.

We evaluated the two strategies with real users. Thirty-six university students from different backgrounds (e.g., arts, humanities, medicine, and engineering) participated in the evaluation. Seventeen participants interacted with a system with the baseline strategy and the other 19 interacted with a system with the learned strategy. Each participant was given a pre-task recognition test to record his/her initial domain knowledge. The experimenter read out a list of technical terms and the participant was asked to point to the domain entities laid out in front of him/her. The participant was then assigned to one of the two systems. Following system instructions, the participant attempted to set up a broadband connection while interacting with the system. After the dialogue had ended, the participant was given a post-task test where the recognition test was repeated. The participant's broadband connection setup was manually examined for task completion (i.e., the percentage of correct connections made in the final setup). The participant was given the task completion results and was then given a user satisfaction questionnaire.

For this experiment, we used human "wizards" for automatic speech recognition and natural language understanding. Each participant utterance was annotated in near-real-time by a wizard for dialogue act information; this was passed to the automated dialogue manager, which responded with a system dialogue action. The wizards were not aware of the referring expression generation strategy used by the system. System utterances were converted to speech by a speech synthesizer (Cereproc) and were played to the participant.

Metrics

In addition to the adaptation accuracy metric mentioned above, we also measured other features of the dialogues: dialogue time (DT), the actual time taken for the participant to complete the task; task completion (TC), the percentage of correct connections in the final setup for each participant; number of turns; and number of clarification requests made by the participant. We also measured the learning effect on the participants as (normalized) learning gain (LG) produced by using unknown jargon expressions. This is calculated using the pre- and post-test scores for the participant domain

Table 7.4. Adaptive referring expression generation: evaluation with real users.
* Statistically significant ($p < 0.05$). ** Statistically significant ($p < 0.001$).
*** Statistically significant ($p < 0.0001$)

Strategy used	Baseline	Learned
Adaptation accuracy (%)	63.91 (± 8.4)	84.72*** (± 4.72)
Learning gain	0.71 (± 0.26)	0.74 (± 0.22)
Dialogue time (mins)	7.86 (± 0.77)	6.98** (± 0.93)
Task completion rate (%)	84.7 (± 14.63)	99.47*** (± 2.29)
No. of turns	30.05 (± 2.33)	28.10* (± 1.85)
No. of clarification requests	4.11 (± 2.36)	2.79 (± 1.43)

knowledge (DK_u) as follows:

$$LG = \frac{Post - Pre}{1 - Pre}$$

The initial knowledge (mean pre-task recognition score) of the two groups of participants was similar (baseline $= 7.35 \pm 1.9$, learned $= 6.57 \pm 2.29$). Hence there is no bias on the user's pre-task score toward any strategy.

We compared the performance of the two strategies using objective parameters and subjective feedback scores. Tests for statistical significance were performed using a Mann–Whitney test for two independent samples (due to the non-parametric nature of the data).

Table 7.4 presents the mean adaptation accuracy (AA), learning gain (LG), dialogue time (DT), and task completion (TC) produced by the two strategies. The learned strategy produced more accurate adaptation than the baseline strategy ($p < 0.0001$). The higher adaptation accuracy (AA) of the learned strategy translates to shorter dialogue time ($p < 0.001$) and higher task completion ($p < 0.0001$) than the baseline strategy. However, there was no significant difference in learning gain (LG).

Users' subjective feedback on the systems did not differ greatly. However, users did notice that it was easier to identify domain objects with the reinforcement learning-based strategy than the baseline strategy ($p < 0.05$).

The learned strategy enabled the system to adapt using the dependencies that it acquired during the training phase. For instance, when the user asked for clarification on a particular jargon referring expression (e.g., *ethernet cable*), the learned strategy used descriptive expressions for related domain objects such as the ethernet light and ethernet socket. Such adaptation across referents enabled the learned strategy to score better than the baseline strategy. Since the agent starts the conversation with no knowledge about the user, it learned to use information-seeking moves (i.e., to use jargon) at appropriate moments. Since it was trained to maximize adaptation accuracy, the agent also learned to restrict such moves and start predicting the user's domain knowledge as soon as possible. By learning to trade off between information-seeking and adaptation, the learned strategy produced faster and more accurate adaptation for users with different domain knowledge levels.

7.4   Adaptive temporal referring expressions

In very recent work a reinforcement learning-based approach has been applied to the generation of temporal referring expressions, such as *The day after tomorrow* or *The same time on the 12th of June* (Janarthanam *et al.*, 2011). This work has been tested in field trials with real users, and we give a brief summary here.

The data-driven approach to temporal expression (TE) generation presented here is in the context of appointment scheduling dialogue systems. Telephone-based automated appointment scheduling systems are currently deployed on a national scale, for example France Telecom's 1013+ service.

The fact that there are multiple ways in which a time can be referred to leads to an interesting NLG problem: how best to realize a temporal expression for a particular individual in a particular context. For example, the following expressions all vary in terms of length, ambiguity, redundant information, and user preference: *next Friday afternoon*, or *Friday next week at the same time*, or *in the afternoon, a week on Friday*.

Temporal expressions can contain two types of references: absolute references such as *Tuesday* and *the 12th of January*, and relative references such as *tomorrow*, *this Tuesday*, and *at the same time*. Generating temporal expressions therefore involves both selecting appropriate pieces of information (date, day, time, month, and week) to present and deciding how to present each one (i.e., via absolute or relative references).

Our objective here is to convey a target appointment slot to users using an expression that is optimal in terms of the tradeoff between understandability, length, and user preference.

7.4.1   Corpus

We used as training data a corpus of 2650 temporal expressions used in a web-based experimental paradigm in which users were presented with a synthesized temporal expression and then asked to rate it (1920 temporal expressions) or to assign it to one or more potentially matching time slots (730 temporal expressions). Participants could hear each temporal expression more than once. We tracked the number of times each participant replayed each temporal expression (all 2650 temporal expressions), and the number of time slots the participant chose to potentially match the temporal expression (730 temporal expressions). For more details about the corpus, see Janarthanam *et al.* (2011).

7.4.2   User simulation

We built a user simulation to simulate the dialogue behavior of users in appointment scheduling conversations. It responds to the temporal expression used by the system to refer to an appointment slot. It responds by either accepting, rejecting, or clarifying the offered slot based on the user's own calendar of their available slots. For instance, the simulated user rejects an offered slot if the user understood the temporal expression but is not available at that time. If it accepts or rejects an offered slot, the user is assumed to understand the temporal expression unambiguously. However, if the simulated user is unable to resolve the appointment slot from the temporal expression, it responds with a clarification request. The simulation therefore responds with a dialogue action $(a_{u,t})$ to temporal expressions based on the system's prior dialogue act $(a_{s,t})$, and the system's temporal expression $(TE_{s,t})$. In addition to $TE_{s,t}$ and $a_{s,t}$, other factors such as distance between the target slot and the current slot (G), the previous slot in context (C), and the user's calendar (Cal) were also taken into account. The model used to generate user dialogue actions aims to maximize the following:

$$P(a_{u,t}|a_{s,t}, TE_{s,t}, G, C, Cal)$$

The user's dialogue action $(a_{u,t})$ is one of {accept_slot, reject_slot, request_clarification}. The probability of clarification requests was calculated as the average of the ambiguity and replay probabilities seen in the corpus.

Reward function

The temporal expression generator was rewarded for each temporal expression that it generated. The reward was given based on tradeoffs between three variables: user preference (UP), length of the temporal expression (L), and clarification request probability (CR). In short, we chose a reward function that penalizes temporal expressions that are long and ambiguous, and that rewards temporal expressions that users prefer. It also indirectly rewards task success by penalizing ambiguous temporal expressions that lead to clarification requests.

7.4.3   Evaluation with real users

The learned temporal expression generation strategy was integrated into an NLG component of a deployed appointment scheduling spoken dialogue system. The learned strategy was activated when the system informed the user of an available time slot. The performance of the learned strategy was compared with that of an existing rule-based adaptive NLG component, which was developed independently for the deployed system. In the rule-based strategy MONTH, DATE, and TIME are always absolute; DAY is relative if the target date is less than three days away (i.e., *today, tomorrow, day after tomorrow*); and WEEK is always relative (i.e., *this week, next week*). All five information units are included in the realization (e.g., *Thursday the 15th July in the afternoon, next week*), although the order is slightly different (DAY-DATE-MONTH-TIME-WEEK).

In this evaluation, participants were asked to make an appointment for an engineer to visit their home. Each participant was given a two-week calendar showing his/her availability and the goal was to arrange an appointment at a time when both the participant and the engineer were available. There were 12 possible scenarios, of different degrees of difficulty, that were evenly rotated across participants and systems.

The system was evaluated by employees at France Telecom and students of partner universities who had never used the appointment scheduling system before. After each scenario, participants were asked to fill out a questionnaire on perceived task success and five user satisfaction questions on a 6-point Likert Scale (Walker *et al.*, 2000). We collected and analyzed 605 dialogues for this study.

The learned strategy showed significant improvement in perceived task success (+23.7%), although no significant difference was observed between the two systems in terms of actual task success (Chi-square test, df $= 1$). Perceived task success is the user's perception of whether they completed the task successfully.

Overall user satisfaction (the average score of all the questions) was significantly higher for the learned strategy (+5%).[7] Dialogues with the learned strategy were significantly shorter with lower call duration in terms of time (-15.7%)[7] and fewer average words per system turn (-23.93%).[7]

The learned strategy consistently results in shorter dialogues across all levels of scenario difficulty. In summary, these results show that using a learned strategy results in shorter dialogues and greater confidence in the user that they have had a successful dialogue.

7.5   Research directions

There are several directions in which research on reinforcement learning-based approaches to NLG for interactive systems can be developed. New research could improve over current methods by treating NLG as a hierarchy of statistical decision processes, and could explore new application domains. For example, some recent work on reinforcement learning for NLG has been done in the domain of generating instructions in virtual environments (Dethlefs and Cuayáhuitl, 2011a,b; Dethlefs *et al.*, 2011).

An interesting challenge for interactive NLG in general is that of "generation under uncertainty" (Lemon *et al.*, 2010; Janarthanam and Lemon, 2011), where language must be generated for users even though there is some uncertainty about their state. This uncertainty can be about their location, their gaze direction and field of view, or even about their goals and preferences. An interesting research direction may be to explicitly represent uncertainty about the generation context using techniques such as Partially Observable Markov Decision Processes (POMDPs).

Future work could also include richer and more complex models of dialogue context, both past and predicted, for optimizing NLG decisions. For example, predicted TTS quality has been explored as a feature (Boidin *et al.*, 2009).

7.6   Conclusions

We have presented a new data-driven method for adaptive NLG in interactive systems, using several case studies: information presentation in spoken dialogue systems and referring expression generation, including generation of temporal referring expressions. At the time of writing, this approach is being actively explored by a variety of

---

[7] Independent two-tailed *t*-test, significant at $p < 0.05$.

researchers (Dethlefs and Cuayáhuitl, 2011b; Dethlefs *et al.*, 2011; Janarthanam *et al.*, 2011; Rieser *et al.*, 2011), and there is closely related work which also explores NLG as a process of maximizing utility (van Deemter, 2009; Golland *et al.*, 2010). Very recent work applies the approach to the generation of utterances during incremental dialogue processing (Dethlefs *et al.*, 2012).

For the information presentation problem we used a statistical optimization framework for content structure planning and attribute selection. Crucially, we used both higher-level features from the dialogue state, and lower-level features from the realizer. This work was the first to apply a data-driven optimization method to this decision space. We found that reinforcement learning-based information presentation strategies significantly improve dialogue task completion for real users compared to conventional, hand-coded presentation prompts. This approach also provides new insights into the nature of the information presentation problem, which has previously been treated as a module following dialogue management with no access to lower-level context features.

In the second case study we showed that user-adaptive referring expression generation policies can also be learned using a reinforcement learning framework and data-driven user simulations. The learned strategy can trade off between adaptive moves and information-seeking moves, to maximize overall adaptation accuracy for a particular user's level of expertise: it starts the conversation with information seeking moves, learns a little about the user, and adapts dynamically as the conversation progresses. We also showed that the learned strategy performs better than a reasonable hand-coded strategy with real users in terms of accuracy of adaptation, dialogue time, task completion, and subjective evaluation.

In the third case study we presented recent work on adaptive generation of temporal referring expressions, using similar techniques. Here the learned strategy showed significant improvement in perceived task success and overall user satisfaction. Dialogues with the learned strategy were also significantly shorter, with lower call duration in terms of time and fewer words per system turn.

It is interesting to note that all three case studies show that an adaptive NLG component significantly contributes to the (perceived or objective) dialogue task success of the system. Thus, data-driven adaptive NLG strategies have "global" effects on overall system performance. The data-driven planning method applied here therefore promises a significant upgrade in the performance of generation modules, and thereby of natural language interaction in general.

Acknowledgments

We would like to thank Simon Keizer, Helen Hastie, and Xingkun Liu for collaboration on the real user evaluations for information presentation and temporal referring expression generation. The research leading to these results has received funding from the European Community's Seventh Framework Programme (FP7) under grant agreement no. 270019 (SpaceBook project), under grant agreement no. 216594 (CLASSiC project, www.classic-project.org), and from the EPSRC, project no. EP/G069840/1.

References

Ai, H. and Litman, D. (2007). Knowledge consistent user simulations for dialog systems. In *Proceedings of the International Conference on Spoken Language Processing (INTERSPEECH)*, pages 2697–2700, Antwerp, Belgium. International Speech Communication Association.

Akiba, T. and Tanaka, H. (1994). A Bayesian approach for user modelling in dialogue systems. In *Proceedings of the International Conference on Computational Linguistics (COLING)*, pages 1212–1218, Kyoto, Japan. International Committee on Computational Linguistics.

Bell, A. (1984). Language style as audience design. *Language in Society*, **13**(2):145–204.

Belz, A. (2007). Probabilistic generation of weather forecast texts. In *Proceedings of Human Language Technologies: The Conference of the North American Chapter of the Association for Computational Linguistics (HLT-NAACL)*, pages 164–171, Rochester, NY. Association for Computational Linguistics.

Boidin, C., Rieser, V., van der Plas, L., Lemon, O., and Chevelu, J. (2009). Predicting how it sounds: Re-ranking dialogue prompts based on TTS quality for adaptive spoken dialogue systems. In *Proceedings of the International Conference on Spoken Language Processing (INTERSPEECH)*, pages 2487–2490, Brighton, UK. International Speech Communication Association.

Cawsey, A. (1993). User modelling in interactive explanations. *User Modeling and User-Adapted Interaction*, **3**(3):221–247.

Chung, G. (2004). Developing a flexible spoken dialog system using simulation. In *Proceedings of the Annual Meeting of the Association for Computational Linguistics (ACL)*, pages 63–70, Barcelona, Spain. Association for Computational Linguistics.

Clark, H. H. and Murphy, G. (1982). Audience design in meaning and reference. In Le Ny, J.-F. and Kintsch, W., editors, *Language and Comprehension*, pages 287–299. North-Holland Publishing Company, Amsterdam, The Netherlands.

Clarkson, P. and Rosenfeld, R. (1997). Statistical language modeling using the CMU-Cambridge toolkit. In *Proceedings of the European Conference on Speech Communication and Technology (EUROSPEECH)*, pages 2707–2710, Rhodes, Greece. International Speech Communication Association.

Cuayáhuitl, H., Renals, S., Lemon, O., and Shimodaira, H. (2005). Human–computer dialogue simulation using hidden Markov models. In *Proceedings of the IEEE workshop on Automatic Speech Recognition and Understanding (ASRU)*, pages 290–295, San Juan, Puerto Rico. Institute of Electrical and Electronics Engineers.

Dale, R. (1989). Cooking up referring expressions. In *Proceedings of the Annual Meeting of the Association for Computational Linguistics (ACL)*, pages 68–75, Vancouver, Canada. Association for Computational Linguistics.

Demberg, V. and Moore, J. (2006). Information presentation in spoken dialogue systems. In *Proceedings of the Conference of the European Chapter of the Association for Computational Linguistics (EACL)*, pages 65–72, Trento, Italy. Association for Computational Linguistics.

Dethlefs, N. and Cuayáhuitl, H. (2011a). Combining hierarchical reinforcement learning and Bayesian networks for natural language generation in situated dialogue. In *Proceedings of the European Workshop on Natural Language Generation (ENLG)*, pages 110–120, Nancy, France. Association for Computational Linguistics.

Dethlefs, N. and Cuayáhuitl, H. (2011b). Hierarchical reinforcement learning and hidden Markov models for task-oriented natural language generation. In *Proceedings of the Annual Meeting of the Association for Computational Linguistics: Human Language Technologies (ACL-HLT)*, pages 654–659, Portland, OR. Association for Computational Linguistics.

Dethlefs, N., Cuayáhuitl, H., and Viethen, J. (2011). Optimising natural language generation decision making for situated dialogue. In *Proceedings of the SIGdial Conference on Discourse and Dialogue (SIGDIAL)*, pages 78–87, Portland, OR. Association for Computational Linguistics.

Dethlefs, N., Hastie, H., Rieser, V., and Lemon, O. (2012). Optimising incremental dialogue decisions using information density for interactive systems. In *Proceedings of the Conference on Empirical Methods in Natural Language Processing and the Conference on Computational Natural Language Learning (EMNLP-CONLL)*, pages 82–93, Jeju Island, Korea. Association for Computational Linguistics.

Eckert, W., Levin, E., and Pieraccini, R. (1997). User modeling for spoken dialogue system evaluation. In *Proceedings of the IEEE workshop on Automatic Speech Recognition and Understanding (ASRU)*, pages 80–87, Santa Barbara, CA. Institute of Electrical and Electronics Engineers.

Frampton, M. and Lemon, O. (2006). Learning more effective dialogue strategies using limited dialogue move features. In *Proceedings of the International Conference on Computational Linguistics and the Annual Meeting of the Association for Computational Linguistics (COLING-ACL)*, pages 185–192, Sydney, Australia. Association for Computational Linguistics.

Georgila, K., Henderson, J., and Lemon, O. (2005). Learning user simulations for information state update dialogue systems. In *Proceedings of the International Conference on Spoken Language Processing (EUROSPEECH)*, pages 893–896, Lisbon, Portugal. International Speech Communication Association.

Golland, D., Liang, P., and Klein, D. (2010). A game-theoretic approach to generating spatial descriptions. In *Proceedings of the Conference on Empirical Methods in Natural Language Processing (EMNLP)*, pages 410–419, Boston, MA. Association for Computational Linguistics.

Henderson, J., Lemon, O., and Georgila, K. (2008). Hybrid reinforcement/supervised learning of dialogue policies from fixed datasets. *Computational Linguistics*, **34**(4):487–513.

Isaacs, E. A. and Clark, H. H. (1987). Reference in conversation between experts and novices. *Journal of Experimental Psychology: General*, **116**(1):26–37.

Isard, A., Oberlander, J., Androutsopoulos, I., and Matheson, C. (2003). Speaking the users' languages. *IEEE Intelligent Systems Magazine: Special Issue "Advances in Natural Language Processing"*, **18**(1):40–45.

Janarthanam, S., Hastie, H., Lemon, O., and Liu, X. (2011). "The day after the day after tomorrow?": A machine learning approach to adaptive temporal expression generation: Training and evaluation with real users. In *Proceedings of the SIGdial Conference on Discourse and Dialogue (SIGDIAL)*, pages 142–151, Portland, OR. Association for Computational Linguistics.

Janarthanam, S. and Lemon, O. (2009). A Wizard of Oz environment to study referring expression generation in a situated spoken dialogue task. In *Proceedings of the European Workshop on Natural Language Generation (ENLG)*, pages 94–97, Athens, Greece. Association for Computational Linguistics.

Janarthanam, S. and Lemon, O. (2010a). Adaptive referring expression generation in spoken dialogue systems: Evaluation with real users. In *Proceedings of the SIGdial Conference on Discourse and Dialogue (SIGDIAL)*, pages 124–131, Tokyo, Japan. Association for Computational Linguistics.

Janarthanam, S. and Lemon, O. (2010b). Learning adaptive referring expression generation policies for spoken dialogue systems. In Krahmer, E. and Theune, M., editors, *Empirical Methods in Natural Language Generation*, pages 67–84. Sp inger, Berlin.

Janarthanam, S. and Lemon, O. (2010c). Learning to adapt to unknown users: Referring expression generation in spoken dialogue systems. In *Proceedings of the Annual Meeting of the Association for Computational Linguistics (ACL)*, pages 69–78, Uppsala, Sweden. Association for Computational Linguistics.

Janarthanam, S. and Lemon, O. (2011). The GRUVE challenge: Generating routes under uncertainty in virtual environments. In *Proceedings of the European Workshop on Natural Language Generation (ENLG)*, pages 208–211, Nancy, France. Association for Computational Linguistics.

Jung, S., Lee, C., Kim, K., Jeong, M., and Lee, G. G. (2009). Data-driven user simulation for automated evaluation of spoken dialog systems. *Computer Speech & Language*, **23**(4): 479–509.

Koller, A. and Petrick, R. (2011). Experiences with planning for natural language generation. *Computational Intelligence*, **27**(1):23–40.

Koller, A. and Stone, M. (2007). Sentence generation as a planning problem. In *Proceedings of the Annual Meeting of the Association for Computational Linguistics (ACL)*, pages 336–343, Prague, Czech Republic. Association for Computational Linguistics.

Lemon, O. (2008). Adaptive natural language generation in dialogue using reinforcement learning. In *Proceedings of the Workshop on the Semantics and Pragmatics of Dialogue (SEMDIAL)*, pages 141–148, London, UK. SemDial.

Lemon, O. (2011). Learning what to say and how to say it: Joint optimisation of spoken dialogue management and natural language generation. *Computer Speech & Language*, **25**(2):210–221.

Lemon, O., Janarthanam, S., and Rieser, V. (2010). Generation under uncertainty. In *Proceedings of the International Conference on Natural Language Generation (INLG)*, pages 255–260, Trim, Ireland. Association for Computational Linguistics.

Mairesse, F. (2008). *Learning to Adapt in Dialogue Systems: Data-driven Models for Personality Recognition and Generation*. PhD thesis, Department of Computer Science, University of Sheffield.

Mairesse, F. and Walker, M. A. (2010). Towards personality-based user adaptation: Psychologically-informed stylistic language generation. *User Modeling and User-Adapted Interaction*, **20**(3):227–278.

Maloor, P. and Chai, J. (2000). Dynamic user level and utility measurement for adaptive dialog in a help-desk system. In *Proceedings of the SIGdial Workshop on Discourse and Dialogue (SIGDIAL)*, pages 94–101, Hong Kong. Association for Computational Linguistics.

Moore, J. D., Foster, M. E., Lemon, O., and White, M. (2004). Generating tailored, comparative descriptions in spoken dialogue. In *Proceedings of the Florida Artificial Intelligence Research Society Conference (FLAIRS)*, pages 917–922, Miami Beach, FL. The Florida Artificial Intelligence Research Society.

Nakatsu, C. (2008). Learning contrastive connectives in sentence realization ranking. In *Proceedings of the SIGdial Workshop on Discourse and Dialogue (SIGDIAL)*, pages 76–79, Columbus, OH. Association for Computational Linguistics.

Paris, C. (1988). Tailoring object descriptions to a user's level of expertise. *Computational Linguistics*, **14**(3):64–78.

Polifroni, J. and Walker, M. (2006). Learning database content for spoken dialogue system design. In *Proceedings of the International Conference on Language Resources and Evaluation (LREC)*, Genoa, Italy. European Language Resources Association.

Polifroni, J. and Walker, M. (2008). Intensional summaries as cooperative responses in dialogue automation and evaluation. In *Proceedings of the Annual Meeting of the Association for Computational Linguistics: Human Language Technologies (ACL-HLT)*, pages 479–487, Columbus, OH. Association for Computational Linguistics.

Reiter, E., Robertson, R., and Osman, L. M. (2003). Lessons from a failure: Generating tailored smoking cessation letters. *Artificial Intelligence*, **144**(1–2):41–58.

Reiter, E., Sripada, S., Hunter, J., and Davy, I. (2005). Choosing words in computer-generated weather forecasts. *Artificial Intelligence*, **167**:137–169.

Rieser, V., Keizer S., Lemon, O., and Liu, X. (2011). Adaptive information presentation for spoken dialogue systems: Evaluation with human subjects. In *Proceedings of the European Workshop on Natural Language Generation (ENLG)*, pages 102–109, Nancy, France. Association for Computational Linguistics.

Rieser, V. and Lemon, O. (2008). Learning effective multimodal dialogue strategies from Wizard of Oz data: Bootstrapping and evaluation. In *Proceedings of the Annual Meeting of the Association for Computational Linguistics: Human Language Technologies (ACL-HLT)*, pages 638–646, Columbus, OH. Association for Computational Linguistics.

Rieser, V. and Lemon, O. (2009). Natural language generation as planning under uncertainty for spoken dialogue systems. In *Proceedings of the Conference of the European Chapter of the Association for Computational Linguistics (EACL)*, pages 683–691, Athens, Greece. Association for Computational Linguistics.

Rieser, V. and Lemon, O. (2011). Learning and evaluation of dialogue strategies for new applications: Empirical methods for optimization from small data sets. *Computational Linguistics*, **37**(1):153–196.

Rieser, V., Lemon, O., and Liu, X. (2010). Optimising information presentation for spoken dialogue systems. In *Proceedings of the Annual Meeting of the Association for Computational Linguistics (ACL)*, pages 1009–1018, Uppsala, Sweden. Association for Computational Linguistics.

Schatzmann, J., Thomson, B., Weilhammer, K., Ye, H., and Young, S. (2007). Agenda-based user simulation for bootstrapping a POMDP dialogue system. In *Proceedings of Human Language Technologies: The Conference of the North American Chapter of the Association for Computational Linguistics (HLT-NAACL)*, pages 149–152, Rochester, NY. Association for Computational Linguistics.

Schatzmann, J., Weilhammer, K., Stuttle, M., and Young, S. (2006). A survey of statistical user simulation techniques for reinforcement-learning of dialogue management strategies. *The Knowledge Engineering Review*, **21**(2):97–126.

Shapiro, D. and Langley, P. (2002). Separating skills from preference: Using learning to program by reward. In *Proceedings of the International Conference on Machine Learning (ICML)*, pages 570–577, Sydney, Australia. The International Machine Learning Society.

Stent, A., Prasad, R., and Walker, M. A. (2004). Trainable sentence planning for complex information presentation in spoken dialog systems. In *Proceedings of the Annual Meeting of the Association for Computational Linguistics (ACL)*, pages 79–86, Barcelona, Spain. Association for Computational Linguistics.

Stent, A., Walker, M., Whittaker, S., and Maloor, P. (2002). User-tailored generation for spoken dialogue: An experiment. In *Proceedings of the International Conference on Spoken Language Processing (INTERSPEECH)*, pages 1281–1284, Denver, CO. International Speech Communication Association.

Stoyanchev, S. and Stent, A. (2009). Concept form adaptation in human–computer dialog. In *Proceedings of the SIGdial Conference on Discourse and Dialogue (SIGDIAL)*, pages 144–147, London, UK. Association for Computational Linguistics.

Sutton, R. S. and Barto, A. G. (1998). *Reinforcement Learning: An Introduction*. MIT Press, Cambridge, MA.

van Deemter, K. (2009). Utility and language generation: The case of vagueness. *Journal of Philosophical Logic*, **38**(6):607–632.

Walker, M., Whittaker, S., Stent, A., Maloor, P., Moore, J., Johnston, M., and Vasireddy, G. (2004). Generation and evaluation of user tailored responses in multimodal dialogue. *Cognitive Science*, **28**(5):811–840.

Walker, M. A., Kamm, C., and Litman, D. (2000). Towards developing general models of usability with PARADISE. *Natural Language Engineering*, **6**(3–4):363–377.

Walker, M. A., Rambow, O., and Rogati, M. (2001). SPoT: A trainable sentence planner. In *Proceedings of the Conference of the North American Chapter of the Association for Computational Linguistics (NAACL)*, Pittsburgh, PA. Association for Computational Linguistics.

Walker, M. A., Stent, A., Mairesse, F., and Prasad, R. (2007). Individual and domain adaptation in sentence planning for dialogue. *Journal of Artificial Intelligence Research*, **30**(1):413–456.

White, M., Rajkumar, R., and Martin, S. (2007). Towards broad coverage surface realization with CCG. In *Proceedings of the Workshop on Using Corpora for NLG: Language Generation and Machine Translation*. Association for Computational Linguistics.

Winterboer, A., Hu, J., Moore, J. D., and Nass, C. (2007). The influence of user tailoring and cognitive load on user performance in spoken dialogue systems. In *Proceedings of the International Conference on Spoken Language Processing (INTERSPEECH)*, pages 2717–2720, Antwerp, Belgium. International Speech Communication Association.

Young, S., Gašić, M., Keizer, S., Mairesse, F., Schatzmann, J., Thomson, B., and Yu, K. (2010). The hidden information state model: A practical framework for POMDP-based spoken dialogue management. *Computer Speech & Language*, **24**(2):150–174.

Young, S., Schatzmann, J., Weilhammer, K., and Ye, H. (2007). The hidden information state approach to dialog management. In *Proceedings of the IEEE International Conference on Acoustics, Speech, and Signal Processing (ICASSP)*, pages IV-149–IV-152, Honolulu, HI. Institute of Electrical and Electronics Engineers.

8 A joint learning approach for situated language generation

Nina Dethlefs and Heriberto Cuayáhuitl

8.1 Introduction

Interactive systems re increasingly **situated**: they have knowledge about the non-linguistic context of the interaction, including aspects related to location, time, and the user (Byron and Fosler-Lussier, 2006; Kelleher *et al.*, 2006; Stoia *et al.*, 2006; Raux and Nakano, 2010; Garoufi and Koller, 2011; Janarthanam *et al.*, 2012). This extra knowledge makes it possible for the Natural Language Generation (NLG) components of these systems to be more **adaptive**, changing their output to suit the larger context. Adaptive NLG systems for situated interaction aim to produce the most effective utterance for each user in each physical and discourse context. At each stage of the generation process (*what to say* or **content selection**, *how to structure content* or **utterance planning**, and *how to express content* or **surface realization**), the best choices depend on the physical and linguistic context, which is constantly changing. Consequently, it is key to successful interaction that adaptive NLG systems constantly monitor the physical environment, the dialogue history, and the user's preferences and behaviors. As the representations of each of these will necessarily be incomplete and error-prone, adaptive NLG systems must also be able to model **uncertainty** in the generation process.

A designer of adaptive N^TLG systems faces at least two challenges. The first challenge is to identify the set of contextual features that are relevant to decision making in a specific generation situation. The second challenge is to develop a method for selecting a (near-)optimal choice in any given situation from a set of competing ones that may initially appear as viable alternatives. Complicating these challenges is the fact that individual generation decisions are tightly interrelated, so the best decision at one stage may easily depend on others.

Consider, for exam le, a system that assists a user in navigating through a virtual environment and solving a number of tasks. During content selection, the system should provide enough information to ensure the success of the user's next move. On the other hand, given that the user needs to process instructions while navigating, the generated instructions should not contain unnecessary or redundant information, and they should be structured in a way that facilitates understanding. Similarly, surface realization choices depend not only on general lexical frequencies, but also on the history of language used in this interaction and/or with this user. Given this interdependence of decisions, the traditional pipeline approach to NLG (which treats generation as an ordered sequence of isolated decisions) is limiting. In this chapter, we present an alternative model that jointly optimizes NLG decisions in a unified framework. We illustrate our model using the GIVE virtual environment, in which a human instruction follower is instructed by an NLG system to perform a sequence of tasks, as part of a situated interaction scenario (Koller *et al.*, 2010a,b).

Related work in NLG for situated interaction has so far focused on finding one best generation strategy to address the task. Stoia *et al.* (2006) proposed to learn decision trees for referring expression generation from human data in a virtual environment very similar to GIVE (Byron and Fosler-Lussier, 2006). They used properties of the spatial context (such as the viewing angle of the instruction follower or their distance from a referent) and the discourse history, and mapped them to noun phrases that humans produced in the same situation. Based on this, they learned how to make various surface realization decisions. Denis (2010) suggested a referring expression generation algorithm for the GIVE task. It is based on Reference Domain Theory (Reboul, 1998; Salmon-Alt and Romary, 2001), which assumes that each act of reference both relies on and modifies the context. Denis views referring expression generation largely as a discrimination process in which the intended referent needs to be singled out from its distractors based on the spatial context and the objects' properties, such as their position and color. Garoufi and Koller (2010) presented a planning algorithm for a GIVE generation system. It makes explicit use of the non-linguistic context to guide the user to positions where unambiguous referring expressions can be generated, and was extended with a maximum entropy model for surface realization in Garoufi and Koller (2011). In this chapter, we will relax the implicit assumption of previous work that one best strategy can be found to address the situated interaction task. Instead, we will argue that learning a set of alternative generation strategies can give the system a richer repertoire of actions to choose from when troubleshooting is needed.

We present a promising computational model for task-oriented NLG based on hierarchical reinforcement learning that learns an optimal generation strategy – unified from multiple sub-strategies – for situated interaction from a limited amount of human data. It learns to balance the tradeoffs arising from two NLG subtasks – content selection and utterance planning. We extend this model with an HMM-based reward function to jointly balance the tradeoffs arising from the previous two NLG subtasks plus the surface realization subtask, in order to optimize long-term interactions rather than isolated utterances. Empirical results show the benefits of this joint model over a sequential generation process.

8.2   GIVE

8.2.1   The GIVE-2 corpus

The GIVE task involves two participants, one instruction giver and one instruction follower, who engage in a "treasure hunt" in a virtual world. To complete the task, the instruction giver guides the instruction follower through the virtual world using navigation instructions and referring expressions. The latter refer to buttons that the instruction

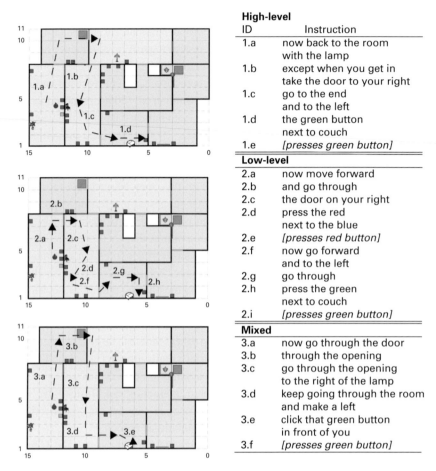

High-level

ID	Instruction
1.a	now back to the room with the lamp
1.b	except when you get in take the door to your right
1.c	go to the end and to the left
1.d	the green button next to couch
1.e	[presses green button]

Low-level

ID	Instruction
2.a	now move forward
2.b	and go through
2.c	the door on your right
2.d	press the red next to the blue
2.e	[presses red button]
2.f	now go forward and to the left
2.g	go through
2.h	press the green next to couch
2.i	[presses green button]

Mixed

ID	Instruction
3.a	now go through the door
3.b	through the opening
3.c	go through the opening to the right of the lamp
3.d	keep going through the room and make a left
3.e	click that green button in front of you
3.f	[presses green button]

Figure 8.1 Example instructions in "high-level," "low-level," and "mixed" instruction modes (all describing the same situation) from the GIVE corpus. The arrows in the maps on the left show the route segment that is described in each instruction. The instruction follower's initial position is indicated by the person in the lower-left room

follower needs to press in a certain sequence in order to unlock a safe and win a trophy. In the GIVE challenge[1] (Koller *et al.*, 2010a,b), the role of instruction giver was taken by an NLG system.

The GIVE-2 corpus (Gargett *et al.*, 2010) is a collection of English and German human–human dialogues collected in a Wizard of Oz study for GIVE. Participants played three games. After the first game, they switched the roles of instruction giver and follower.

Figure 8.1 displays instruction giving sequences from three GIVE-2 dialogues for the same situation. There are fundamental differences in content selection (the level of abstraction in the instructions and referring expressions), utterance planning (the

[1] http://www.give-challenge.org/

amount of information in each instruction), and surface realization. The first set of instructions adopts a "high-level" strategy: the follower is guided by explicit reference to places that are known from the dialogue history, and the structure of the environment is used to construct references to paths, doors, and rooms. The second set of instructions adopts a "low-level" strategy, giving the follower simple commands about basic actions such as turning, going straight, or changing orientation. The last set of instructions uses a "mixed" strategy: while it makes use of visual information to abstract away from low-level actions, there are no references to the dialogue history.

8.2.2 Natural language generation for GIVE

By examining the dialogues in the GIVE-2 corpus, we identified general NLG strategies for GIVE. In terms of content selection, the system should be able to choose a level of abstraction – "low-level," "high-level," or "mixed." The system should also be able to vary the amount of detail included in instructions depending on the complexity of the spatial setting as well as on the dialogue history. With regard to utterance planning, the system should be able to make information packaging decisions – for example, to present a set of instructions in one utterance or across several utterances, linked by a conjunction or as a temporal sequence, with a marked or unmarked theme. Finally, the system should be able to make appropriate lexical and syntactic choices depending on the physical situation and dialogue history.

Figure 8.2 contrasts the dynamics of two possible architectures, a traditional sequential architecture and the joint one proposed in this chapter. In the traditional architecture, a game starts with information about the spatial setting (objects in the virtual world, e.g., rooms, doors) and the dialogue history being sent to the content selection component. Based on this information, the system constructs a first scaffold of a semantic form, which is passed on to the utterance planning component. Here, the utterance is

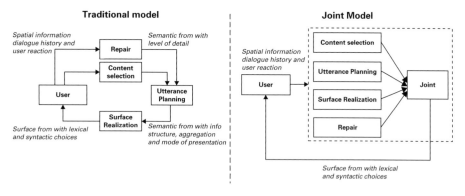

Figure 8.2   Contrasting architectures of an NLG system with the subtasks of content selection, repair, utterance planning, and surface realization. The traditional model on the left treats NLG tasks in sequence; the model on the right (developed in this chapter) treats them jointly. After each generation cycle, a user reaction is observed.

organized in terms of sentential and information structure, and passed on to surface realization. Finally, a string of words is generated and presented to the instruction follower. The system then observes the follower's reaction, and either repairs the previous utterance (if the follower appears lost or confused), or generates the next one. In the joint architecture, there is no sequential order on decision making. Since NLG decisions are treated interdependently, the system works in a unified way by considering all relevant variables jointly and then presenting the chosen utterance to the user.

The tight coupling of the giver and follower's behavior is a key feature of situated NLG. It allows the system to constantly monitor the progress of the interaction and adapt its strategy as soon as the instruction follower shows signs of confusion or hesitation. Since the follower needs to process system utterances online during the course of the interaction, the NLG system faces a tradeoff between generating instructions that are informative and ones that are easy to understand. For example, in content selection, to minimize the chances of the follower getting lost, it seems wise to be specific. On the other hand, unnecessary detail may make an utterance long and hard to comprehend. In utterance planning a goal may be to be efficient, and generate only a few utterances. However, if there is a lot of content to communicate, breaking it up into more and smaller chunks can facilitate comprehension. In surface realization, the system may choose frequently used words and phrases on the assumption that these are easier to understand. Alternatively, the system may use less-frequent words and phrases if the follower responds to them more accurately. The decisions made for any one NLG task may affect the instruction follower and the options available to other NLG tasks. Since the data in the GIVE-2 corpus represent only the output of entire NLG "systems," it makes sense to model the different NLG tasks jointly, rather than in isolation, to be able to balance their tradeoffs.

8.2.3   Data annotation and baseline NLG system

To obtain training data for building our NLG model, we annotated a subset of GIVE-2 dialogues using the features shown in Table 8.1. These features are organized into the following groups: (a) features describing the type of utterance; (b) features capturing the spatial environment; (c) features describing the navigation level and content; (d) referring expression (RE) features; and (e) features about the follower's understanding of the instructions.

We also used the annotated data to create a simple but sensible rule-based baseline NLG system for the GIVE task:

1. Always use a low-level navigation strategy (to avoid overestimating the follower's prior knowledge).
2. Always use destination instructions if the goal is visible (for efficient instruction giving).
3. If the current utterance type is not a repair, then mention the referent's color. Else, use the referent's color only if it is a discriminating feature.
4. If a distractor is mentioned, always mention its color.

Table 8.1. Annotation scheme for the GIVE corpus grouped by feature types

Type	ID	Feature	Values
Utterance	f_1	utterance_type	confirm, repair, RE, navigation
	f_2	within_dialogue_history	true, false
Environment	f_3	all_rooms_known	true, false
	f_4	discriminating_color_referent	true, false
	f_5	discriminating_color_distractor	true, false
	f_6	leaving_room	true, false
	f_7	number_of_distractors	1, 2, 3, 4, 5 or more
	f_8	number_of_landmarks	1, 2, 3, 4, 5 or more
	f_9	is_visible_and_near_object	true, false
Navigation	f_{10}	navigation_level	high_level, low_level, mixed
	f_{11}	navigation_content	destination, "straight", orientation path, direction
RE	f_{12}	referent_color_mentioned	true, false
	f_{13}	distractor_color_mentioned	true, false
	f_{14}	distractor_mentioned	true, false
	f_{15}	landmark_mentioned	true, false
	f_{16}	spatial_relation	left, right, above, below, next_to
User	f_{17}	user_confusions	0, 1, 2 or more
	f_{18}	user_reaction	perform_desired_action, request_help, wait, perform_undesired_action

5. Mention a distractor, if one is present and the referent's color is not a discriminating feature.
6. Mention a landmark, if one is present, the referent's color is not a discriminating feature, and there is no suitable distractor present.
7. Repeat the previous utterance, if the follower requests help. After three unsuccessful attempts, rephrase the utterance.
8. Use a conjunction for pairs of instructions and include temporal markers for more than three instructions.

8.3   Hierarchical reinforcement learning for NLG

8.3.1   An example

Consider one of the GIVE NLG tasks: generate referring expressions for situations where the instruction follower has to press a button (the **referent**) in order to open a door. We want to create a learning agent for this task; the agent should learn when to mention the referent's color, when to include a distractor, when to mention a landmark, and when to mention the referent's spatial position relative to the follower (e.g., left, right, in front). To design such an agent, we need to define a **state space**, represent-ing the agent's knowledge of the task, and a set of **actions**, representing the agent's

(a) Flat state–action space

Feature	Values	Description
f_0	true,false	Is the referent's color discriminating?
f_1	none,one,few,many	How many distractors are present?
f_2	none,one,few,many	How many landmarks are present?
f_3	true,false	Is the referent's spatial position discriminating?
f_4	null,yes,no	Has color been mentioned?
f_5	null,yes,no	Has a distractor been mentioned?
f_6	null,yes,no	Has a landmark been mentioned?
f_7	null,yes,no	Has a spatial position been mentioned?

Action	Description
a_0	Mention referent color
a_1	Do not mention referent color
a_2	Include distractor
a_3	Do not include distractor
a_4	Include landmark
a_5	Do not include landmark
a_6	Include spatial position
a_7	Do not include spatial position

(b) Hierarchical state–action space

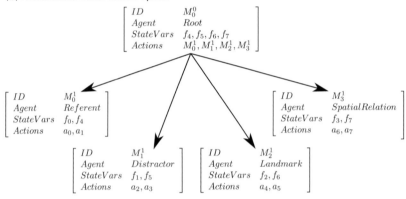

Figure 8.3   Flat and hierarchical state–action spaces for a simple learning agent for an example referring expression generation task. While the flat agent has 41 472 state–action pairs, the hierarchical agent has only 420 state–action pairs. In the hierarchical case, a parent subtask can invoke child subtasks. When they terminate, control is returned to the caller.

capabilities. An example state space is given by features $f_0, \ldots, f_7$ in Figure 8.3, and an example action set is provided by actions $a_0, \ldots, a_7$. The goal states are defined when all feature values (used as high-level representations for surface realization) are non-null. For example, the goal state $f_0 = true, f_1 = one, f_2 = one, f_3 = true, f_4 = yes, f_5 = no, f_6 = no, f_7 = yes$ corresponds to the referring expression *Press the blue button on your left*. Given the state–action space, the agent can learn a referring expression generation strategy using reinforcement learning provided that it receives feedback, or **reward**, after exploring a state–action pair.

Notice that the number of decisions for this simple example grows quickly as we add more state features and actions to the agent's state–action space. Bigger state–action spaces mean slower learning (requiring more training data) with scalability depending on the available computational resources. We can reduce the state–action space by factoring it using reasonable heuristics. The size of the flat state–action space for our example (shown at the top of Figure 8.3) is $|S \times A| = \sum_i |f_i| \times |A| = 41\,472$. If we factor the state–action space into subtasks (as shown at the bottom of Figure 8.3), the size becomes $|S \times A| = \sum_{i,j} |S_j^i| \times |A_j^i| = 420$. This dramatic reduction of the size of the state–action space is produced by (a) dividing the state space into smaller state spaces, (b) ignoring unnecessary state features in each smaller state space, and (c) reducing the action set in each smaller state space. The hierarchical state–action space, essential to scale up to more complex problems, must be designed carefully to limit any accuracy losses in the final solution (in comparison to the non-hierarchical setting).

8.3.2 Reinforcement learning with a flat state–action space

The concept of reinforcement learning (RL) for NLG can be described briefly as follows: given a set of generation states that capture the system's knowledge (e.g., the features in Table 8.1), a set of actions that represent the system's capabilities (e.g., "mention the color of a referent" or "choose a high-level navigation strategy"), and a reward function that rewards each action the agent takes, an optimal generation strategy chooses actions for each state that lead to the highest possible reward. Reinforcement learning is a powerful method to use whenever we do not know the best strategy to address a problem, because we can define a learning agent to find a (near-)optimal strategy automatically by trial and error.

One fundamental model for a reinforcement learning agent is the **Markov Decision Process** (MDP). An MDP can be defined as a four-tuple $< S, A, T, R >$. S is a set of states, A is a set of actions, $T(s', a, s)$ is a probabilistic state transition function that determines the next state s' from the current state s and the performed action a, and $R(s'|s, a)$ is a reward function that specifies the reward that an agent receives for taking action a in state s. The goal of an MDP is to find an optimal policy π^* that maximizes the reward for each visited state, according to $\pi^*(s) = argmax_{a \in A} Q^*(s, a)$, where Q is the expected reward for executing action a in state s and then following the policy π^*. For learning small-scale NLG policies using flat reinforcement learning, we can use algorithms such as SARSA (Sutton, 1996) or Q-learning (Watkins, 1989; Sutton and Barto, 1998). See Janarthanam and Lemon (2010) and Rieser and Lemon (2011) for additional applications of reinforcement learning to NLG.

8.3.3 Reinforcement learning with a hierarchical state–action space

Reinforcement learning systems with large state spaces are affected by a problem referred to as **the curse of dimensionality**: the fact that state spaces grow exponentially with the number of features they take into account. When the state space grows too large the agent will not be able to find an optimal policy for a task. The best one can do in such situations is to provide an approximate solution. One way to do this is to use a divide-and-conquer approach: divide the task represented by the state space into several subtasks, each of which has its own (smaller) state space and the solution to each of which can therefore be found more easily.[2] In other words, learn a hierarchy of policies for related subtasks, rather than learning a single policy for the whole task.

We denote the hierarchy of reinforcement learning agents by $M = \{M_j^i\}$, where the indexes i and j identify an individual agent in the hierarchy. The execution sequence of subtasks is also learned. In our approach, each agent in the hierarchy is defined by a discrete time semi-MDP (SMDP) $< S_j^i, A_i^j, T_j^i, R_j^i >$, which is similar to an MDP but includes actions that last a variable amount of time. S_j^i is the set of states for the agent; A_i^j is its set of actions, and T_j^i is the state transition function that determines the next state s' from the current state s and the performed action a. The reward function $R_j^i(s', \tau | s, a)$ is defined slightly differently than for MDPs due to the hierarchical setting. It specifies the reward that an agent receives for taking an action a in state s lasting τ time steps (Dietterich, 2000). Actions in A_i^j can be either primitive single actions, or composite actions, each of which corresponds to a subtask modeled by a SMDP. Primitive actions yield single rewards, while composite actions yield cumulative rewards. The goal of each SMDP is to find an optimal policy that maximizes the reward for each visited state, according to $\pi^{*i}_j(s) = argmax_{a \in A} Q^{*i}_j(s, a)$, where $Q_j^i(s, a)$ specifies the expected cumulative reward for executing action a in state s and then following π^{*i}_j. For learning hierarchical NLG policies, we use the HSMQ-learning algorithm (Dietterich, 2000). See Cuayáhuitl (2009, p. 92) for an application of HSMQ-learning to spoken dialogue management, and Dethlefs and Cuayáhuitl (2010) for an application to NLG.

In the next section, we present a hierarchical reinforcement learning agent for the GIVE task.

8.4   Hierarchical reinforcement learning for GIVE

8.4.1   Experimental setting

The state–action space

To define the agent's state space, we started with the annotation features introduced in Table 8.1, which define our content selection features. We extended this feature set with the features shown in Table 8.2, which define our utterance planning and surface realization features. The utterance planning features include information about how to present a set of instructions, while the surface realization features include information for choosing among competing surface form candidates that can realize a semantic concept.

For the agent's action space, we use the values that semantic features can take as specified in Table 8.1 for content selection, and as specified in Table 8.2 for utterance planning and surface realization. The concrete state–action sets for our hierarchical NLG agent are shown in Table 8.3. The size of the flat state–action space (considering

---

[2] Other ways of dealing with the curse of dimensionality are based on function approximation (Denecke *et al.*, 2004; Henderson *et al.*, 2005; Pietquin *et al.*, 2011; Cuayáhuitl *et al.*, 2012).

Table 8.2. Features describing utterance planning and surface realization decisions involved in the GIVE task

Type	ID	Feature	Values
Utterance Planning	f_{19}	aggregation	conjunctive, sequential, none
	f_{20}	info_structure	marked_theme, unmarked_theme
	f_{21}	number_of_instructions	1, 2, 3 or more
	f_{22}	presentation	joint, incremental
	f_{23}	temporal_markers	include, dont_include
Surface Realization	f_{24}	detail	direction, path, destination
	f_{25}	destination_preposition	into, in, to, towards, until
	f_{26}	destination_relatum	container_like, point_like
	f_{27}	destination_verb	go, keep_going, walk, continue, get, return, you_need, you_want
	f_{28}	direction_preposition	to (your), to (the)
	f_{29}	direction_verb	bear, turn, go, make, take, hang, move
	f_{30}	orientation	90/180/360 degrees, around, round
	f_{31}	orientation_verb	turn, keep_going, you, look
	f_{32}	path_preposition	down, along, through, across, out, past
	f_{33}	path_relatum	path_like, tunnel_like, container_like, space_like, point_like
	f_{34}	path_verb	go, continue, walk, keep_going, follow, pass
	f_{35}	straight_verb	go, keep_going, move, proceed
	f_{36}	re_determiner	the, this, that
	f_{37}	re_spatial_relation	AP, PP, none
	f_{38}	re_type	it, button, one
	f_{39}	re_verb	push, press, click, click_on, hit, choose

all combinations of feature values) is 10^{23}, which makes it a large-scale decision-making problem.

The hierarchy of agents

An NLG system for the GIVE task needs to give navigation instructions and make references to objects, preferably by taking the spatial context, the dialogue history, and the instruction follower into account. Navigation instructions require decisions about whether to use a low-level, high-level, or mixed instruction-giving strategy, and whether to use a destination, direction, orientation, path, or "straight" instruction. In referring to objects, the system typically needs to decide whether to mention distractors or landmarks, and which attributes of an object to mention. To design a hierarchical reinforcement learning agent that learns to make these decisions, we organize the NLG tasks in the form of a hierarchy as shown in Figure 8.4.

Every generation sequence starts with the root agent M_0^0, which decides whether to generate a reference to an object or a navigation instruction. Its state–action space is defined by the features and actions shown in Table 8.3. If a reference is to be generated, control is passed to agent M_0^1, which can either repair a previous referring expression

Table 8.3. State–action spaces for our hierarchical NLG learning agent

Subtask	State variables	Actions
M_0^0	f_1, f_2, f_{17}, f_{18}	confirm, M_0^1, M_1^1
M_0^1	$f_4, f_5, f_7, f_8, f_9, f_{12}, f_{13},$ $f_{14}, f_{15}, f_{16}, f_{17}, f_{18}$	M_0^2, M_0^3, include referent color, don't include referent color, include distractor, don't include distractor, include landmark, don't include landmark, include spatial relation, M_1^2
M_1^1	$f_3, f_6, f_{11}, f_{17}, f_{18}$	$M_0^2, M_2^2, M_3^2, M_1^2$
M_0^2	$f_{19}, f_{20}, f_{21}, f_{22}, f_{23}$	use conjunctive structure, use sequential structure, use temporal markers, don't use temporal markers, present instructions together, present instructions incrementally, marked theme, unmarked theme
M_1^2	f_1, f_2, f_7, f_{18}	repeat, paraphrase, switch navigation strategy
M_2^2	$f_3, f_6, f_{10}, f_{11}, f_{17}, f_{18}$	$M_1^3, M_2^3, M_3^3, M_4^3, M_5^3$
M_3^2	$f_3, f_6, f_{10}, f_{11}, f_{17}, f_{18}$	$M_1^3, M_2^3, M_3^3, M_4^3, M_5^3$
M_0^3	$f_{36}, f_{37}, f_{38}, f_{39}$	push, press, click, click on, hit, choose, it, button, one, the, this, that, AP, PP, none
M_1^3	$f_{24}, f_{25}, f_{26}, f_{27}$	direction, path, destination, into, in, to, towards, until, go, keep going, walk, continue, get, return
M_2^3	$f_{24}, f_{28}, f_{29}, f_{30}$	direction, path, destination, to (your), to (the), bear, turn, go, make, take, hang, move
M_3^3	$f_{24}, f_{26}, f_{30}, f_{31}$	direction, path, destination, 90/180/360 degrees, around, round, turn, keep going, look
M_4^3	$f_{24}, f_{26}, f_{32}, f_{33}, f_{34}$	direction, path, destination, down, along, through, follow, across, out, past, go, continue, walk, keep going, pass
M_5^3	$f_{24}, f_{30}, f_{33}, f_{35}$	direction, path, destination, go, keep going, move, proceed

(in case the previous one was unsuccessful) by calling agent M_1^2, or generate a new referring expression by calling agent M_0^3. If the decision is made to repair a previous referring expression, it can be repeated or paraphrased or more detail can be added. If the decision is to generate a new referring expression, surface realization decisions need to be made.

If the root agent decides to generate a navigation instruction, it calls agent M_1^1. This agent decides whether to repair a previous unsuccessful navigation instruction by calling agent M_1^2, or generate a new navigation instruction. If the decision is to repair, the previous utterance can either be repeated or paraphrased or more detail can be added, or the navigation strategy can be switched (from high-level to low-level, for example). If the decision is to generate a new instruction, the agent can either choose a low-level navigation strategy by calling agent M_2^2, or a high-level navigation strategy by calling agent M_3^2. Mixed strategies emerge by alternating the two choices. M_2^2 and M_3^2 select content for the navigation instruction: destination, direction, orientation, path, or "straight." Control is then transferred to agents that make surface realization decisions, i.e., agents $M_{0...5}^3$. Before agents M_0^1 (reference) and M_1^1 (navigation) terminate and return control to the root agent, however, some final utterance planning decisions need to be made. Control is therefore transferred to agent M_0^2. This can happen before or after surface realization decisions are made. Once the root agent has reached its goal state, the process terminates and an instruction is presented to the user. A special characteristic of our hierarchical NLG agent is that it combines the decisions of content selection, utterance planning, and surface realization. In this way, we learn one joint policy for all tasks in the hierarchy and achieve a joint treatment, with interrelated decision making, for different NLG tasks. Training and testing the resulting policy will reveal whether this yields any benefits or improvement over isolated optimizations for each individual task.

The simulated environment

Since a reinforcement learning agent needs a large number of interactions to learn an optimal policy, we used a **simulation** to train it. Our simulator simulates properties of (a) the spatial context, represented by features $f_3, \ldots, f_9$; (b) the dialogue context, represented by features f_1, f_2, and f_{21}; and (c) the instruction follower, represented by feature f_{17}. These features are simulated using a unigram language model trained on part of the annotated GIVE-2 corpus for worlds 1 and 2 (data for world 3 was used for testing; see Section 8.4.2).

In addition, we simulated the instruction follower's reaction to an instruction. We used two Naive Bayes classifiers to simulate the following reactions of the instruction follower: perform desired action, perform undesired action, wait, and request help. Only the first reaction indicates a successful instruction. The first classifier was used for utterances that refer to objects; its feature vector included features $f_4, f_5, f_7, f_8, f_9, f_{12}, f_{13}, f_{14}, f_{15}, f_{16}$, and f_{17}. The second classifier was used for navigation instructions, and included features f_3, f_6, f_{10}, f_{11}, and f_{17}. The classifiers were also trained from the part of the annotated GIVE-2 corpus for worlds 1 and 2, and achieved an average accuracy of 82% in a ten-fold cross-validation experiment.

The reward function

We define the reward function for our hierarchy of agents to reward them for generating the shortest possible instructions exhibiting consistency (of navigation strategy). To ensure that the learned navigation instructions are informative, the root agent can only reach its final goal state when the instruction follower has executed the task. We use the reaction classifiers described above to observe whether this is the case. We use the following rewards: 0 for reaching the goal state, -2 for an already invoked subtask or primitive action, $+1$ for generating instruction u consistent (in terms of navigation strategy chosen) with instruction u_{-1}, and -1 otherwise.

We trained all policies for 100 000 episodes using the following parameters: (a) the **step-size parameter** α (one per agent), which indicates the learning rate, was initiated with 1 and then reduced over time by $\alpha = 1/1 + t$, where t is the time step; (b) the **discount rate**, which indicates the relevance of future rewards in relation to immediate rewards, was set to 0.99; and (c) the probability of a random action ϵ was set to 0.01. See Sutton and Barto (1998) for details about the use of these parameters.

8.4.2   Experimental results

Learned policy

In terms of content selection in referring expressions, the agent learns to mention a referent's color, if the color helps to discriminate it from other buttons. In this case, using a color is preferred over using a spatial relation. If color is not a discriminating factor, it may either be used or not be used with equal likelihood, but if distractors are present, the referent may be identified either in relation to them or to a close-by landmark. When repairing a referring expression, a paraphrase with more detail is preferred over a repetition.

When generating navigation instructions, a high-level strategy is preferred because it can lead to more efficient interactions. If the instruction follower has shown signs of confusion more than twice, though, the agent switches permanently to a low-level strategy. When repairing a high-level navigation instruction, switching to a low-level strategy is preferred if not already being used; otherwise a paraphrase with more detail is preferred.

In terms of utterance planning, the agent prefers presentations in instalments for three or more instructions (that is, one instruction per turn), and joint presentation (with a conjunction) otherwise. It varies its use of temporal markers.

In terms of surface realization, the agent learns to generate grammatical utterances. This is the only variable it considers.

An example interaction using our learned NLG policy is shown in Table 8.4. In this interaction, the agent starts with a low-level navigation strategy, and then switches to a high-level one. When the instruction follower gets confused, the system temporarily switches back to a low-level strategy. As far as referring expression generation, the agent first attempts to locate the referent by reference to a distractor (the blue button), and then, as the instruction follower requests help, repairs with a paraphrase that uses a landmark (the chair) instead. While, at first glance, the content selection and utterance planning decisions that were learned seem fine, surface realization looks odd. Since the agent's only restriction is to generate grammatical utterances, it has no language model on which it can base its decisions, and therefore no conception of the frequency with which individual phrases occur in the GIVE-2 corpus. The result is that low-frequency terms are chosen with high frequency (for example, *proceed* in utterances 2 and 10, or *move* in utterances 4 and 6, or *hang* in utterance 8). In the realization of the direction of movement, the agent continuously prefers *to the left / right* over the simpler *left / right*. In the referring expression in utterance 8, the agent left the head noun of the referent (the button) unrealized. This is an available choice according to the GIVE-2

Table 8.4. Dialogue 1: Sample interaction demonstrating the learned NLG policy including referring expression generation, a mixed navigation strategy, and two different repair strategies (switching the navigation strategy and paraphrasing a referring expression)

Turn	Conversant	Utterance
1	USR	Action: starts game
2	SYS	*First, turn 90 degrees left. Then proceed forward.*
3	USR	Reaction: perform desired action
4	SYS	*OK. Move to the left at the end of the hall.*
5	USR	Reaction: wait
6	SYS	*Move to the right and move down the hallway.*
7	USR	Reaction: perform desired action
8	SYS	*Now hang left and press the red beside the blue.*
9	USR	Reaction: request help
10	SYS	*Proceed to the red button beside the chair. Press it.*
11	USR	Reaction: perform desired action
12	SYS	*Well done.*

data, but a very infrequent one, and it confused the instruction follower in this interaction. It therefore seems necessary to equip the agent with a language model that can provide better motivation for its surface realization decisions. Nevertheless, before we do this, we were interested in how humans would rate our instructions (with normalized surface realization choices) against the hand-crafted baseline we presented in Section 8.2.3.

In a human evaluation of our learned agent (Dethlefs *et al.*, 2011), participants rated the helpfulness of three different instructions (human, learned, and baseline) for a spatial navigation scenario in the GIVE task. They rated the instructions generated by the learned agent significantly higher than the baseline instructions ($p < 0.002$), and similarly to human instructions (human instructions were rated slightly higher, but the difference was not statistically significant). These results confirm our hypothesis that hierarchical reinforcement learning (with interrelated decision making) can outperform a rule-based baseline with isolated decision making. The main weakness of the learned agent was with regard to surface realization. In the next section, we show how the agent's generation space – the set of all possible surface forms it can generate – can be formalized as a Hidden Markov Model and used to inform the agent's learning process for surface realization decisions.

8.5 Hierarchical reinforcement learning and HMMs for GIVE

8.5.1 Hidden Markov models for surface realization

In the evaluation of our first learned agent, we noticed that its main weakness was surface realization: the agent was choosing low-frequency lexical items with high frequency and, as a result, produced language that sounded unnatural. We wanted it

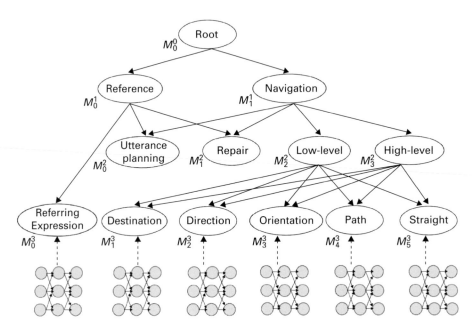

Figure 8.4   Hierarchy of NLG reinforcement learning agents, some using HMM-based reward functions

to learn to make more "human-like" surface realization decisions. Since the reward function in a reinforcement learning agent has a strong impact on the learned behavior, we induced a reward function for surface realization from a statistical language model trained on a corpus of human-produced language (Dethlefs and Cuayáhuitl, 2011b).

The concept of representing the generation space of a surface realizer as a Hidden Markov Model, or HMM (Rabiner, 1989), can be roughly defined as the inverse of part of speech (POS) tagging. A POS-tagger has the task of mapping an observation sequence of words onto a hidden sequence of part of speech tags. In our scenario, the task is to map a sequence of semantic symbols onto a hidden sequence of surface realizations. Figure 8.5 provides an illustration of this idea. We treat the states of the HMM as corresponding to individual words or phrases and sequences of states $i_0, \ldots, i_n$ as corresponding to larger phrases or sentences. An observation sequence $o_0, \ldots, o_m$ consists of a set of semantic symbols (features f_{24} to f_{39} in Table 8.2) that are specific to the content of the instruction (destination, direction, orientation, path or "straight," referring expression). Each of these symbols has an associated observation likelihood $b_i(o_t)$, which gives the probability of observing symbol o in state i at time t.

We used six different HMMs, one for each instruction type. The HMMs were trained over the GIVE-2 corpus in two steps. First, we used the ABL algorithm (van Zaanen, 2000), which aligns strings based on minimum edit distance, and induces a context-free grammar (CFG) from the aligned examples. In this way, we were able to obtain six CFGs, one per instruction type. All grammar rules were of one of two forms, either $r \to \alpha\beta$ or $r \to \alpha, \beta, \ldots$. In the first rule form, r denotes the root node (the sentence node) and α and β correspond to non-terminal symbols realizing a sentence. In the second

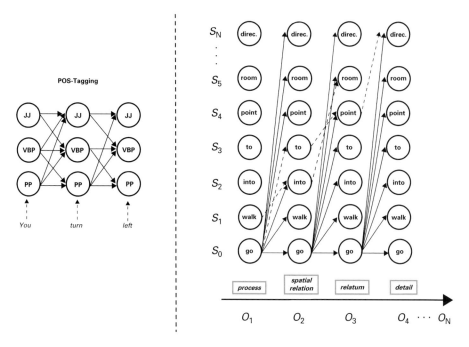

Figure 8.5 (Left): A simple example Hidden Markov Model for part of speech tagging, where an observation sequence of words is mapped onto a hidden sequence of states/part of speech tags. (Right): A Hidden Markov Model for surface realization, where an observation sequence of semantic symbols is mapped onto a sequence of states corresponding to surface realizations.

rule form, r is a semantic constituent and α, β, etc. are alternative surface realizations for the semantic concept denoted by r. We mapped all terminal symbols that realize a specific non-terminal symbol into separate HMM states. For example, if the grammar learned a rule $verbs \rightarrow turn, go, bear, \ldots$, then we grouped *turn*, *go*, and *bear* into a separate HMM state each. When all HMMs had been constructed, we trained them on the GIVE-2 corpus. Their transition and emission probabilities were estimated using the Baum–Welch algorithm (Baum *et al.*, 1970).

For related work on using graphical models for surface realization, see Mairesse *et al.* (2010) for an approach using Bayesian Networks within an Active Learning framework, and Dethlefs and Cuayáhuitl (2011a) for an application of Bayesian Networks within a hierarchical reinforcement learning framework. Dethlefs and Cuayáhuitl (2012) provide a comparison between HMMs and Bayesian Networks. Both directions are interesting for context-aware statistical language generation.

8.5.2 Retraining the learning agent

To retrain the agent so that it learns to make more "human-like" surface realization decisions, while still balancing the tradeoffs of interrelated decision making among content selection, utterance planning, and surface realization, we added a reward of

$P(w_0 \ldots w_n)$ to the original reward function (presented in Section 8.4.1). Whenever the agent has generated a word sequence $w_0 \ldots w_n$, it receives $P(w_0 \ldots w_n)$ as a reward. This corresponds to the likelihood of observing the sequence in the data. Note that this new reward only applies to surface realization agents M_0^3 through M_5^3; all other agents use the original reward function to optimize interaction length at maximal task success using a consistent navigation instruction giving strategy. Using this new reward function, we trained the agent for 150 000 episodes using the same simulated environment and training parameters as before.

8.5.3   Results

We have made two claims in this chapter: (a) that due to their interrelated nature, decisions of content selection, utterance planning, and surface realization should all be optimized in a joint fashion in order to reach optimal performance of NLG for situated interaction; and (b) that HMMs can prove beneficial in supporting the NLG agent's learning process for surface realization decisions. To test both claims, we evaluated the success of our learned agent and its similarity with human-generated navigation instructions.

Results using simulation

After retraining the agent using the HMM-based surface realization modifications presented above, we found that the content selection and utterance planning policies had not changed. In terms of surface realization, the agent now learned to choose a (short) most likely sequence when the user had a low information need (showed no signs of confusion or being lost), but to include more information when this was not the case, even if this meant choosing a less likely surface form. To measure the performance of different NLG tasks using simulation, we measured the average rewards of different policies over all training episodes. We compared our learned policies against the following baselines: (ISO) isolated decision making for all NLG tasks; (CS+UP) joint content selection and utterance planning with isolated surface realization; (CS+SR) joint content selection and surface realization with isolated utterance planning; (UP+SR) joint utterance planning and surface realization with isolated content selection; and (CS+UP+SR) fully joint decision making. This comparison reveals that fully joint decision making outperforms all other policies. An absolute comparison of the average rewards (rescaled from 0 to 1) of the last 1000 training episodes of each policy shows that CS+UP+SR improves over ISO by 71%. It improves over CS+UP by 29%, CS+SR by 26%, and UP+SR by 13%. These differences are significant at $p < 0.01$.

Results for humanlikeness

To evaluate how similar our generated instructions are to human-generated instructions from the GIVE-2 corpus, we compare probability distributions of the automatically generated and human-generated instructions using the symmetric Kullback–Leibler

(KL) divergence (the lower the value the better), a bigram version of the method presented in Cuayáhuitl *et al.* (2005). We compared the following pairs of instructions:

1. Two human data sets ("Real1" and "Real2") that were obtained by splitting the GIVE-2 corpus.
2. The combination of "Real1" and "Real2" ("Real") against the instructions generated by the HMM-based hierarchical learned policy ("Learned").
3. "Real" against a greedy baseline surface realization policy that always chooses the most likely surface form ("Greedy"); this behavior corresponds to using only HMMs for surface realization.
4. "Real" against a baseline that picks randomly among surface forms as long as the result is grammatical ("Random"); this baseline corresponds to the learned policy without the HMMs.

Unsurprisingly, the two sets of human instructions are most similar to each other, obtaining a value of 1.77. The "Learned" instructions score second best with a value of 2.80, followed by "Greedy" with a value of 4.34, and finally "Random" with a value of 10.06.

Example dialogues

Tables 8.5 and 8.6 compare dialogues generated (using hierarchical reinforcement learning with HMM-based surface realization) using isolated and joint decision making, respectively, for the dialogue first shown in Table 8.4.

Table 8.5 shows a dialogue generated with the rule-based baseline formulated in Section 8.2.3. Most noticeably, instructions are long and no flexible repair strategy is used. This is due to isolated utterance planning and content selection choices. At two points, the instruction follower becomes confused. The first time (utterances 2–5), the follower had difficulties comprehending the system's long utterance and had to receive it twice before being successful. The second confusion (utterances 6–13) is more serious, since the follower has trouble identifying the intended referent and needs to wait through three repetitions before the utterance is rephrased. In addition, surface realization choices (which in the baseline system only aim to be grammatical) are made without reference to a language model or the dialogue history, resulting in the choice of infrequent words. Finally, the chain of awkward referring expressions (referring to the referent first using a *one* reference, then eliding the head noun and only last using the noun *button*) in utterances 6, 8, and 10 is caused by isolated content selection and surface realization choices.

In contrast, the utterances in Table 8.6 were generated using the joint reinforcement learning policy with HMM-based surface realization. While content selection and utterance planning behaviors do not change from the dialogue in Table 8.4, this time the surface realization choices were optimized using the HMM-based language model.

Table 8.5. Dialogue 2: Sample interaction demonstrating the baseline NLG system for the same situation as in Table 8.4. The system uses low-level navigation and repeats unsuccessful utterances as a main repair strategy.

Turn	Conversant	Utterance
1	USR	Action: starts game
2	SYS	*Turn 90 degrees to the left and move to the front to the end of the hall.*
3	USR	Reaction: wait
4	SYS	*Turn 90 degrees to the left and move to the front to the end of the hall.*
5	USR	Reaction: perform desired action
6	SYS	*Now proceed to the red one left of the blue and click the button.*
7	USR	Reaction: request help
8	SYS	*Proceed to the red one left of the blue and click the button.*
9	USR	Reaction: perform undesired action
10	SYS	*Proceed to the red one left of the blue and click the button.*
11	USR	Reaction: wait
12	SYS	*Turn slightly to the left and hit the red one beside the door.*
13	USR	Reaction: perform desired action
14	SYS	*Well done.*

Table 8.6. Dialogue 3: Sample interaction demonstrating the learned NLG policy from Table 8.4, this time with an HMM-based reward function for surface realization.

Turn	Conversant	Utterance
1	USR	Action: starts game
2	SYS	*Turn 90 degrees left and go straight.*
3	USR	Reaction: perform desired action
4	SYS	*OK. Now go left at the end of the hall.*
5	USR	Reaction: wait
6	SYS	*Turn right and head down the hallway.*
7	USR	Reaction: perform desired action
8	SYS	*Now bear left and press the red button on the right.*
9	USR	Reaction: request help
10	SYS	*Go to the red button beside the chair and click it.*
11	USR	Reaction: perform desired action
12	SYS	*Well done.*

8.6   Discussion

In this chapter, we argued for joint optimization of different NLG tasks for situated interaction, especially content selection, utterance planning, and surface realization, in order to take advantage of their interrelated nature and strong interdependencies. We showed that these tasks are interrelated in several ways and their treatment can affect the comprehensibility and humanlikeness of instruction giving in a navigation task, and the troubleshooting capabilities of an NLG system. We demonstrated experimentally that joint optimization of NLG tasks can lead to a significant increase in performance when compared with models that were optimized in isolation.

We used hierarchical reinforcement learning to optimize joint decision making for NLG in a fairly large and complex situated interaction domain. Reinforcement learning agents, presented with some knowledge of the task and the ultimate goal the system should achieve, are able to discover (near-)optimal strategies using trial-and-error search. This makes them a very useful tool to use when we have limited knowledge of the exact best way to approach a problem, but we have a very concrete idea of what we would like the final outcome to look like.

Of course other approaches are conceivable. One alternative is example-based NLG, which recycles utterances from a corpus of human–human interactions (e.g., DeVault *et al.*, 2008; Stent *et al.*, 2008). Unfortunately, for NLG a corpus cannot be considered a gold standard as much as in other NLP tasks (Belz and Reiter, 2006; Scott and Moore, 2007): the fact that a phrase does not occur in a corpus does not mean that it is unacceptable, or less good than phrases that do occur. This means that a system should also be able to make sensible generalizations over a corpus.

Another alternative approach may be purely statistical models such as n-gram-based language models, which are capable of some generalization. Unfortunately, such models do not take contextual cues into account to a sufficient degree, leading to little or no adaptation.

A third alternative approach is supervised learning, in which an NLG strategy is learned from a training corpus. Unfortunately, supervised learning requires large amounts of labeled training data (Rieser and Lemon, 2011). In addition, supervised learning-based systems may sometimes produce unpredictable behavior in circumstances that differ substantially from the ones they were trained on (Levin *et al.*, 2000). For example, if exposed to unseen spatial environments or instruction followers that behave differently from those in the training data, a supervised learning-based system may find itself unable to adapt. See Lemon (2008) for more details of the advantages and drawbacks of statistical and trainable approaches to NLG.

Related work has also shown the benefits of treating interrelated decisions in a joint fashion. Angeli *et al.* (2010) treat content selection and surface realization in a joint fashion using a log-linear classifier, which allows each decision to depend on all decisions made previously in the generation process. Their results show that treating decisions jointly achieves better results than treating them in isolation. They do not mention their system's performance on unseen data, however. Similarly, Lemon (2011) uses flat reinforcement learning to jointly optimize dialogue management and NLG decisions for information presentation in the domain of restaurant recommendations. The system learns to balance the tradeoffs of keeping an interaction short and obtaining more information about the user's restaurant preferences, so as to always present information when this is most advantageous. Cuayáhuitl and Dethlefs (2011b) use hierarchical reinforcement learning for the joint optimization of spatial behaviors and dialogue behaviors in a learning agent that gives route instructions to users in an unfamiliar environment. By optimizing both behaviors jointly, their agent learns to plan routes by taking individual users and their dialogue history into account. It learns to guide them past landmarks that they are already familiar with and to avoid locations where they may likely get lost. These recent investigations show that jointly optimized policies outperform policies optimized in isolation.

8.7   Conclusions and future work

Traditionally, NLG has been divided into a number of subtasks – such as content selection, utterance planning, and surface realization – and these subtasks have been dealt with in a sequential and isolated fashion. However, NLG subtasks are highly interrelated; furthermore, in interactive scenarios factors including the dialogue partner, the physical situation, and the dialogue history may affect all NLG subtasks. A better way of approaching NLG for interaction is joint optimization. Hierarchical reinforcement learning is a joint optimization method that has a number of benefits: (a) it is able to learn a (near-)optimal strategy automatically from a limited amount of human data; (b) if there is more than one optimal strategy, it learns the alternative(s) and can switch between them whenever the need arises; and (c) through its divide-and-conquer approach, it can handle even large and complex domains and is not restricted to artificial toy world scenarios. In this chapter we presented a case study in the use of hierarchical reinforcement learning for NLG in the form of a system that interactively produces navigation instructions for a human in a virtual environment. The results of an evaluation of this system showed that joint optimization of NLG subtasks gave a significant increase in system performance compared to isolated optimization.

An important challenge facing researchers at the intersection of dialogue and NLG is that of developing robust and sophisticated adaptive systems for real-world applications. Such systems must adapt quickly and flexibly to individual users and situations during the course of an interaction, while at the same time dealing with uncertainty concerning the user's preferences and prior knowledge and the dialogue context. This means that systems will need to observe their environment and user carefully to detect dynamic and sudden changes and be able to adapt their own behavior and generation strategies accordingly (see Cuayáhuitl and Dethlefs, 2011a,b). A promising way to achieve this is by learning in real time during the course of the interaction. In addition, systems need scalable techniques for dealing with uncertainty during an interaction so that they can make optimal decisions even with incomplete knowledge. To date, such techniques do not exist, but see Gašić et al. (2011) for some first advances. Systems either deal with uncertainty using computationally intensive methods (e.g., Williams, 2006; Thomson, 2009; Young et al., 2010) or can learn in real time but assume complete knowledge (no uncertainty).

Another challenge for the joint optimization of NLG subtasks is multimodal NLG. Multimodal NLG involves more constraints and more degrees of freedom than unimodal NLG, so is harder to model. However, it has broad applicability for human–robot interaction and embodied conversational agents.

A further promising direction for dialogue and NLG researchers consists of designing hierarchies of learning agents automatically – automatically learning the structure of a task. In our work, we have assumed that the task structure is specified manually, but this does not guarantee an optimal design. Automatic acquisition of task models would be useful for the application of hierarchical reinforcement learning to NLG and for dialogue systems in general.

For all of these outlined challenges, machine learning methods, such as reinforcement learning combined with supervised and/or unsupervised learning, are likely to play an important role in inducing system behavior from human data. Essential insights can further be gained from empirical research on human cognition and linguistic behavior and by an active data-driven development cycle between computational model building and empirical studies.

Acknowledgments

The research leading to this work was done while the authors were at the University of Bremen, Germany, and the Transregional Collaborative Research Center SFB/TR8 for Spatial Cognition. Funding by the German Research Foundation (DFG) and the European Commission's FP7 projects ALIZ-E (ICT-248116) and PARLANCE (287615) is gratefully acknowledged.

References

Angeli, G., Liang, P., and Klein, D. (2010). A simple domain-independent probabilistic approach to generation. In *Proceedings of the Conference on Empirical Methods in Natural Language Processing (EMNLP)*, pages 502–512, Boston, MA. Association for Computational Linguistics.

Baum, L. E., Petrie, T., Soules, G., and Weiss, N. (1970). A maximization technique occurring in the statistical analysis of probabilistic functions of Markov chains. *The Annals of Mathematical Statistics*, **41**(1):164–171.

Belz, A. and Reiter, E. (2006). Comparing automatic and human evaluation of NLG systems. In *Proceedings of the Conference of the European Chapter of the Association for Computational Linguistics (EACL)*, pages 313–320, Trento, Italy. Association for Computational Linguistics.

Byron, D. and Fosler-Lussier, E. (2006). The OSU Quake 2004 corpus of two-party situated problem-solving dialogs. Technical Report OSU-CISRC-805-TR57, Ohio State University.

Cuayáhuitl, H. (2009). *Hierarchical Reinforcement Learning for Spoken Dialogue Systems*. PhD thesis, University of Edinburgh.

Cuayáhuitl, H. and Dethlefs, N. (2011a). Optimizing situated dialogue management in unknown environments. In *Proceedings of the International Conference on Spoken Language Processing (INTERSPEECH)*, pages 1009–1012, Florence, Italy. International Speech Communication Association.

Cuayáhuitl, H. and Dethlefs, N. (2011b). Spatially-aware dialogue control using hierarchical reinforcement learning. *ACM Transactions on Speech and Language Processing*, **7**(3):5:1–5:26.

Cuayáhuitl, H., Korbayová, I. K., and Dethlefs, N. (2012). Hierarchical dialogue policy learning using flexible state transitions and linear function approximation. In *Proceedings of the International Conference on Computational Linguistics (COLING)*, pages 95–102, Mumbai, India. International Committee on Computational Linguistics.

Cuayáhuitl, H., Renals, S., Lemon, O., and Shimodaira, H. (2005). Human–computer dialogue simulation using hidden Markov models. In *Proceedings of the IEEE workshop on Automatic Speech Recognition and Understanding (ASRU)*, pages 290–295, San Juan, Puerto Rico. Institute of Electrical and Electronics Engineers.

Denecke, M., Dohsaka, K., and Nakano, M. (2004). Fast reinforcement learning of dialogue policies using stable function approximation. In *Proceedings of the International Joint Conference on Natural Language Processing (IJCNLP)*, pages 1–11, Hainan Island, China. Association for Computational Linguistics.

Denis, A. (2010). Generating referring expressions with reference domain theory. In *Proceedings of the International Workshop on Natural Language Generation (INLG)*, pages 27–36, Trim, Ireland. Association for Computational Linguistics.

Dethlefs, N. and Cuayáhuitl, H. (2010). Hierarchical reinforcement learning for adaptive text generation. In *Proceedings of the International Workshop on Natural Language Generation (INLG)*, pages 37–46, Trim, Ireland. Association for Computational Linguistics.

Dethlefs, N. and Cuayáhuitl, H. (2011a). Combining hierarchical reinforcement learning and Bayesian networks for natural language generation in situated dialogue. In *Proceedings of the European Workshop on Natural Language Generation (ENLG)*, pages 110–120, Nancy, France. Association for Computational Linguistics.

Dethlefs, N. and Cuayáhuitl, H. (2011b). Hierarchical reinforcement learning and hidden Markov models for task-oriented natural language generation. In *Proceedings of the Annual Meeting of the Association for Computational Linguistics: Human Language Technologies (ACL-HLT)*, pages 654–659, Portland, OR. Association for Computational Linguistics.

Dethlefs, N. and Cuayáhuitl, H. (2012). Comparing HMMs and Bayesian Networks for surface realisation. In *Proceedings of the Conference of the North American Chapter of the Association for Computational Linguistics: Human Language Technologies (NAACL-HLT)*, pages 636–640, Montréal, Canada. Association for Computational Linguistics.

Dethlefs, N., Cuayáhuitl, H., and Viethen, J. (2011). Optimising natural language generation decision making for situated dialogue. In *Proceedings of the SIGdial Conference on Discourse and Dialogue (SIGDIAL)*, pages 78–87, Portland, OR. Association for Computational Linguistics.

DeVault, D., Traum, D., and Artstein, R. (2008). Practical grammar-based NLG from examples. In *Proceedings of the International Workshop on Natural Language Generation (INLG)*, pages 77–85, Salt Fork, OH. Association for Computational Linguistics.

Dietterich, T. G. (2000). Hierarchical reinforcement learning with the MAXQ value function decomposition. *Journal of Artificial Intelligence Research*, **13**:227–303.

Gargett, A., Garoufi, K., Koller, A., and Striegnitz, K. (2010). The GIVE-2 corpus of giving instructions in virtual environments. In *Proceedings of the International Conference on Language Resources and Evaluation (LREC)*, Valletta, Malta. European Language Resources Association.

Garoufi, K. and Koller, A. (2010). Automated planning for situated natural language generation. In *Proceedings of the Annual Meeting of the Association for Computational Linguistics (ACL)*, pages 1573–1582, Uppsala, Sweden. Association for Computational Linguistics.

Garoufi, K. and Koller, A. (2011). Combining symbolic and corpus-based approaches for the generation of successful referring expressions. In *Proceedings of the European Workshop on Natural Language Generation (ENLG)*, pages 121–131. Nancy, France. Association for Computational Linguistics.

Gašić, M., Jurcicek, F., Thomson, B., Yu, K., and Young, S. (2011). On-line policy optimisation of spoken dialogue systems via live interaction with human subjects. In *Proceedings of the IEEE workshop on Automatic Speech Recognition and Understanding (ASRU)*, pages 312–317, Waikoloa, HI. Institute of Electrical and Electronics Engineers.

Henderson, J., Lemon, O., and Georgila, K. (2005). Hybrid reinforcement/supervised learning for dialogue policies from Communicator data. In *Proceedings of the Workshop on Knowledge and Reasoning in Practical Dialogue Systems (KRPDS)*, pages 68–75, Edinburgh, Scotland. International Joint Conference on Artificial Intelligence.

Janarthanam, S. and Lemon, O. (2010). Learning adaptive referring expression generation policies for spoken dialogue systems. In Krahmer, E. and Theune, M., editors, *Empirical Methods in Natural Language Generation*, pages 67–84. Springer, Berlin.

Janarthanam, S., Lemon, O., and Liu, X. (2012). A web-based evaluation framework for spatial instruction-giving systems. In *Proceedings of the Annual Meeting of the Association for Computational Linguistics (ACL)*, pages 49–54, Jeju Island, Korea. Association for Computational Linguistics.

Kelleher, J. D., Kruijff, G.-J. M., and Costello, F. J. (2006). Incremental generation of spatial referring expressions in situated dialog. In *Proceedings of the International Conference on Computational Linguistics and the Annual Meeting of the Association for Computational Linguistics (COLING-ACL)*, pages 745–752, Sydney, Australia. Association for Computational Linguistics.

Koller, A., Striegnitz, K., Byron, D., Cassell, J., Dale, R., and Moore, J. D. (2010a). The first challenge on generating instructions in virtual environments. In Krahmer, E. and Theune, M., editors, *Empirical Methods in Natural Language Generation*, pages 328–352. Springer LNCS, Berlin, Germany.

Koller, A., Striegnitz, K., Gargett, A., Byron, D., Cassell, J., Dale, R., Moore, J., and Oberlander, J. (2010b). Report on the second NLG challenge on generating instructions in virtual environments (GIVE-2). In *Proceedings of the International Conference on Natural Language Generation (INLG)*, pages 243–250, Trim, Ireland. Association for Computational Linguistics.

Lemon, O. (2008). Adaptive natural language generation in dialogue using reinforcement learning. In *Proceedings of the Workshop on the Semantics and Pragmatics of Dialogue (SEMDIAL)*, pages 141–148, London, UK. SemDial.

Lemon, O. (2011). Learning what to say and how to say it: Joint optimisation of spoken dialogue management and natural language generation. *Computer Speech & Language*, **25**(2):210–221.

Levin, E., Pieraccini, R., and Eckert, W. (2000). A stochastic model of human–machine interaction for learning dialog strategies. *IEEE Transactions on Speech and Audio Processing*, **8**(1):11–23.

Mairesse, F., Gašić, M., Jurčíček, F., Keizer, S., Thomson, B., Yu, K., and Young, S. (2010). Phrase-based statistical language generation using graphical models and active learning. In *Proceedings of the Annual Meeting of the Association for Computational Linguistics (ACL)*, pages 1552–1561, Uppsala, Sweden. Association for Computational Linguistics.

Pietquin, O., Geist, M., Chandramohan, S., and Frezza-Buet, H. (2011). Sample-efficient batch reinforcement learning for dialogue management optimization. *ACM Transactions on Speech and Language Processing (Special Issue on Machine Learning for Robust and Adaptive Spoken Dialogue Systems)*, **7**(3):7.

Rabiner, L. R. (1989). Tutorial on hidden Markov models and selected applications in speech recognition. *Proceedings of the IEEE*, **77**(2):257–286.

Raux, A. and Nakano, M. (2010). The dynamics of action corrections in situated interaction. In *Proceedings of the SIGdial Conference on Discourse and Dialogue (SIGDIAL)*, pages 165–174, Tokyo, Japan. Association for Computational Linguistics.

Reboul, A. (1998). A relevance theoretic approach to reference. In *Proceedings of the Relevance Theory Workshop*, pages 45–50, Luton, UK. University of Luton.

Rieser, V. and Lemon, O. (2011). Learning and evaluation of dialogue strategies for new applications: Empirical methods for optimization from small data sets. *Computational Linguistics*, **37**(1):153–196.

Salmon-Alt, S. and Romary, L. (2001). Reference resolution within the framework of cognitive grammar. In *Proceedings of the International Colloquium on Cognitive Science*, Donostia – San Sebastian, Spain. Institute for Logic, Cognition, Language and Information (ILCLI) and the Dept. of Logic and Philosophy of Science of the University of the Basque Country.

Scott, D. and Moore, J. (2007). An NLG evaluation competition? Eight reasons to be cautious. In *Proceedings of the NSF Workshop on Shared Tasks and Comparative Evaluation in Natural Language Generation*, Arlington, VA. National Science Foundation.

Stent, A., Bangalore, S., and Di Fabbrizio, G. (2008). Where do the words come from? Learning models for word choice and ordering from spoken dialog corpora. In *Proceedings of the IEEE International Conference on Acoustics, Speech, and Signal Processing (ICASSP)*, pages 5037–5040, Las Vegas, NV. Institute of Electrical and Electronics Engineers.

Stoia, L., Shockley, D. M., Byron, D., and Fosler-Lussier, E. (2006). Noun phrase generation for situated dialogs. In *Proceedings of the International Conference on Natural Language Generation (INLG)*, pages 81–88, Sydney, Australia. Association for Computational Linguistics.

Sutton, R. (1996). Generalization in reinforcement learning: Successful examples using sparse coarse coding. In Touretzky, D., Mozer, M., and Hasselmo, M., editors, *Advances in Neural Information Processing Systems*, pages 1038–1044. MIT Press, Cambridge, MA.

Sutton, R. S. and Barto, A. G. (1998). *Reinforcement Learning: An Introduction*. MIT Press, Cambridge, MA.

Thomson, B. (2009). *Statistical Methods for Spoken Dialogue Management*. PhD thesis, University of Cambridge.

van Zaanen, M. (2000). Bootstrapping syntax and recursion using alignment-based learning. In *Proceedings of the International Conference on Machine Learning (ICML)*, pages 1063–1070, Stanford, CA. The International Machine Learning Society.

Watkins, C. (1989). *Learning from Delayed Rewards*. PhD thesis, Kings College, University of Cambridge.

Williams, J. D. (2006). *Partially Observable Markov Decision Processes for Spoken Dialogue Management*. PhD thesis, University of Cambridge.

Young, S., Gašić, M., Keizer, S., Mairesse, F., Schatzmann, J., Thomson, B., and Yu, K. (2010). The hidden information state model: A practical framework for POMDP-based spoken dialogue management. *Computer Speech & Language*, **24**(2):150–174.

Part IV

Engagement

9   Data-driven methods for linguistic style control

François Mairesse

9.1   Introduction

Modern spoken language interfaces typically ignore a fundamental component of human communication: human speakers tailor their speech and language based on their audience, their communicative goal, and their overall personality (Scherer, 1979; Brennan and Clark, 1996; Pickering and Garrod, 2004). They control their linguistic style for many reasons, including social (e.g., to communicate social distance to the hearer), rhetorical (e.g., for persuasiveness), or task-based (e.g., to facilitate the assimilation of new material). As a result, a close acquaintance, a politician, or a teacher are expected to communicate differently, even if they were to convey the same underlying meaning. In contrast, the style of most human–computer interfaces is chosen once for all at development time, typically resulting in cold, repetitive language, or **machinese**. This chapter focuses on methods that provide an alternative to machinese by learning to control the linguistic style of computer interfaces from data.

Natural Language Generation (NLG) differs from other areas of natural language processing in that it is an under-constrained problem. Whereas the natural language understanding task requires learning a mapping from linguistic forms to the corresponding meaning representations, NLG systems must learn the reverse one-to-many mapping and choose among all possible realizations of a given input meaning. The criterion to optimize is often unclear, and largely dependent on the target application. Hence it is important to identify relevant control dimensions – i.e., linguistic styles – to optimize the generation process based on the application context and the user.

Previous work has looked at a large range of control dimensions. The earliest work on stylistic control dates back to the PAULINE system (Hovy, 1988), which focused on general aspects of pragmatics manually derived from a set of rhetorical goals, such as the distance between the speaker and the hearer, or the subjective connotations of the utterance. For example, the rhetorical goals of low formality, high force, and high partiality produce a "no-nonsense" effect. Similarly, Green and DiMarco (1996) used handcrafted **stylistic grammars** to control dimensions such as clarity, concreteness, and dynamism during the language generation process. Both approaches show that stylistic control requires deep models of how context affects language production. For example, the production of "no-nonsense" language typically requires consistent generation decisions during content selection, syntactic aggregation, and lexical choice. This complexity makes it difficult for NLG systems to scale from one domain to another, let alone from one style to another. A recent line of research has therefore looked at methods for learning to control linguistic style from labeled data.

Most previous work on data-driven NLG uses the **overgenerate and rank** paradigm (OR), in which a large number of candidate utterances are ranked using a statistical model trained to maximize an objective function, e.g., the likelihood of the utterance in a corpus (Langkilde and Knight, 1998), average utterance quality ratings according to users (Stent *et al.*, 2002; Walker *et al.*, 2004), individual stylistic preferences (Walker *et al.*, 2007), or the personality of weblog authors (Isard *et al.*, 2006). In contrast, we only know of two existing methods for data-driven stylistic control *without* overgeneration. Belz (2009) modeled individual stylistic differences by training probabilistic context-free grammars (PCFGs) on weather forecasts from different authors. While PCFGs can be easily be trained from a parsed corpus, their expressiveness is limited. Paiva and Evans (2005) trained linear regression models to predict stylistic scores from utterances produced by an arbitrary generator. At generation time, this method requires searching for the input generation decisions yielding the stylistic scores that are closest to the target scores. While these methods are promising alternatives to the OR approach, neither was evaluated as to whether humans can successfully perceive the intended stylistic variations.

In this chapter we describe the traditional OR approach to stylistic control, and compare it with two novel methods that do not require any overgeneration phase. As a proof of concept, we focus on the most general set of control dimensions characterizing human behavior. That is, we characterize linguistic style in terms of the **personality traits** conveyed by the generated utterances. Many psychology studies refer to personality in terms of the **Big Five** traits, which were shown to explain the most variance when clustering large sets of adjective descriptors (Norman, 1963; Goldberg, 1990). The Big Five traits consist of extraversion, emotional stability, agreeableness, conscientiousness, and openness to experience. For the purposes of this chapter, a target linguistic style is therefore defined as a point within a five-dimensional space; however, the methods presented here can be applied to any other set of stylistic dimensions.

In the next section we briefly describe PERSONAGE, a language generation system that implements a large range of generation parameters shown to affect the perception of personality. In Section 9.3 we present two methods for learning to control PERSONAGE from data, (a) by using the traditional OR approach and (b) by learning **parameter estimation models** predicting generation decisions from stylistic targets. In Section 9.4 we present BAGEL, a novel statistical NLG framework that has the potential to greatly improve the scalability of NLG systems with stylistic controls, by learning **factored language models** from semantically aligned data. Finally, in Section 9.5 we discuss our results and important challenges for stylistic control in language generation.

9.2   PERSONAGE: personality-dependent linguistic control

PERSONAGE is an NLG system that provides recommendations and comparisons of restaurants. It follows the standard NLG pipeline architecture (Reiter and Dale, 2000).

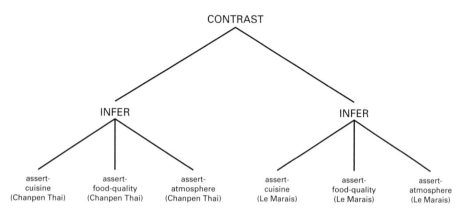

CONTRAST

INFER   INFER

assert-	assert-	assert-	assert-	assert-	assert-
cuisine	food-quality	atmosphere	cuisine	food-quality	atmosphere
(Chanpen Thai)	(Chanpen Thai)	(Chanpen Thai)	(Le Marais)	(Le Marais)	(Le Marais)

Figure 9.1   Content plan tree comparing the cuisine, food quality, and atmosphere of two restaurants (Chanpen Thai and Le Marais)

The main difference between PERSONAGE and other language generators is that, at each level of the pipeline, it uses a large number of generation parameters motivated by studies on the linguistic indicators of personality. PERSONAGE's input is a high-level *communicative goal* (e.g., to compare two restaurants) which is converted into a content plan tree combining elementary propositions using relations from Rhetorical Structure Theory (Mann and Thompson, 1988), such as the one illustrated in Figure 9.1. Parameters at this level can affect the verbosity of the output utterance; for example by controlling the number of restaurant attributes in the content plan tree, or whether positive restaurant attributes are to be emphasized. Propositions in the content plan tree are then associated with syntactic templates from a handcrafted generation dictionary. Propositions may be aggregated by selecting sentence planning operations that use syntactic aggregation constraints over syntactic templates (Stent *et al.*, 2004). For example, the contrast rhetorical relation in Figure 9.1 can be associated with the sentence planning operation *however*, or the infer relation with one that embeds one of the input propositions within a relative clause. Once the full syntactic tree has been aggregated, PERSONAGE implements a large range of syntactic transformations to convey pragmatic effects; for example, inserting hedging expressions (e.g., *basically*), in-group markers (e.g., *buddy*), tag questions (e.g., *isn't it?*), filled pauses (e.g., *err*), mild expletives, stuttering, and many other transformations summarized in Table 9.3. The final content words are determined by the lexical choice component, by choosing among a set of WordNet synonyms based on their length, frequency of use, and strength. Finally, the aggregated syntactic tree is converted to a surface string using the RealPro surface realizer (Lavoie and Rambow, 1997). PERSONAGE's 67 generation parameters are illustrated in Tables 9.1, 9.2, and 9.3. The reader is referred to Mairesse (2008) and Mairesse and Walker (2010) for implementation details and pointers to the psycholinguistic studies motivating each parameter.

We have experimented with several methods to generate language conveying a target personality (e.g., highly extrovert). Our first method simply derives parameter values by

Table 9.1. PERSONAGE's continuous generation parameters (I)

Parameter	Description
Content planning	
VERBOSITY	Control the number of propositions in the utterance
RESTATEMENTS	Paraphrase an existing proposition, e.g., *Chanpen Thai has great service, it has fantastic waiters*
REPETITIONS	Repeat an existing proposition
CONTENT POLARITY	Control the polarity of the propositions expressed, i.e., referring to negative or positive attributes
REPETITION POLARITY	Control the polarity of the restated propositions
CONCESSION	Emphasize one attribute over another, e.g., *even if Chanpen Thai has great food, it has bad service*
CONCESSION POLARITY	Determine whether positive or negative attributes are emphasized
POLARIZATION	Control whether the expressed polarity is neutral or extreme
POSITIVE CONTENT FIRST	Determine whether positive propositions – including the claim – are uttered first
Pragmatic markers	
SUBJECT IMPLICITNESS	Make the restaurant implicit by moving the attribute to the subject, e.g., *the service is great*
STUTTERING	Duplicate the first letters of a restaurant's name, e.g., *Ch-ch-anpen Thai is the best*
PRONOMINALIZATION	Replace occurrences of the restaurant's name by pronouns
Lexical choice	
LEXICON FREQUENCY	Control the average frequency of use of each content word, according to frequency counts from the British National Corpus
LEXICON WORD LENGTH	Control the average number of letters of each content word
VERB STRENGTH	Control the strength of the verbs, e.g., *I would suggest* vs. *I would recommend*

hand based on findings from the psycholinguistic literature. For example, findings that high extroverts produce more content in their utterance suggest a high VERBOSITY parameter value (Furnham, 1990; Pennebaker and King, 1999), whereas findings that neurotics produce more non-verbal hesitations suggest a high FILLED PAUSE parameter value (Scherer, 1979). Such parameter settings were previously derived for each end of each Big Five trait. A human evaluation has shown that each of these ten parameter settings produces utterances with recognizable personality (Mairesse and Walker, 2007, 2010). This is quite remarkable given that the psycholinguistic findings on which PERSONAGE's parameters are based focus on a large range of textual or spoken genres, using either self-reports or observer reports of personality. Yet those findings were shown to carry over to a specific generation domain (restaurant recommendations and comparisons), through a single turn, and without any speech cues or dialogue context.

While these results suggest that PERSONAGE's stylistic coverage is large enough for personality modeling, this method is unlikely to scale to other stylistic dimensions,

Table 9.2. PERSONAGE's continuous generation parameters (II)

Parameter	Description
Syntactic template selection	
SELF-REFERENCES	Control the number of first person pronouns
SYNTACTIC COMPLEXITY	Control syntactic complexity (syntactic embedding)
TEMPLATE POLARITY	Control the connotation of the claim, i.e., whether positive or negative affect is expressed
Aggregation operations	
PERIOD	Leave two propositions in their own sentences, e.g., *Chanpen Thai has great service. It has nice decor.*
RELATIVE CLAUSE	Aggregate propositions with a relative clause, e.g., *Chanpen Thai, which has great service, has nice decor*
With CUE WORD	Aggregate propositions using *with*, e.g., *Chanpen Thai has great service, with nice decor*
CONJUNCTION	Join two propositions using a conjunction, or a comma if more than two propositions
MERGE	Merge the subject and verb of two propositions, e.g., *Chanpen Thai has great service and nice decor*
Also CUE WORD	Join two propositions using *also*, e.g., *Chanpen Thai has great service, also it has nice decor*
CONTRAST - CUE WORD	Contrast two propositions using *while, but, however*, or *on the other hand*, e.g., *While Chanpen Thai has great service, it has bad decor. Chanpen Thai has great service, but it has bad decor*
JUSTIFY - CUE WORD	Justify a proposition using *because, since*, or *so*, e.g., *Chanpen Thai is the best, because it has great service*
CONCEDE - CUE WORD	Concede a proposition using *although, even if*, or *but/though*, e.g., *Although Chanpen Thai has great service, it has bad decor, Chanpen Thai has great service, but it has bad decor though*
MERGE WITH COMMA	Restate a proposition by repeating only the object, e.g., *Chanpen Thai has great service, nice waiters*
OBJECT ELLIPSIS	Restate a proposition after replacing its object by an ellipsis, e.g., *Chanpen Thai has …, it has great service*

especially if there are no existing studies relating those dimensions to linguistic behavior. Furthermore, handcrafting generation decisions limits the granularity of the stylistic control to a small set of predefined styles, e.g., high introvert or low emotional stability. Given the complexity of the pragmatic effects found in human language, it would be desirable for a language generator to be able to interpolate between different styles along continuous scales, rather than being limited to a few points along those scales. As the human effort required to build such models is prohibitively large, in this chapter we focus on methods for learning them from data.

Table 9.3. PERSONAGE's binary generation parameters.

Binary parameter	Description
Content planning	
REQUEST CONFIRMATION	Begin the utterance with a confirmation of the restaurant's name, e.g., *did you say Chanpen Thai?*
INITIAL REJECTION	Begin the utterance with a mild rejection, e.g., *I'm not sure*
COMPETENCE MITIGATION	Express the speaker's negative appraisal of the hearer's request, e.g., *everybody knows that …*
Pragmatic markers	
NEGATION	Negate a verb by replacing its modifier by its antonym, e.g., *Chanpen Thai doesn't have bad service*
SOFTENER HEDGES	Insert syntactic elements (*sort of, kind of, somewhat, quite, around, rather, I think that, it seems that, it seems to me that*) to mitigate the strength of a proposition, e.g., *Chanpen Thai has kind of great service* or *It seems to me that Chanpen Thai has rather great service*
EMPHASIZER HEDGES	Insert syntactic elements (*really, basically, actually, just*) to strengthen a proposition, e.g., *Chanpen Thai has really great service* or *Basically, Chanpen Thai just has great service*
ACKNOWLEDGMENTS	Insert an initial backchannel (*yeah, right, ok, I see, oh, well*), e.g., *Well, Chanpen Thai has great service*
FILLED PAUSES	Insert syntactic elements expressing hesitancy (*like, I mean, err, mmhm, you know*), e.g., *I mean, Chanpen Thai has great service, you know* or *Err... Chanpen Thai has, like, great service*
EXCLAMATION	Insert an exclamation mark, e.g., *Chanpen Thai has great service!*
EXPLETIVES	Insert a swear word, e.g., *the service is damn great*
NEAR EXPLETIVES	Insert a near-swear word, e.g., *the service is darn great*
TAG QUESTION	Insert a tag question, e.g., *the service is great, isn't it?*
IN-GROUP MARKER	Refer to the hearer as a member of the same social group, e.g., *pal, mate,* or *buddy*

9.3   Learning to control a handcrafted generator from data

In this section we investigate two methods for training PERSONAGE on personality-annotated utterances. The first one is based on the "overgenerate and rank" (OR) framework used in most previous work on statistical NLG, whereas the second one learns parameter estimation models predicting generation decisions directly from target stylistic scores.

9.3.1   Overgenerate and rank

As the OR framework has been widely used for statistical NLG (Langkilde and Knight, 1998; Walker *et al.*, 2002; Stent *et al.*, 2004; Isard *et al.*, 2006), we use it for stylistic control by (a) generating candidate utterances by randomly varying PERSONAGE's parameters, and (b) selecting the utterance yielding the closest personality score to the target according to a statistical regression model. The resulting statistical generator is referred to as PERSONAGE-OR.

The statistical scoring models are trained on a corpus of personality-annotated utterances. The corpus consists of utterances generated by randomly varying PERSONAGE's generation decisions. Most expressions of linguistic variation – e.g., style, emotion, mood, and personality – can only be measured subjectively; thus a major advantage of the Big Five framework is that it offers standard questionnaires validated over time by the psychology community (John *et al.*, 1991; Costa and McCrae, 1992; Gosling *et al.*, 2003). The data was therefore annotated by asking native speakers of English to rate the personality conveyed by each utterance, by completing the Ten-Item Personality Inventory (TIPI) in Figure 9.2 (Gosling *et al.*, 2003). The answers were averaged over all judges to produce a rating for each trait ranging from 1 (e.g., highly neurotic) to 7 (e.g., very stable). Three judges rated 320 utterances for extroversion, and two judges rated 160 utterances for the other four traits. These utterances were generated from 20 communicative goals (40 for extroversion). Thus our final dataset consists of total of 480 utterances and their personality ratings averaged over all judges.

The main advantage of the OR approach is that it can model features that are emergent properties of multiple generation decisions, such as the utterance's length, instantiations of word categories, or specific phrase patterns. We trained regression models to predict the average personality score of an utterance from generation decision features, n-gram features, and content analysis features. We first computed the 67 generation decisions made for the parameters in Tables 9.1, 9.2, and 9.3. As n-gram frequency counts were shown to correlate with the author's personality (Gill and Oberlander, 2002), we also computed unigram, bigram, and trigram counts after replacing restaurant names, cuisine types, and numbers with generic labels. Our content analysis features were derived from

"Basically, Flor De Mayo isn't as bad as the others. Obviously, it isn't expensive. I mean, actually, its price is 18 dollars."

I see the speaker as...

1. Extroverted, enthusiastic	Disagree strongly	1 ○ 2 ○ 3 ○ 4 ○ 5 ○ 6 ○ 7 ○ Agree strongly
2. Reserved, quiet	Disagree strongly	1 ○ 2 ○ 3 ○ 4 ○ 5 ○ 6 ○ 7 ○ Agree strongly
3. Critical, quarrelsome	Disagree strongly	1 ○ 2 ○ 3 ○ 4 ○ 5 ○ 6 ○ 7 ○ Agree strongly
4. Dependable, self-disciplined	Disagree strongly	1 ○ 2 ○ 3 ○ 4 ○ 5 ○ 6 ○ 7 ○ Agree strongly
5. Anxious, easily upset	Disagree strongly	1 ○ 2 ○ 3 ○ 4 ○ 5 ○ 6 ○ 7 ○ Agree strongly
6. Open to new experiences, complex	Disagree strongly	1 ○ 2 ○ 3 ○ 4 ○ 5 ○ 6 ○ 7 ○ Agree strongly
7. Sympathetic, warm	Disagree strongly	1 ○ 2 ○ 3 ○ 4 ○ 5 ○ 6 ○ 7 ○ Agree strongly
8. Disorganized, careless	Disagree strongly	1 ○ 2 ○ 3 ○ 4 ○ 5 ○ 6 ○ 7 ○ Agree strongly
9. Calm, emotionally stable	Disagree strongly	1 ○ 2 ○ 3 ○ 4 ○ 5 ○ 6 ○ 7 ○ Agree strongly
10. Conventional, uncreative	Disagree strongly	1 ○ 2 ○ 3 ○ 4 ○ 5 ○ 6 ○ 7 ○ Agree strongly

Figure 9.2 Online version of the TIPI used in our experiments (Gosling *et al.*, 2003), adapted to the evaluation of personality in generated utterances

handcrafted dictionaries typically used for studying psycholinguistic properties of texts, in order to provide information about the author or the speaker. We first extracted features based on the Linguistic Inquiry and Word Count (LIWC) dictionary (Pennebaker *et al.*, 2001), which consists of 88 word categories including both syntactic (e.g., ratio of pronouns) and semantic information (e.g., positive emotion words). Pennebaker and King (1999) and Mehl *et al.* (2006) previously found significant correlations between these features and each of the Big Five personality traits, and they were also previously used for personality recognition (Mairesse and Walker, 2007). We also extracted features based on the MRC psycholinguistic database (Coltheart, 1981), which contains numerous statistics for over 150 000 words, including estimates of the age of acquisition, concreteness, frequency of use, and familiarity. Finally, to reduce data sparsity, we filtered out features that do not correlate significantly with personality ratings with a coefficient above 0.1.

The learning algorithm that is the most appropriate for predicting linguistic style or personality from the above features is still an open research question. We therefore compare four algorithms with a baseline model that returns the mean personality score in the training data (Base). The baseline model returns a constant value, which is equivalent to selecting a candidate utterance at random. We experimented with a linear regression model (LR), an M5′ regression tree (M5R), a piecewise-linear M5′ model tree (M5), and a support vector machine model with a radial-basis function kernel (SVM_r).

In order to evaluate the potential issues with using the OR approach for stylistic control, the errors made by PERSONAGE-OR are broken down into two components: **modeling error** due to incorrect assumptions made by the rescoring model, and **sampling error** due to the partial stylistic coverage of the models and the limited number of pre-generated utterances.

Modeling error results from the inaccuracy of the rescoring model, i.e., assuming that the candidate utterance set contains an utterance matching the target style exactly, is the model likely to select it? For regression models, this can be evaluated by computing how well the predicted score correlates with the true score of the utterance. Table 9.4 shows the average correlation between the models' predicted scores and the judges' average ratings over ten 10-fold cross-validations, using the best performing learning algorithm for each trait.

Results show that all regression models significantly outperform the baseline for all traits, with correlations ranging from 0.19 to 0.54. Extroversion and agreeableness are predicted the most accurately, whereas openness to experience is the hardest trait to predict. Given the difficulty of the personality recognition task, and the fact that the randomly generated utterances are likely to convey conflicting cues, such correlations are relatively high. They are indeed higher than most correlations observed for personality cues in natural text or conversations (Pennebaker and King, 1999; Mehl *et al.*, 2006). However, are they large enough to convey personality reliably in a generation system? This depends largely on the the sampling error resulting from the overgeneration phase, which represents a major issue with the OR approach.

Table 9.4. Pearson's correlation coefficients between the ratings and the predictions of selection models trained on the best performing feature sets (Content = Gen + LIWC + MRC). All results are averaged over ten 10-fold cross-validations. Best results for each trait are in bold. • = statistically significant improvement over the mean value baseline ($p < 0.05$).

Trait	Features	Base	LR	M5R	M5	SVM$_r$
Extroversion	LIWC	0.00	**0.45** •	0.21 •	0.45 •	0.35 •
Emotional stability	Gen	0.00	0.19 •	0.18 •	0.25 •	**0.32** •
Agreeableness	Gen	0.00	0.29 •	**0.54** •	0.43 •	0.49 •
Conscientiousness	Content	0.00	0.13 •	0.14 •	**0.25** •	0.11 •
Openness to experience	LIWC	0.00	**0.19** •	0.09 •	0.16 •	0.16 •

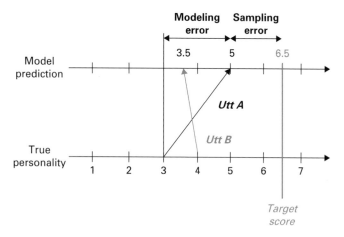

Figure 9.3   Modeling error and sampling error for a candidate utterance set of size 2. Utterance A is selected because its predicted score is closer to the target, yielding a sampling error of 1.5.

   Sampling error is observed if the candidate utterance set does not contain an utterance predicted to have the target style. We define this error as the distance between the target personality score and the predicted score of the closest utterance in the candidate utterance set. Given perfect scoring models, the sampling error would be the only error component of the overall generator. Figure 9.3 illustrates sampling error for a small candidate utterance set.

   As most previous work on statistical NLG only focuses on one end of the selection scale (e.g., naturalness), sampling error is typically ignored (Langkilde-Geary, 2002; Walker *et al.*, 2002). Sampling error can be estimated by generating candidate utterances for a single communicative goal, and computing the distance between the target score and the predicted score of the candidate utterance whose score is closest to the target, without knowing the true score. Figure 9.4 shows the distribution of the sampling error for different numbers of candidate utterances and different target scores, averaged over ten randomly selected communicative goals, and using the best performing models from Table 9.4.

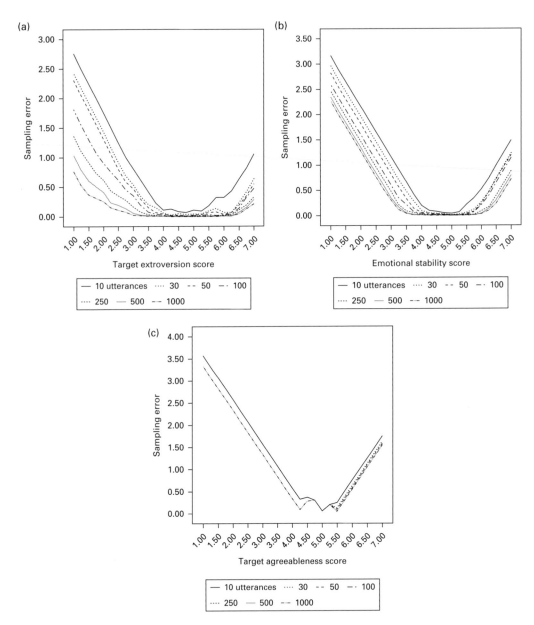

Figure 9.4 PERSONAGE-PE's parameter estimation framework. (a) Extroversion (linear regression with LIWC), (b) emotional stability (SVM with generation features), (c) agreeableness (regression tree with generation features).

The first noticeable result is that the sampling error is the lowest in the middle of the scale, especially for emotional stability and agreeableness. This is a consequence of the fact that randomly generated utterances are rarely rated as extreme, which has two implications: (a) small candidate utterance sets are less likely to contain examples of extreme utterances; and (b) models trained on random utterances are unlikely to learn to predict extreme scores due to training data sparsity. The latter affects the granularity of the model, i.e., whether the model can output scores covering the full range of the target stylistic dimension. It is important to note that models differ in the way they extrapolate over sparse regions of their feature space. The linear regression model in Figure 9.4(a) and the SVM regression model in Figure 9.4(b) have similar error distribution for small candidate utterance sets. However, the coverage of the linear regression model improves with the utterance set size, whereas the coverage of the SVM regression model only improves marginally above 250 utterances. This phenomenon is even more strikingly depicted for the agreeableness regression tree in Figure 9.4(c), which despite being the best performing model only outputs values between 4 and 5.5.

Whereas modeling error is likely to decrease with more training data and more complex models, sampling error remains a fundamental issue for the application of the OR approach to stylistic modeling, because increasing the number of candidate utterances only reduces the error marginally. In the next section we present a second approach to data-driven stylistic control, which avoids the need for an overgeneration phase and is more robust to sampling error.

9.3.2 Parameter estimation models

While the scoring models of the OR approach learn to map generation decisions to target stylistic dimensions, **parameter estimation models** (PEMs) learn the reverse mapping, i.e., they predict individual generation decisions given target stylistic scores. Once the generation decisions have been predicted by the models, the generator is called once for all to generate the target style (Figure 9.5). The resulting generator is referred to as PERSONAGE-PE.

The parameters used in PERSONAGE-PE are estimated using either regression or classification models, depending on whether they are continuous (e.g., verbosity) or binary (e.g., exclamation). As in Section 9.3.1, the continuous generation parameters in Tables 9.1 and 9.2 are modeled with a linear regression model (LR), an M5′ model tree (M5), and a model based on support vector machines (SVM). Binary generation parameters in Table 9.3 are modeled using classifiers that predict whether the parameter should be enabled or disabled. We tested a Naive Bayes classifier (NB), a C4.5 decision tree (J48), a nearest neighbor classifier using one neighbor (NN), the Ripper rule-based

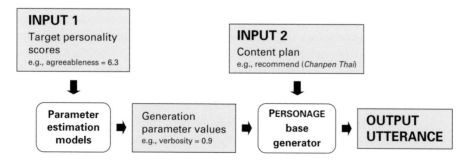

Figure 9.5   PERSONAGE-PE's parameter estimation framework

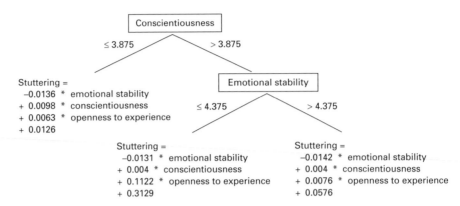

Figure 9.6   M5′ model tree predicting the STUTTERING parameter

learner (JRIP), the AdaBoost boosting algorithm (ADA), and a support vector machine with a linear kernel (SVM). Parameter estimation models were trained on the same data as the scoring models of PERSONAGE-OR. The final learning algorithm for each parameter was chosen based on a 10-fold cross-validation. Figure 9.6 and Table 9.6 illustrate the best performing models for the STUTTERING and EXCLAMATION parameters, and Table 9.5 shows example utterances produced when using the parameter settings predicted by all PEMs.

Although there has been considerable work on the expression of various stylistic effects (e.g., Hovy, 1988; DiMarco and Hirst, 1993; Paiva and Evans, 2005; Isard *et al.*, 2006), there have only been a few attempts to evaluate whether the intended style is perceived correctly by human judges (Traum *et al.*, 2003; Porayska-Pomsta and Mellish, 2004). We evaluated PERSONAGE-PE by generating utterances targeting different personality scores, which were then rated by judges using the questionnaire in Figure 9.2 (Gosling *et al.*, 2003). A total of 24 judges rated 50 utterances, each targeting an extreme value for two traits (either 1 or 7 out of 7) and neutral values for the remaining three traits (4 out of 7). The goal is for each utterance to project multiple traits on a continuous scale. To limit the experiment's duration, only the two traits with extreme target values were evaluated for each utterance. The judges were also asked to evaluate the naturalness of the generated utterances.

The evaluation results in Table 9.7 show that PERSONAGE-PE can successfully convey extroversion, producing a 0.45 correlation between the target extroversion and the subjects' ratings (significant at $p < 0.01$). Emotional stability, agreeableness, and openness to experience ratings also correlate strongly with the target scores, with correlations of 0.39, 0.36, and 0.17 respectively (all significant at $p < 0.01$). Additionally, Table 9.7 shows that the magnitude of the correlation increases when considering the perception of a hypothetical average subject – i.e., smoothing individual variation by averaging the ratings over all 24 judges – producing a correlation r_{avg} of up to 0.80 for extraversion. These correlations are unexpectedly high; in corpus analyses, significant correlations as low as 0.05 to 0.15 are typically observed between averaged personality ratings and linguistic markers (Pennebaker and King, 1999; Mehl *et al.*, 2006).

Table 9.5. Example outputs from PERSONAGE-PE for a comparison (#1) and a recommendation (#2), with the average judges' personality (Pers) and naturalness (Nat) scores. Ratings are on a scale from 1 to 7, with 1 = very low (e.g., introvert) and 7 = very high (e.g., extrovert).

#	Trait and target		Pers	Nat	Output utterance
1.a	Extroversion	high 6.90	4.42	4.79	*Radio Perfecto's price is 25 dollars but Les Routiers provides adequate food. I imagine they're alright!*
	Agreeableness	high 7.00	4.94		
1.b	Emotional stability	high 7.00	5.35	5.04	*Let's see, Les Routiers and Radio Perfecto... You would probably appreciate them. Radio Perfecto is in the East Village with kind of acceptable food. Les Routiers is located in Manhattan. Its price is 41 dollars.*
	Conscientiousness	high 6.21	5.21		
1.c	Agreeableness	low 1.28	4.04	4.58	*Les Routiers and Radio Perfecto are outstanding. Radio Perfecto's price is 25 dollars, isn't it? The food is acceptable, also it's located in the East Village. Err... although Les Routiers offers adequate food, it's damn costly.*
	Openness to experience	high 7.00	4.52		
2.a	Agreeableness	low 1.92	3.65	3.21	*Err... you would probably appreciate Trattoria Rustica, wouldn't you? It's in Manhattan, also it's an Italian restaurant. It offers poor ambience, also it's quite costly.*
	Extroversion	low 1.73	4.02		
2.b	Emotional stability	low 1.97	4.13	4.50	*Trattoria Rustica isn't as bad as the others. Err... even if it's costly, it offers kind of adequate food, alright? It's an Italian place.*
	Openness to experience	low 1.00	3.85		
2.c	Agreeableness	low 1.00	3.27	3.58	*Trattoria Rustica is the only eating place that is any good. Err... it's located in Manhattan. This restaurant is an Italian place with poor ambience. It's bloody costly, even if this eating house has friendly waiters you see?*
	Openness to experience	low 1.33	3.94		

These results show that the PEM approach offers a viable alternative to the OR method, without any costly overgeneration phase. However, it still requires a pre-existing generator that is expressive enough to hit all regions of the target stylistic space. In the next section we describe a method for learning such a generator entirely from a corpus of labeled data that has both semantic and stylistic annotations.

Table 9.6. AdaBoost model predicting the exclamation parameter. Given input trait values, the model outputs the class yielding the largest sum of weights for the rules that return that class.

Rules	Weight
if extroversion > 6.42 then enabled else disabled	1.81
if extroversion > 4.42 then enabled else disabled	0.38
if extroversion ≤ 6.58 then enabled else disabled	0.22
if extroversion > 4.71 then enabled else disabled	0.28
if extroversion > 5.13 then enabled else disabled	0.42
if extroversion ≤ 6.58 then enabled else disabled	0.14
if extroversion > 4.79 then enabled else disabled	0.19
if extroversion ≤ 6.58 then enabled else disabled	0.17

Table 9.7. Pearson's correlation coefficient r and mean absolute error e between the target personality scores and the 480 judges' ratings (20 ratings per trait for 24 judges); the third measure is the correlation between the target scores and the 20 ratings averaged over all judges. ● = statistically significant correlation ($p < 0.05$), ○ $p = 0.07$ (two-tailed).

Trait	r	r_{avg}	e
Extroversion	0.45 ●	0.80 ●	1.89
Emotional stability	0.39 ●	0.64 ●	2.14
Agreeableness	0.36 ●	0.68 ●	2.38
Conscientiousness	−0.01	−0.02	2.79
Openness to experience	0.17 ●	0.41 ○	2.51

9.4 Learning a generator from data using factored language models

While PEMs show promising results, they have two major flaws: (a) they require an existing expressive generator, and (b) they assume that parameters are conditionally independent given the target style. Predicting multiple parameter values simultaneously while taking their interdependencies into account can be treated as a structured prediction problem. This type of problem has recently received a lot of attention in the machine learning community. Structured prediction methods aim at learning to predict elements in large output spaces, such as the set of all possible generation decision values. Typically, structured prediction methods make use of the structure of the output to find good candidates efficiently, e.g., by using dynamic programming to find the optimal sequence in Markov chains, as used for speech recognition or part-of-speech tagging (Rabiner, 1989; Tsochantaridis *et al.*, 2004). Unfortunately, the structure behind generation decisions is typically complex and generator-dependent, so finding optimal parameter values is likely to require an exhaustive search, as in the OR paradigm.

In order to use structured prediction methods for stylistic NLG, one must identify a model for representing language and input semantics such that (a) the model can be

Table 9.8. Example utterance with aligned semantic stacks. The `inform` communicative goal is the root concept of the semantic tree, indicating that the user is providing some information.

Char Sue	*is a*	*Chinese*	*restaurant*	*in the*	*center of town*
Charlie Sue		Chinese	restaurant		center
name		food	type	area	area
inform	`inform`	inform	inform	`inform`	inform
$t = 1$	$t = 2$	$t = 3$	$t = 4$	$t = 5$	$t = 6$

learned from semantically labeled data; (b) such data can be collected easily; (c) the model structure allows for efficient search algorithms to find good candidates; and (d) the model is expressive enough to convey the large range of styles found in human language. If we cast the NLG task as a search over factored language models (FLMs) (Bilmes and Kirchhoff, 2003), from which the optimal realization phrase sequence can be found efficiently using Viterbi-style dynamic programming (Mairesse *et al.*, 2010), we can satisfy these requirements. Each semantically labeled training utterance is split into atomic phrases, and each phrase is associated with a distinct semantic stack as illustrated in Table 9.8. The semantically aligned utterances are used to train FLMs predicting: (a) the most likely semantic stack sequence given the input communicative goal; and (b) the most likely realization phrase sequence given the optimal stack sequence. Each FLM can be seen as a smoothed dynamic Bayesian network in which each time frame corresponds to a phrase and a semantic stack. Our BAGEL generator combines the two FLMs to generate unseen semantic inputs, by learning from a corpus of semantically annotated utterances. BAGEL's semantic ordering model uses a trigram stack model, and the realization phrase model is conditioned on the previous phrase and the neighboring stacks, as illustrated in Figure 9.7. BAGEL has been shown to produce utterances that are indistinguishable from human utterances in a relatively constrained tourist information domain,[1] even for inputs that were not entirely seen during training (Mairesse *et al.*, 2010).

While BAGEL is trained to maximize the likelihood of the data – i.e., favoring frequent phrase sequences – it can be extended to model stylistic variation explicitly. A first approach would be to condition the predicted phrase on stylistic variables as well as on the semantics. However, whether the increase in vocabulary size would affect performance remains to be evaluated. An alternative approach models stylistic variation as a separate adaptation task, by learning to transform a canonical neutral utterance into arbitrary target styles. Decoupling the problem in such a way is likely to reduce data sparsity issues, especially for learning to insert local stylistic markers such as those in Table 9.3. While a similar approach has been successfully used for speaker-dependent speech recognition (Gales and Woodland, 1996), it has not yet been applied to discrete spaces.

---

[1] There was no statistical difference in naturalness over 200 utterances according to an independent sample t-test ($p < 0.05$).

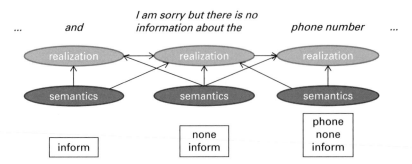

Figure 9.7   Dynamic Bayesian network for BAGEL's realization model. The semantic stack variable is observed during decoding.

Overall, training a generator with large stylistic control capabilities entirely from data remains an open research challenge; however, we believe that FLMs offer a promising tradeoff between expressiveness and tractability.

9.5   Discussion and future challenges

In this chapter we have explored three different approaches to data-driven language generation under stylistic constraints. Both the OR and the PEM approaches can learn to tailor the output of an existing handcrafted generator. The OR method – the most widely used – can make use of a large set of utterance features to predict stylistic scores, whereas PEMs only take generation decisions into account. Furthermore, the OR method can model dependencies between features (e.g., *kind of* together with an exclamation mark might indicate extroversion), whereas PEMs currently predict generation decisions independently of each other. However, despite those drawbacks, we have seen that PEMs can produce recognizable personality, without the large computational cost required by an overgeneration phase. While one could argue that computational cost is not important given the large amount of parallel processing power available, we showed in Section 9.3.1 that sampling error decreases only marginally with the number of pre-generated utterances, suggesting that the OR paradigm for stylistic control inherently lacks scalability.

The third approach to data-driven language generation under stylistic constraints involves training a generator entirely from semantically and stylistically annotated data, by casting the NLG task as a structured prediction problem over all phrase sequences in the data, which can be solved using FLMs. Whereas the OR method amounts to an exhaustive search over the realization space, the modeling assumptions made by an FLM-based generator enable it to find optimal realizations efficiently. Furthermore, this approach presents the major advantage of not requiring any handcrafting beyond the definition of the domain's semantics, which greatly reduces development effort. Hence we believe that future NLG research should focus on extending search-based probabilistic methods to model linguistic style explicitly.

Regardless of the model used, there remains a fundamental issue with statistical NLG: the lack of semantically annotated data. Given that – by definition – extreme displays of linguistic style (e.g., high extroversion) are uncommon in traditional corpora, the problem is even more severe for stylistic modeling. Semantically annotated data is difficult to collect in the first place, since annotators need to be familiar with a specific semantic labeling scheme, for which there is currently no consensus. We believe that crowdsourcing offers a real opportunity for collecting large amounts of data, as we were able to successfully train BAGEL on utterances collected by asking untrained annotators to generate utterances matching a simple predicate/argument semantic description (Mairesse *et al.*, 2010). This method could easily be extended to include stylistic elements, e.g., by modifying an utterance to convey an arbitrary style. We have also demonstrated that active learning can be used to optimize the data collection process, by iteratively annotating the utterances that are generated with a low confidence by BAGEL's probabilistic model (Mairesse *et al.*, 2010). Active learning methods are likely to be highly beneficial for stylistic modeling, given the sparsity of the data characterizing extreme styles.

Another issue with linguistic style modeling is the lack of taxonomies that relate one style to another; i.e., generating a new style requires collecting a whole new dataset. As the Big Five personality traits represent the most important dimensions of variation of human behavior according to psychology studies (Norman, 1963; Goldberg, 1990), other dimensions could be approximated as a linear combination of those traits. In practice, because of the loss of information resulting from the characterization of an individual using only five traits, it is not clear yet whether interpolating these dimensions produces convincing stylistic variation for new dimensions. Additionally, some personality dimensions manifest themselves primarily through means other than language (e.g., openness to experience), hence focusing purely on linguistic behavior might reveal a different set of major variation dimensions. Research in psycholinguistics is likely to be useful to identify a unified framework for modeling all components of human stylistic variation in a principled way.

Perhaps the most important future challenge for stylistic modeling in NLG is identifying areas in which it can make a difference, i.e., for which pre-programmed machinese is not optimal for conveying information. As most users are now familiar with the way computer interfaces behave, they might see computers that produce human-like language as surprising, or even disturbing, especially if the computer fails to convey the desired style accurately. However, some research suggests that users prefer interacting with machines conveying the same personality type as theirs (Reeves and Nass, 1996), and that the generation of human-like language reduces the user's cognitive load, resulting in better task performance (Campana *et al.*, 2004). Additional research shows that the expression of politeness (a behavior related to personality) by automatic tutoring systems can positively affect student learning (Wang *et al.*, 2008). While stylistic modeling is a fascinating topic in itself, these findings suggest that customizing the style of human–computer interfaces can be beneficial in many applications. Given the pervasiveness of such interfaces in our daily lives, the potential for customization through user modeling is immense.

Acknowledgments

This work was done while the author was at the University of Sheffield, working under the supervision of Marilyn Walker.

References

Belz, A. (2009). Prodigy-METEO: Pre-alpha release notes. Technical Report NLTG-09-01, Natural Language Technology Group, CMIS, University of Brighton.

Bilmes, J. and Kirchhoff, K. (2003). Factored language models and generalized parallel backoff. In *Proceedings of the Human Language Technology Conference of the North American Chapter of the Association for Computational Linguistics (HLT-NAACL)*, pages 4–6, Edmonton, Canada. Association for Computational Linguistics.

Brennan, S. E. and Clark, H. H. (1996). Conceptual pacts and lexical choice in conversation. *Journal of Experimental Psychology*, **22**(6):1482–1493.

Campana, E., Tanenhaus, M. K., Allen, J. F., and Remington, R. W. (2004). Evaluating cognitive load in spoken language interfaces using a dual-task paradigm. In *Proceedings of the International Conference on Spoken Language Processing (INTERSPEECH)*, pages 1721–1724, Jeju Island, Korea. International Speech Communication Association.

Coltheart, M. (1981). The MRC psycholinguistic database. *Quarterly Journal of Experimental Psychology*, **33A**:497–505.

Costa, P. T. and McCrae, R. R. (1992). *NEO PI-R Professional Manual*. Psychological Assessment Resources, Odessa, FL.

DiMarco, C. and Hirst, G. (1993). A computational theory of goal-directed style in syntax. *Computational Linguistics*, **19**(3):451–499.

Furnham, A. (1990). Language and personality. In Giles, H. and Robinson, W. P., editors, *Handbook of Language and Social Psychology*, pages 73–95. Wiley, New York, NY.

Gales, M. and Woodland, P. (1996). Mean and variance adaptation within the MLLR framework. *Computer Speech & Language*, **10**(4):249–264.

Gill, A. J. and Oberlander, J. (2002). Taking care of the linguistic features of extraversion. In *Proceedings of the Annual Conference of the Cognitive Science Society (CogSci)*, pages 363–368, Fairfax, Virginia. Cognitive Science Society.

Goldberg, L. R. (1990). An alternative "description of personality": The Big-Five factor structure. *Journal of Personality and Social Psychology*, **59**(6):1216–1229.

Gosling, S. D., Rentfrow, P. J., and Swann, W. B. (2003). A very brief measure of the Big-Five personality domains. *Journal of Research in Personality*, **37**(6):504–528.

Green, S. J. and DiMarco, C. (1996). Stylistic decision-making in natural language generation. In Adorni, G. and Zock, M., editors, *Trends in Natural Language Generation: An Artificial Intelligence Perspective*, volume 1036, pages 125–143. Springer LNCS, Berlin, Germany.

Hovy, E. (1988). *Generating Natural Language under Pragmatic Constraints*. Lawrence Erlbaum Associates, Hillsdale, NJ.

Isard, A., Brockmann, C., and Oberlander, J. (2006). Individuality and alignment in generated dialogues. In *Proceedings of the International Conference on Natural Language Generation (INLG)*, pages 25–32, Sydney, Australia. Association for Computational Linguistics.

John, O. P., Donahue, E. M., and Kentle, R. L. (1991). The Big Five inventory–versions 4a and 54. Technical report, Berkeley: University of California, Institute of Personality and Social Research.

Langkilde, I. and Knight, K. (1998). Generation that exploits corpus-based statistical knowledge. In *Proceedings of the Annual Meeting of the Association for Computational Linguistics and the International Conference on Computational Linguistics (COLING-ACL)*, pages 704–710, Montréal, Canada. Association for Computational Linguistics.

Langkilde-Geary, I. (2002). An empirical verification of coverage and correctness for a general-purpose sentence generator. In *Proceedings of the International Conference on Natural Language Generation (INLG)*, pages 17–24, Arden Conference Center, NY. Association for Computational Linguistics.

Lavoie, B. and Rambow, O. (1997). A fast and portable realizer for text generation systems. In *Proceedings of the Applied Natural Language Processing Conference (ANLP)*, pages 1–7, Washington, DC. Association for Computational Linguistics.

Mairesse, F. (2008). *Learning to Adapt in Dialogue Systems: Data-driven Models for Personality Recognition and Generation*. PhD thesis, Department of Computer Science, University of Sheffield.

Mairesse, F., Gašić, M., Jurčíček, F., Keizer, S., Thomson, B., Yu, K., and Young, S. (2010). Phrase-based statistical language generation using graphical models and active learning. In *Proceedings of the Annual Meeting of the Association for Computational Linguistics (ACL)*, pages 1552–1561, Uppsala, Sweden. Association for Computational Linguistics.

Mairesse, F. and Walker, M. A. (2007). PERSONAGE: Personality generation for dialogue. In *Proceedings of the Annual Meeting of the Association for Computational Linguistics (ACL)*, pages 496–503, Prague, Czech Republic. Association for Computational Linguistics.

Mairesse, F. and Walker, M. A. (2010). Towards personality-based user adaptation: Psychologically informed stylistic language generation. *User Modeling and User-Adapted Interaction*, **20**(3):227–278.

Mann, W. C. and Thompson, S. A. (1988). Rhetorical structure theory: Toward a functional theory of text organization. *Text*, **8**(3):243–281.

Mehl, M. R., Gosling, S. D., and Pennebaker, J. W. (2006). Personality in its natural habitat: Manifestations and implicit folk theories of personality in daily life. *Journal of Personality and Social Psychology*, **90**(5):862–877.

Norman, W. T. (1963). Toward an adequate taxonomy of personality attributes: Replicated factor structure in peer nomination personality rating. *Journal of Abnormal and Social Psychology*, **66**(6):574–583.

Paiva, D. S. and Evans, R. (2005). Empirically-based control of natural language generation. In *Proceedings of the Annual Meeting of the Association for Computational Linguistics (ACL)*, pages 58–65, Ann Arbor, MI. Association for Computational Linguistics.

Pennebaker, J. W., Francis, M. E., and Booth, R. J. (2001). Linguistic inquiry and word count. Available from http://www.liwc.net/. Accessed on 11/24/2013.

Pennebaker, J. W. and King, L. A. (1999). Linguistic styles: Language use as an individual difference. *Journal of Personality and Social Psychology*, **77**(6):1296–1312.

Pickering, M. and Garrod, S. (2004). Toward a mechanistic psychology of dialogue. *Behavioral and Brain Sciences*, **27**(2):169–226.

Porayska-Pomsta, K. and Mellish, C. (2004). Modelling politeness in natural language generation. In *Proceedings of the International Conference on Natural Language Generation (INLG)*, pages 141–150, Brockenhurst, UK. Springer.

Rabiner, L. R. (1989). Tutorial on hidden Markov models and selected applications in speech recognition. *Proceedings of the IEEE*, **77**(2):257–286.

Reeves, B. and Nass, C. (1996). *The Media Equation*. University of Chicago Press, Chicago, IL.

Reiter, E. and Dale, R. (2000). *Building Natural Language Generation Systems*. Cambridge University Press, Cambridge, UK.

Scherer, K. R. (1979). Personality markers in speech. In Scherer, K. R. and Giles, H., editors, *Social Markers in Speech*, pages 147–209. Cambridge University Press, Cambridge, UK.

Stent, A., Prasad, R., and Walker, M. A. (2004). Trainable sentence planning for complex information presentation in spoken dialog systems. In *Proceedings of the Annual Meeting of the Association for Computational Linguistics (ACL)*, pages 79–86, Barcelona, Spain. Association for Computational Linguistics.

Stent, A., Walker, M., Whittaker, S., and Maloor, P. (2002). User-tailored generation for spoken dialogue: An experiment. In *Proceedings of the International Conference on Spoken Language Processing (INTERSPEECH)*, pages 1281–1284, Denver, CO. International Speech Communication Association.

Traum, D., Fleischman, M., and Hovy, E. (2003). NL generation for virtual humans in a complex social environment. In *Working Papers of the AAAI Spring Symposium on Natural Language Generation in Spoken and Written Dialogue*, pages 151–158, Stanford, CA. AAAI Press.

Tsochantaridis, I., Hofmann, T., Joachims, T., and Altun, Y. (2004). Support vector learning for interdependent and structured output spaces. In *Proceedings of the International Conference on Machine Learning (ICML)*, Banff, Canada. The International Machine Learning Society, Association for Computing Machinery.

Walker, M., Whittaker, S., Stent, A., Maloor, P., Moore, J. D., Johnston, M., and Vasireddy, G. (2002). Speech-Plans: Generating evaluative responses in spoken dialogue. In *Proceedings of the International Conference on Natural Language Generation (INLG)*. Association for Computational Linguistics.

Walker, M., Whittaker, S., Stent, A., Maloor, P., Moore, J., Johnston, M., and Vasireddy, G. (2004). Generation and evaluation of user tailored responses in multimodal dialogue. *Cognitive Science*, **28**(5):811–840.

Walker, M. A., Stent, A., Mairesse, F., and Prasad, R. (2007). Individual and domain adaptation in sentence planning for dialogue. *Journal of Artificial Intelligence Research*, **30**(1):413–456.

Wang, N., Johnson, W. L., Mayer, R. E., Rizzo, P., Shaw, E., and Collins, H. (2008). The politeness effect: Pedagogical agents and learning outcomes. *International Journal of Human-Computer Studies*, **66**(2):98–112.

10 Integration of cultural factors into the behavioral models of virtual characters

Birgit Endrass and Elisabeth André

10.1 Introduction

The design and implementation of **embodied virtual agents** that converse with human users and/or other agents using natural language is an important application area for Natural Language Generation (NLG). While traditional NLG systems focus on information provision (the transformation of content from a knowledge base into natural language for providing information to readers), embodied conversational agents also need to exhibit human-like qualities. These qualities range from the integration of verbal and non-verbal communicative behaviors – such as facial expressions, gestures, and speech – to the simulation of social and emotional intelligence. To be believable, embodied conversational agents should not all show the same behavior. Rather, they should be realized as individuals that portray personality and culture in a convincing manner.

Whereas a number of approaches exist to tailor system behavior to **personality** (for example, see the approach by Mairesse and Walker (2008) for NLG and the approach by Hartmann *et al.* (2005) for gesture generation), few researchers have so far taken up the challenge of modeling the influences of **culture** on communicative behavior. Even when communication partners speak the same language, irritations and misunderstandings can arise due to cultural differences in what people say and how they say it. For example, Germans tend to get straight to business and be rather formal, while casual small talk is more common in the United States. As a consequence, Americans might perceive Germans as rather reserved and distant, and Germans might feel rather uncomfortable about sharing private thoughts with Americans they have just met. Furthermore, cultural misconceptions can arise from differences in non-verbal communication; gestures or body postures that are common in one culture do not necessarily convey the same meaning in another culture. An example is the American *ok* gesture (bringing the thumb and the index finger together to form a circle). While it means *ok* in American culture, it is considered an insult in Italy and is a symbol for money in Japanese culture. The performance of gesture can also vary across cultures; for example, while gesturing expressively is considered a sign of engagement in some cultures, it is regarded as inappropriate in others.

Possible culture-related misunderstandings are sometimes not even recognized. If communication partners tend to take a common basis of cultural knowledge for granted, they may interpret each other's behaviors in their own culture-specific ways. Behaviors may be decoded wrongly, ignored, or interpreted as deliberate misconduct. Consequently, people might be confronted with rejection without knowing why, which in turn can lead to frustration.

In a similar manner, computer-based systems can be misunderstood: the programmer may use culture-specific indicators, which may be misperceived by the user. This could easily happen with a programmer and a user from different cultural backgrounds. For example, Marcus and Alexander (2007) described different perceptions of a homepage by users with different cultural backgrounds. Regarding virtual agents, these differences can play an even more crucial role. Since virtual agents try to simulate human behavior in a natural manner, social background should be taken into account. If cultural factors are overlooked in the implementation of communicative behaviors for a virtual agent, this may affect user acceptance.

Iacobelli and Cassell (2007), for example, investigated the perceptions of ethnic differences in verbal and non-verbal behaviors by virtual agents. They changed the behaviors of the virtual agent to match behaviors of different ethnicities, but used a constant, racially ambiguous appearance for the virtual agent. They tested children's perceptions of the ethnic identity of the virtual agent. Children were able to correctly assign the virtual agent to different ethnicities by behavior, and they engaged with the virtual agent that behaviorally matched their own ethnicity in a promising way for educational applications.[1]

If culture is not explicitly considered in the modeling process of a virtual agent, the character nevertheless will show a certain cultural background – usually that of the programmer since he or she is the one who judges the naturalness of the agent's behavior. For example, if the programmer is a member of a Western culture where direct eye contact is considered to indicate honesty, the character will probably maintain a lot of eye contact. However, a user from a different cultural background, e.g., a member of an Asian culture, might judge this behavior as impolite.

Integrating culture into the behavioral models of virtual agents is a challenging task. We need to adapt both the content and the form of an agent's utterances to particular cultures. For example, an agent might choose different topics in smalltalk and use different discourse markers to indicate politeness depending on the culture it represents. It might also vary the amount and quality of gestures depending on its assigned cultural background.

In this chapter, we investigate approaches to the generation of culture-specific behaviors for virtual agents. We first present an overview of culture models from the social sciences; these identify different levels of adaptation we will use as the basis for a computational approach to the generation of culture-specific behaviors. Findings from the social sciences are often described in tendencies rather than as concrete rules of behavior that could be implemented for a computational model, but data collections can augment findings from the research literature and help identify

---

[1] This effect can be explained by the **similarity principle** (Byrne, 1971), which states that interaction partners who perceive themselves as being similar are more likely to like each other. This phenomenon happens partly unconsciously (Chartrand and Bargh, 1999).

typical behavior patterns that have not been described in sufficient detail. We introduce a hybrid (theory-based and corpus-driven) approach to integrating social factors into the behavioral models of virtual agents. We exemplify our approach for German and Japanese cultures.

10.2   Culture and communicative behaviors

Culture plays a crucial, though often unrecognized, role in the perception and selection of communicative behaviors. But what exactly is culture and how can we distinguish different cultures and their behaviors? Most people instinctively know what the term **culture** means; however, it is hard to formalize and explain what drives people to feel they belong to a certain culture. In this section, we survey different explanations of the notion of culture, including different levels of culture, dichotomies that distinguish cultures, and dimensional models of culture.

10.2.1   Levels of culture

We first look at theories that use **layers** to describe the influence of culture on human behavior. These layers highlight, among other things, that culture not only determines behavioral differences on the surface but also works on the cognitive level.

Culture can be seen as one social factor that influences a whole group of people. It is, however, hard to formalize the extent to which each of these aspects determines an individual's behavior. Hofstede (2001) referred to the collection of factors that influence human behavior as a mental program unique to each person. This so-called "software of the mind" can be categorized into three layers: human nature, culture, and personality (see Figure 10.1), where

- **Human nature** represents the universal level in an individual's mental program and contains physical and basic psychological functions.
- **Personality** is the level that is specific to the individual.
- **Culture** is the middle layer. Culture is specific to the group and the environment, ranging from the domestic circle, through the neighborhood and workplace, to the country.

Enormous efforts have been made to incorporate two of these three layers into virtual agents. Human nature was included through **embodiment** and the simulation of credible verbal and non-verbal behaviors. Virtual characters simulate people's physical natures in more and more sophisticated ways, using natural-seeming speech and non-verbal behaviors such as gestures and body postures. Basic psychological functions have also been integrated into virtual agents; for example, implementations of the ability to express **emotions** and act accordingly are described in Gratch *et al.* (2002) and Aylett *et al.* (2005). There has also been research on the expression of **personality** in virtual agents, for example, Rist *et al.* (2003) and Kang *et al.* (2008). However, culture has come into focus only recently.

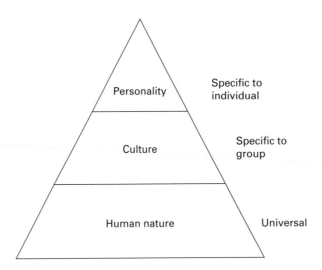

Figure 10.1   Hofstede's levels of uniqueness in an individual's mental program (Hofstede, 2001)

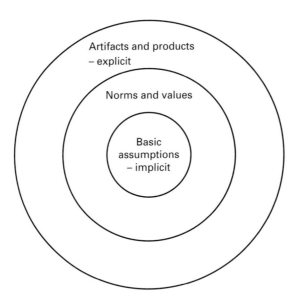

Figure 10.2   Implicit and explicit layers of culture (Trompenaars and Hampden-Turner, 1997)

Culture itself can also be described in terms of layers. For example, Trompenaars and Hampden-Turner (1997) distinguished between implicit and explicit layers of culture (see Figure 10.2). The explicit layer contains observable realities such as language, clothes, buildings, or food. The middle layer consists of norms and values that are reflected in the explicit layer. While norms are related to a group's sense of right and wrong, values are associated with the sense of good and bad. Thus, norms determine how people think they *should* behave and values determine the way people *wish* to behave. The innermost layer of culture contains basic assumptions that have vanished from conscious questioning and become self-evident. When basic assumptions are questioned, the result is confusion or even annoyance, as for example when one asks an American or Dutch person why he or she thinks that all people are equal.

In sum, implicit and explicit layers of culture can be distinguished, with explicit layers consisting of things that can be observed in reality and implicit layers containing internal values and basic assumptions. For the simulation of cultural behaviors in virtual agents, the explicit layer is thus of special interest since this layer holds observable differences in verbal and non-verbal behavior. Implicit layers of culture could also be taken into account, for example when building a cognitive model to control the behavior of virtual agents.

10.2.2 Cultural dichotomies

Cultural **dichotomies** categorize different cultural groups. Most dichotomies focus on one aspect of culture, for example different perceptions of time, and describe prototypical behaviors for the groups that are being distinguished. Cultural dichotomies seem well suited for computational modeling since concrete distinctions between cultures are clearly stated, and are described in terms of behavioral differences.

Hall (1966), for example, distinguished so-called **high-contact** and **low-contact** cultures, that present behavioral differences in proxemics (the use of space) and haptics (the use of gesture). Ting-Toomey (1999) characterized the varieties of these cultural groups in more detail. According to her, features of high-contact cultures include direct facing, frequent direct eye contact, close interaction, and a rather loud voice, whereas features of low-contact cultures include indirect facing, greater interpersonal distance, little or no touching, indirect glances, and a soft or moderate voice.

Hall (1966) also distinguished between **high-context** and **low-context** cultures. In high-context cultures little information is explicitly encoded in verbal messages and communication relies heavily on physical context and non-verbal clues. Thus, interlocutors are expected to "read between the lines" in order to decode the whole meaning of a verbal message. In contrast, in low-context cultures the content of most communications is explicitly encoded. The speaker is thus expected to construct clear messages that can be understood easily without reference to context and non-verbal cues. Most Western cultures are low-context cultures, whereas most Asian cultures are high-context cultures.

Another dichotomy analyzed by Ting-Toomey (1999) is the distinction between **monochronic** and **polychronic** cultures. One behavioral pattern described for monochronic cultures is that members tend to do one thing at a time. Most Western cultures are in the monochronic group. Most Asian cultures belong to the polychronic group; members of these cultures prototypically tend to do several things at a time. Generalizing these behavioral patterns, members of Western cultures tend to finish one thing before starting another, while it is more common in Asian cultures to switch back and forth between tasks.

10.2.3   Hofstede's dimensional model and synthetic cultures

Although culture is often described in terms of abstract behavioral tendencies, there are approaches that define cultures as points in multi-dimensional attribute spaces. These **dimensional models** constitute an excellent starting point for building behavior models for virtual agents, since they lend themselves to implementation.

An example of a dimensional model for culture is given by Hofstede (2001, 2011). His model is based on a broad empirical survey, covering more than 70 countries (however, only the largest 40 countries have been analyzed in detail). The culture of each country is captured as a set of scores along five dimensions,[2] power distance, individualism, masculinity, uncertainty avoidance, and long-term orientation, where:

- **Power distance** captures the extent to which an unequal distribution of power is accepted by the less powerful members of a culture. Scoring high on this dimension indicates a high level of inequality of power and wealth within the society. A low score, on the other hand, indicates greater equality within social groups, including government, organizations, and families.
- **Individualism** captures the degree to which individuals are integrated into groups. On the individualist end ties between individuals are loose, and everybody is expected to take care of himself or herself. On the collectivist end, people form strong, cohesive in-groups.
- **Masculinity** captures the distribution of roles between the genders. The two extremes are masculine and feminine; masculine values include assertiveness and competitiveness, while feminine values include empathy and negotiation.
- **Uncertainty avoidance** captures tolerance for uncertainty and ambiguity. The extent to which a member of the culture feels uncomfortable or comfortable in an unknown situation is a key factor of this dimension. Uncertainty-avoiding cultures try to minimize the possibility of unknown situations and stick to laws and rules, whereas uncertainty-accepting cultures are more tolerant of different opinions.
- **Long-term orientation**: long-term orientation is associated with thrift and perseverance, whereas short-term oriented cultures show respect for tradition, fulfilling of social obligations, and saving "face."

In addition to positioning each culture in this five-dimensional space, Hofstede explains how the dimensional scores of a culture impact the behavior of its members (Hofstede, 2011).

A so-called **synthetic culture** (Hofstede *et al.*, 2002) is a thought experiment in the characteristics of a hypothetical culture at the extreme of a single cultural dimension. For each synthetic culture Hofstede *et al.* (2002) defined a profile that contains the culture's values, core distinction, and key elements as well as words with a positive or negative connotation. The individualistic synthetic culture, for example, has the core value "individual freedom" and the core distinction is the distinction between the self

---

[2] Originally, Hofstede used a four-dimensional model. The fifth dimension, long-term orientation, was added later in order to better model Asian cultures. So far this dimension has been applied to 23 countries.

and others. Key elements are statements such as "Honest people speak their mind," "Laws and rights are the same for all," or "Everyone is supposed to have a personal opinion on any topic." These key elements are golden rules for appropriate behavior in this culture and explain the way in which members of this culture would think. Words with a positive connotation in the extreme individualistic culture include self, "do your own thing," self-respect, dignity, I, me, pleasure, adventure, guilt, and privacy. Words with a negative connotation include harmony, obligation, sacrifice, tradition, decency, honor, loyalty, and shame. For the collectivistic synthetic culture the connotations of these words would be the other way round. Hofstede *et al.* (2002) also defined stereotypical behaviors for members of synthetic cultures. Extreme individualists, for example, are described as verbal, self-centered, defensive, tending to be loners, and running from one appointment to the next.

Although synthetic cultures are a valuable tool, no existing culture is exclusively influenced by one dimension. For example, the US American culture scores high on the individualist dimension, but also scores high on the masculinity dimension. A combination of these two dimensions explains the culture better than either in isolation.

10.3   Levels of cultural adaptation

Culture determines not only *what* we communicate, but also *how* we communicate it. In addition, different aspects of human communicative behavior are influenced by people's cultural background, including verbal communications, posture, eye gaze, and gesture. Culturally adaptable embodied virtual agents should consequently include culture in models for every aspect of communicative behavior. In this section, we review computational approaches to culture-specific adaptation of the behavior of virtual agents in terms of both the content and form of communication, and communication management.

10.3.1   Culture-specific adaptation of context

Language, as the main medium of human communication, is the most obvious barrier to intercultural communication. However, even for speakers of the same language, verbal behavior can vary vastly across cultures. For example, members of different cultures will tend to choose culturally relevant topics and use culture-specific scripts when discussing those topics.

Isbister *et al.* (2000) describe a culturally adaptable conversational agent for smalltalk conversations. The agent, a so-called "Helper Agent," in the appearance of a dog, virtually joins users of a chat room. The agent plays the role of party host, interacting with the human interlocutors when their conversation stagnates by introducing prototypical smalltalk topics. The agent distinguishes between **safe** topics (e.g., weather, music) and **unsafe** topics (e.g., religion, money). According to Isbister *et al.* (2000), the categorization of a topic as safe or unsafe varies by culture. However, their agent does not consider the cultural background of the human interlocutors when introducing conversational topics.

Yin *et al.* (2010) present conversational agents that tailor their conversational scripts based on the assigned cultural background. Two different virtual agents were designed, one representing a member of an Anglo-American culture and the other resembling a member of a Hispanic culture. The cultural background of each virtual agent was expressed in their physical appearance and physical context (the appearance of the flat in the background of the agent). In addition, each agent used different conversational scripts. While the Anglo-American agent focuses on the interlocutor's well-being, the Hispanic agent shows interest in the interlocutor's family and friends. Thus the content of conversations varies with simulated cultural background.

Another example of a culturally adaptable conversational agent is the Tactical Language Training System (TLTS) presented in Johnson *et al.* (2004). This system is designed to help soldiers acquire cross-cultural skills through interaction in a game-like virtual environment while learning a foreign language. The soldier can interact with the system using speech, and through the selection of culture-specific gestures. This system is interesting because the content of both verbal and non-verbal communicative behaviors is culturally specific.

10.3.2   Culture-specific adaptation of form

Another important aspect of culture-specific communication is the form and performance of conversational behaviors.

Generation of culture-specific wording

Personality has been considered as a parameter of linguistic variation (see, for example, Mairesse and Walker, 2008), but only a small amount of research has considered culture, and that mainly for **politeness behaviors**. Most noteworthy is the work of House, who has performed a series of contrastive German–English discourse analyses over the past twenty years (e.g., House, 2006). Among other things, she observed that Germans tend to be more direct and more self-referential, and to resort less frequently to using verbal routines such as "how-are-you's" (and consequently tend to interpret such phrases literally).

Alexandris and Fotinea (2004) investigated the role of discourse particles as politeness markers to inform the design of a Greek speech technology application for the tourist domain. They performed a study in which evaluators had to rank different variations of written dialogues according to their perceived degree of naturalness and acceptability. The study revealed that dialogues in modern Greek with discourse particles emphasizing approval of the user are perceived as friendlier and more natural, while dialogues without any discourse particles, or containing discourse particles that perform other functions, were perceived as unnatural. The authors regard these findings as culture-specific elements of the Greek language.

Johnson *et al.* (2005) investigated the potential benefits of politeness in a tutoring system. By examining interactions between a real tutor and his students, they identified a set of politeness behavior templates, which they implemented in their system. Their system automatically selects different tutorial strategies, which use different politeness behaviors, depending on the type of expected face threat (Brown and Levinson, 1987) to the student.

Wu and Miller (2010) present another example of a tutoring system that incorporates culture-specific politeness behaviors. In their system, a user can interact with the virtual agent by selecting phrases and/or gestures from an interactive phrasebook. The system then calculates the appropriateness of the selected action(s) from the perspective of a member of a Middle Eastern culture based on the social relationship between the interlocutors, and categorizes the action(s) as polite, nominal, or rude.

Generation of culture-specific gestures and postures

An interesting aspect of culture-specific behavior is the **expressivity** of non-verbal behaviors. How someone performs a gesture can sometimes be as crucial for the observer's perception as the gesture itself. For example, a hand gesture may be small (involving only the fingers, hand, and wrist) and performed near the torso; or it may be large (involving the whole arm) and performed near the face. These types of difference in dynamic variation can be described according to expressivity parameters, including spatial extent, speed, power, fluidity, repetivity, and overall activation (see Pelachaud, 2005). The spatial extent parameter, for example, describes the arm's extent toward the torso. The fluidity parameter captures the degree of continuity between consecutive gestures, while the repetivity parameter holds information about the repetition of the stroke. The overall activation parameter captures the number of gestures performed.

Expressivity is a function of both individual and social factors, including personality, emotional state, and culture. Lipi *et al.* (2009) present a cross-culture corpus study of the expressivity of body postures using Hofstede's model of cultural dimensions. Each body posture was analyzed in terms of several behavioral parameters, including spatial extent, rigidness, mirroring, frequency, and duration. Then a Bayesian network model was trained on this data and used to predict non-verbal expressions appropriate for specific cultures.

10.3.3   Culture-specific communication management

When people communicate, they do not typically need to think about the management of their conversation; **turn taking** and **grounding** behaviors are produced automatically. However, these communication management behaviors are also realized in culturally specific ways. According to Ting-Toomey (1999), several types of culture-specific **regulators**, including vocalics, kinesics, and oculesics, control the flow of a conversation. Vocalics are verbal feedback signals, such as the English *uh-huh*, silences, and interruptions. Kinesics include hand gestures and body postures, while oculesics include eye gaze and face orientations. Regulators are learned at a very young age and are thus used at a very low level of awareness. Ting-Toomey (1999) states that the indiscriminate use of regulators in cross-cultural communication often causes misunderstandings and frustration. However, people from contrasting cultures may not be able to name the reason for their distress since regulators are used subconsciously.

There are a number of computational approaches to modeling verbal regulators for turn taking, (e.g., Sato *et al.*, 2002; Sidner, 2004; Jonsdottir *et al.*, 2008). However, hardly any effort has been made so far to integrate culture as a parameter in these models. An exception is the work of Endraß *et al.* (2009, 2010), who focus on culture-specific aspects of communication management for dialogues with culturally adaptable virtual agents.

10.4   Approaches to culture-specific modeling for embodied virtual agents

In principle, there are two types of approach that might be taken to integrate aspects of culture into the behavioral models of virtual agents: **bottom-up** and **top-down**. The top-down approach is model-driven: descriptions of culture-related differences in behavior are extracted from the research literature and transformed into computational models that can be incorporated into the virtual agents' behavior repertoires. By contrast, the bottom-up approach is data-driven: corpora of human communicative behaviors are analyzed to identify culture-specific behavioral tendencies, and these are then integrated into the behavioral models of the virtual agents.

10.4.1   Top-down approaches

The most well-known system that aims at simulating culture-specific behaviors in virtual agents is the Tactical Language Training System (TLTS; Johnson *et al.*, 2004; Johnson and Valente, 2009). The TLTS is used to teach functional verbal communication skills in foreign languages and cultures, including Iraqi, Dari, Pashto, and French. The TLTS virtual environment is a virtual village, where users can gain communicative skills by interacting with virtual agents to perform everyday activities, such as asking for directions or buying food. The user's goal is to learn how to communicate in culturally appropriate ways. The user is represented by an avatar in the virtual world; the user speaks for the avatar and can choose gestures by selecting them from a menu. The virtual agents represent different cultural backgrounds, and incorporate culture-specific communication behaviors that are derived from the research literature; that is, the TLTS uses a top-down approach to simulating culture-specific communication. Figure 10.3 shows a screenshot of the TLTS in an Iraqi environment.

Mascarenhas *et al.* (2009) present another top-down approach to simulating culture-specific communication. Instead of simulating existing national cultures, they simulate Hofstede's synthetic cultures (Hofstede *et al.*, 2002). This work focuses on culture-specific **rituals**, symbolic social activities carried out in a predetermined fashion. Groups of virtual agents representing different synthetic cultures perform rituals in a culturally inflected manner. Figure 10.4 shows the virtual agents in a dinner ritual. Characters in the left picture, representing a low-power culture, rush to the table, while characters in the right picture wait for the elder to sit first, as they represent a high-power

Figure 10.3   A user, represented by an avatar, interacting with a member of a different culture in the Tactical Language Training System (from Johnson *et al.* (2004); used with permission)

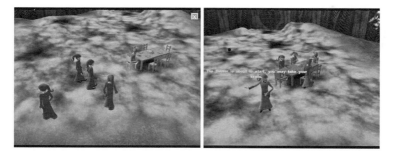

Figure 10.4   A group of virtual agents interacting in different culture-specific rituals (from Mascarenhas *et al.* (2009); used with permission)

culture. In an evaluation, participants were asked to observe the agents' performances of culture-specific rituals and then categorize each group of agents using a set of adjectives. Participants identified significant behavioral differences between the groups of agents, and were able to relate these differences to culture.

10.4.2   Bottom-up approaches

Abstract theoretical models of culture-specific behaviors are useful for understanding cultural differences and implementing heuristics for virtual agents. However, more

Figure 10.5   A frame from a video corpus displaying the emotions anger and despair (left) and simulation with a virtual agent (right) (from Martin *et al.* (2005); used with permission)

accurate and natural simulations of culture-specific behaviors may be obtained using data-driven approaches.

Kipp *et al.* (2007a,b) describe a data-driven approach to simulating non-verbal behaviors that are performed by a virtual agent to resemble a certain speaker. The approach requires an annotated **multimodal corpus**, from which gestural profiles and animation lexicons are extracted for each speaker. Conversational gestures are synthesized for the virtual agent at runtime by an animation engine for any given input text in the style of the particular speaker. Kipp *et al.* illustrated their approach using data from two talk-show hosts, Jay Leno and Marcel Reich-Ranicki. In an evaluation, human observers were able to correctly assign the virtual agent's behavior to the speaker whose data was used to train the model.

Martin *et al.* (2005) use a similar data-driven approach to simulate emotional behaviors. Video recordings of human speakers are manually annotated to identify multimodal emotional behaviors and capture their expressivity parameters. The annotations for each individual human speaker form a behavioral profile, which is used by an animation engine to resynthesize emotional behaviors for performance by a virtual agent. Figure 10.5 left shows a frame from a video recording where a woman talks about a recent trial in which her father was kept in jail. Figure 10.5 right shows the corresponding virtual agent, displaying anger fading into despair. Martin *et al.* (2005) state that they focus on modeling visible behaviors, rather than internal models that describe the motivations for those behaviors.

10.5   A hybrid approach to integrating culture-specific behaviors into virtual agents

In this section, we present a **hybrid approach** to simulating culture-specific communication behaviors. Like Martin *et al.* (2005) and Kipp *et al.* (2007b), we aim to model visible behaviors – the explicit layer of culture, rather than the implicit layers. However, we focus on behaviors that are generalizable across a culture, rather than individual behavioral differences. We use an iterative, top-down then bottom-up, approach. First, we explore the research literature to find descriptions of stereotypical behaviors for different cultures. Then, we validate and ground these descriptions in empirical data through annotation of multimodal corpora. Finally, we extract computational rules from the annotated data. To summarize this combination of top-down and bottom-up approaches, the social sciences tell us *what* aspects of communication behaviors might be relevant to culture-specific models, and the empirical data tells us *how* differences in these aspects manifest themselves. We exemplify our approach for German (Western) and Japanese (Asian) cultures, since the differences between Western and Asian cultures are supposed to be large.

10.5.1 Cultural profiles for Germany and Japan

The two cultures of Germany and Japan are very different. Table 10.1 summarizes some of these differences as identified in the research literature. Germany is a Western culture, while Japan is an Asian culture. In terms of cultural dichotomies (see Section 10.2.2) the following distinctions can be made:

- Germany is a medium-contact culture while Japan is a low-contact culture (Ting-Toomey, 1999). This means that standing close to or touching the conversation partner should not be very common in either culture, but would be more acceptable in Germany than in Japan.
- Germany is a very low-context culture, while Japan is a very high-context culture (Ting-Toomey, 1999). This means that German speakers will tend to be very direct and explicit, while Japanese speakers will rely more on context and non-verbal behaviors.
- Germany is a monochronic culture, while Asian cultures, including that of Japan, are polychronic (Hall and Hall, 1987). This means that in Germany clock time is important, and tasks are solved sequentially. In Japan, on the other hand, the notion of time is more relative, and people may be involved in several tasks simultaneously.

Table 10.1 also summarizes Hofstede's five-dimension analysis of German and Japanese cultures. Germany is a low-power-distance (egalitarian), individualistic, and long-term oriented culture, while Japan is a high-power-distance, collectivistic, and short-term oriented culture.

10.5.2 Behavioral expectations for Germany and Japan

The top-down categorization of German and Japanese cultures from the research literature can be used as a basis to state expectations for behavioral differences. Following the discussion in Section 10.3, we will list expectations for cultural-specific adaptations to (1) communication content, (2) the form of communication behaviors, and (3) communication management. We do not attempt to cover every culture-specific variation in these three aspects, but focus on one or two adaptations for each one that are likely

Table 10.1. Summary of culture profiles for Germany and Japan

Classification	Germany	Japan
regional	Western	Asian
contact dichotomy	medium-contact	low-contact
context dichotomy	low-context	high-context
time dichotomy	monochronic	polychronic
power distance	low power distance	high power distance
individualism	individualistic	collectivistic
masculinity	masculine	masculine
uncertainty avoidance	avoiding	avoiding
long-term orientation	long-term	short-term

Table 10.2. Summary of behavioral expectations for Germany and Japan

Behavioral aspect	Germany	Japan
topic selection	more private	less private
pauses	avoided	consciously used
overlaps	uncommon	common during feedback
nonverbal expressivity	more expressive	less expressive

to highlight the contrasts between German and Japanese cultures. Table 10.2 summa-rizes our expectations for behavioral differences between German and Japanese cultures based on their behavioral profiles.

Communication content

For communication content, we focus on casual smalltalk such as one might make when first meeting someone. We chose this domain, since meeting someone for the first time is a fundamental interaction in everyday communications, occurring in every culture as well as in cross-cultural encounters – so much so, that it is the topic of the first chapter of most language learning books. Smalltalk is often thought of as neutral, non-task-oriented conversation about safe topics, where no specific goals need to be achieved. But smalltalk serves important social purposes. Schneider (1988) categorizes topics that might occur in smalltalk conversations into three groups, with the choice of topic depending on the social context:

- The **immediate situation** group holds topics that are elements of the so-called *frame* of the situation. In order to explain the idea of a frame, Schneider (1988) uses a smalltalk situation that takes place at a party. Possible topics within a party frame could be the atmosphere, drinks, music, participants, or food.
- The **external situation** or "supersituation" group holds topics in the larger context of the immediate situation. This category is the least limited of the three. Topics within this category could be the latest news, politics, sports, movies, or celebrities.

- The **communication situation** group holds topics pertaining to the conversation participants themselves. Topics in this category could include hobbies, family, and career.

However, Schneider (1988) only considered Western cultures in his studies and did not look at topic selection in other cultures. Thus, these groups do not necessarily hold true for other cultures. According to Isbister *et al.* (2000), the categorization of smalltalk topics as **safe** or **unsafe** varies with cultural background. If the distinction into safe and unsafe topics varies with culture, we expect that the overall choice of topic categories is also dependent on culture. For example, in Table 10.1, Germany was categorized as a low-context culture, while Japan was categorized as a high-context culture. Ting-Toomey (1999) describes people belonging to high-context cultures as having a lower "public self" than people belonging to low-context cultures. A typical behavioral pattern for members of high-context cultures is not to reveal too much personal information during a first-time meeting. Consequently, we might expect topics from the communication situation group to be more common in German smalltalk conversations than in Japanese smalltalk conversations.

Communication form

The degree of **expressivity** of non-verbal behaviors is known to vary by culture (see Section 10.3.2). In particular, the individualism dimension in Hofstede's model is related to the expression of emotions and acceptable emotional displays in a culture (Hofstede *et al.*, 2010). For communication form, we focus on the expression of emotions. In individualistic cultures such as Germany, it is more acceptable to publicly display emotions than in collectivistic cultures such as Japan (Ekman, 2001). This suggests that non-verbal behaviors may be performed more expressively in German conversations than in Japanese ones. We expect the display of emotions by German speakers to more obviously affect the expressivity of gestures in a way that increases parameters such as speed, power, or spatial extent. This should also be the case for body postures as the whole body can be used to express emotions.

Communication management

In Section 10.3.3, we discussed how **regulators** (such as verbal feedback, silences, gestures, posture, and eye gaze) are used to manage the flow of a conversation. For communication management, we focus on the use of **silence** and on **overlapping speech**. As summarized in Table 10.1, Japan is a collectivistic culture. In these cultures silence can occur in conversations without creating tension. In addition, pauses can be a crucial feature of conversations in collectivistic cultures. These observations do not hold true for individualistic cultures such as Germany's (Hofstede *et al.*, 2010). Rather, in German culture silences are likely to create tension and thus pauses in speech are avoided if possible.

The power dimension in the model of Hofstede *et al.* (2010) may also influence the flow of communication. High-power-distance cultures are described as verbal, soft-spoken, and polite, and in these cultures interpersonal synchrony is much more important than in low-power-distance cultures, whose members tend to talk freely in any social context (Ting-Toomey, 1999). One way to achieve interpersonal synchrony in a conversation is by giving verbal feedback. This **feedback** often occurs while the conversational partner has the floor, creating overlapping speech segments. We expect this to occur more often in the Japanese culture due to their higher score on the power distance dimension.

10.5.3   Formalization of culture-specific behavioral differences

As we discussed previously, culture-specific behavioral tendencies identified from the research literature can be hard to formalize for use in computational models. The hypothesis that there should be more pauses in Japanese conversations than in German ones, for example, is not quantified: it does not indicate how many pauses we should include in the virtual agent's behavior or how long these pauses should last. To get a deeper insight into these issues and to obtain some quantitative data, we analyzed the video corpus recorded for the Cube-G project (Rehm *et al.*, 2007). This corpus was recorded in Germany and Japan and includes three prototypical interaction scenarios (a first-time meeting, a negotiation, and a conversation with status differences). Altogether, more than 20 subjects from each of the two cultures participated, and about 20 hours of video material were collected. Participants interacted with actors to ensure that they had not met before and that all scenarios lasted for about the same time. For the first-time meeting, participants were asked to get acquainted with their conversational partner in order to be better able to solve later tasks together. The analysis described in this section focuses on this first-time meeting scenario, which lasted for around 5 minutes for each subject. The analysis includes all the German videos and about half of the Japanese videos due to translation issues.

The video corpus was annotated using the Anvil tool (Kipp, 2001) for the behavioral tendencies summarized in Table 10.2. In particular:

- Topics that occurred in the conversations were categorized as pertaining to immediate, external, or communication situations, as described in Section 10.5.2.
- Expressivity was annotated for every gesture, taking into account the parameters power, speed, spatial extent, repetition, and fluidity (see Section 10.3.2). We also labeled body postures.
- Pauses and overlaps were calculated as segments on the time line where either neither of the participants spoke, or both participants spoke at the same time.

Analysis of topic selection

We compared the frequencies of the three topic categories within first-time meeting conversations involving German and Japanese participants. Topics pertaining to immediate and external communication situations occurred significantly more often in the Japanese conversations than in the German ones ($p = 0.014$ for immediate situations and $p = 0.036$ for external situations), while topics covering the communication situation occurred significantly more often in the German conversations ($p = 0.035$;

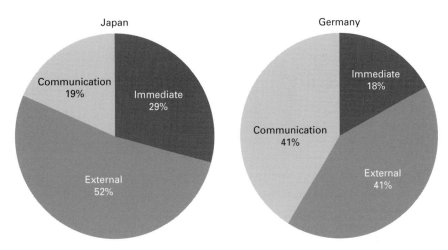

Japan   Germany

Figure 10.6 Average distribution of topic categories during smalltalk conversations recorded in Japan and Germany (Endraß *et al.*, 2011b)

Endraß *et al.*, 2011b). We also calculated the average percentage distribution of topic categories in the German and Japanese conversations, shown in Figure 10.6. This distribution is in line with our top-down expectations and provides a possible target topic distribution for a computational model of culture-specific topic choice.

Analysis of non-verbal expressivity

Using a 7-point scale for each parameter, we annotated the expressivity of non-verbal behaviors in each conversation for each of the parameters power, speed, spatial extent, repetition, and fluidity. We found significant differences between the two cultures for all parameters (ANOVA, $p < 0.01$). Figure 10.7 (left) shows the average ratings of the expressivity parameters for German and Japanese conversational participants. Gestures were performed faster and more powerfully in the German conversations than in the Japanese ones, and German participants used more space for their gestures compared to Japanese participants. Gestures were also performed more fluently in the German conversations, and the stroke of a gesture was repeated less in the Japanese conversations (Endraß *et al.*, 2011a).

We used the coding scheme described in Bull (1987) to label the type/shape of body postures. Postures most frequently observed in the German conversations were folding arms in front of the trunk and putting hands in the pockets of trousers, while the most frequent postures in the Japanese conversations were joining both hands in front of the body and putting hands behind the back. It is notable that postures frequently observed in the German conversations were rated higher in spatial extent and lower in rigidity (on a 7-point scale), compared to postures frequently observed in the Japanese conversations. In addition, our analysis indicates that the Japanese conversational participants remained in the same posture longer, engaged in more frequent mirroring, took up less space, and displayed a more rigid posture in comparison to the German participants (see Figure 10.7, right).

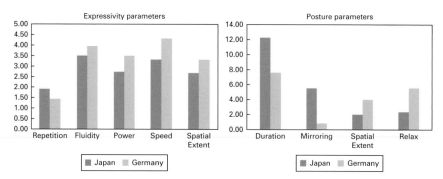

Figure 10.7   Ratings of expressivity (left) and posture parameters (right) averaged over participants (Endraß *et al.*, 2011a)

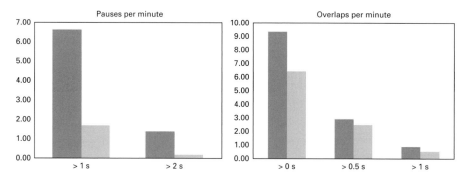

Figure 10.8   Pauses (left) and overlaps in speech (right) per minute, averaged over participants (Endraß *et al.*, 2011a)

Analysis of pauses and overlaps

We counted pauses in conversations that lasted for more than 1 second (short pauses) and pauses that lasted for more than 2 seconds (longer pauses); we discarded very brief pauses, such as those used for breathing. We found more pauses in conversations involving Japanese participants than in those involving German participants; these differences were significant both for short pauses (two-tailed *t*-test, $p < 0.001$) and for longer pauses ($p < 0.001$). In conversations involving German participants, there were on average 7.1 short pauses and 1.3 longer pauses. By contrast, in conversations with Japanese participants, there were on average 31 short pauses and 8.4 long pauses. Figure 10.8 (left) shows the distribution of short and longer pauses that were found on average per minute in each video.

We counted, for each conversation, the total amount of overlapping speech, as well as the number of overlaps lasting for more than 0.5 seconds (short overlaps) and the number lasting for more than 1 second (longer overlaps). In line with our predictions, there was significantly more overlapping speech in Japanese conversations ($p = 0.04$); however, the differences in numbers of short overlaps and numbers of long overlaps were not significant. The average occurrences of overlapping speech per subject per minute for the two cultures are shown in Figure 10.8 (right). We observed an average of 6 overlaps per minute in conversations with German participants, and of 9 overlaps per minute in conversations with Japanese participants (Endraß *et al.*, 2011a).

10.5.4 Computational models for culture-specific conversational behaviors

For the generation of culture-specific dialogues, a distributed planning system was implemented based on the SHOP planner (University of Maryland, 2005). This distributed approach allows us to implement autonomous agents that can act proactively on their internal goals, as well as react to the ongoing conversation. Dialogue planning consists of triggering appropriate speech acts given the dialogue history, the agent's goals, and other agent-specific features, here including cultural features.

Computational model for topic selection

In each conversational agent's knowledge base a personal motivation is provided for each available topic. This motivation represents the agent's internal drive to talk about the topic, e.g., because of an increased personal interest in the topic or a general desire to talk a lot. In our analysis of human–human conversations, we found that Japanese participants prefer topics from the immediate and external situation, while German participants will also select topics from the communication situation (see Figure 10.6). To implement a culture-specific computational model for topic selection, we also set cultural thresholds for each topic based on the topic category and the selected culture for each agent. So, a more "Japanese" agent might have a high personal motivation to talk about children, but a high cultural threshold against discussing communication situation topics, while a more "German" agent, similarly motivated to discuss children, would have a lower cultural threshold for the topic. In order for an agent to initiate discussion of a topic, its personal motivation would have to exceed its cultural threshold.

Computational model for non-verbal expressivity

We used a Bayesian network model for culture-specific non-verbal expressivity. Using this approach, culture as a nondeterministic concept can be modeled in an intuitive manner and without giving up a certain amount of variability that is necessary to ensure that an agent is perceived as an individual. In the network, the causality of cultural background and corresponding most probable behavior is modeled according to Hofstede's dimensional model (see Section 10.2.3). The Bayesian network was implemented using the GeNie modeling environment (Decision Systems Laboratory, 2007), and incorporates nodes reflecting the culture, the dimensions of the culture, and the aspects of gestural expressivity that are influenced by each dimension. Each gesture is divided into three phases: preparation, stroke, and retraction. The preparation and retraction phases are used for animation blending, while in the stroke phase the actual gesture is performed. The stroke phase of a gesture is customized for different parameters and degrees of gestural expressivity. For example, the repetition parameter can be varied by playing the stroke phase several times, while the speed parameter controls the speed of performance of each repetition. To vary spatial extent, the stroke phase is blended with a neutral hand position close to the trunk of the virtual agent.

In some cases, culture-specific behaviors cannot be generated by customizing culture-neutral behaviors because culture is reflected by specific (usually emblematic) gestures and postures that need to be accurately executed in order not to be misunderstood. To account for this fact, the library of non-verbal behaviors was extended with behaviors extracted directly from the corpus. For example, a prototypical posture for a German agent would be putting their hands in the pockets of their trousers, or folding their arms in front of the trunk, while a prototypical posture for a Japanese agent would be folding both hands in front of the body.

10.5.5   Simulation

The environment for our virtual agents is the Virtual Beergarden scenario from http://mm-werkstatt.informatik.uni-augsburg.de/projects/AAA. We designed culture-specific virtual agents that had a prototypical German or Japanese appearance, and spoke with a German or Japanese synthesized voice. Figure 10.9 shows examples of prototypical characters in the Virtual Beergarden scenario.

Each virtual agent has a dialogue planner, and access to a verbal knowledge base containing smalltalk sentences relating to different topics and speech acts that can be triggered by the planner. Each agent also has access to over 40 different animations of non-verbal behavior, including gestures and body postures. Some behaviors are culture-specific (e.g., a bow for a Japanese greeting, a hand wave for a German greeting). Other behaviors can be exhibited by an agent of any cultural background, but are realized in a culture-specific way using our Bayesian model for non-verbal expressivity. For behavioral examples of our virtual agents, please see our website at http://www.hcm-lab.de/projects/animations/.

Figure 10.9   Prototypical male (left) and female characters (right) resembling members of German and Japanese cultures

10.5.6   Evaluation

We conducted two evaluations of the perception of our virtual agents, one focusing on verbal behavior (Endraß *et al.*, 2011b) and the other on communication management and non-verbal behavior (Endraß *et al.*, 2011a). We did this to reduce dependencies between the different types of behavior that might influence participants. The evaluation of verbal behavior did not include variations in non-verbal behaviors such as gestures, while in the evaluation of non-verbal behavior the agents spoke gibberish. In each evaluation, we asked participants to observe and rate generated dialogues. In the studies, the virtual agents' appearance was left constant to resemble the observers' cultural background and only the culture-specific behavioral model was changed. Thus, German participants observed Western-looking characters and Japanese participants watched Asian-looking characters with varying behavior.

In line with previous work (e.g., Byrne, 1971), we hypothesized that our participants would prefer agents whose behavior reflected their own cultural background. That is, we expected German participants to prefer German-acting agents, while we expected Japanese participants to prefer Japenese-acting agents.

First we look at the results of the verbal behavior study. German participants found the German version of the smalltalk dialogues significantly more appropriate and interesting than the Japanese version, thought they would have quite liked to join the conversation, and thought that the agents got along with each other better. Japanese participants found the Japanese version of the smalltalk dialogues significantly more appropriate and interesting than the German version, and thought that the agents got along with each other better.

In the second study, we measured differences in the usage of pauses, in instances of overlapping speech, in gestural speed, in the spatial extent of gestures, and in postures. Each aspect was tested in isolation. For German participants, we found statistically significant preferences for the German version of the dialogues for overlapping speech and spatial extent of gestures. For all other aspects, we observed trends toward a preference for the more German behaviors, although none was statistically significant. For Japanese participants, we found statistically significant preferences for the Japanese version of the dialogues for postures. For most other aspects, we observed trends toward a preference for the more Japanese behaviors. However, for pauses in speech and overlapping speech we observed trends toward a preference for the less Japanese behaviors. One reason for this outcome might be that the dialogues in this evaluation involved gibberish rather than actual speech. Since the Japanese version of the dialogues contained both more pauses and more overlaps in speech, but lacked linguistic content, participants might have chosen the "safe" solution.

In sum, our evaluation studies suggest that human observers of different cultural backgrounds seem to prefer agent behaviors designed to reflect their own cultural background in most cases. We conclude that the integration of culture-specific features into the behavioral models of virtual agents can make them more believable and acceptable to users.

10.6   Conclusions

In this chapter, we examined the role of culture in the expression of conversational behaviors. Through an analysis of the research literature, we demonstrated that culture can influence the expression of conversational behaviors at the content, form, and communication management levels. We then presented a hybrid approach to the generation of culture-specific behaviors for embodied virtual agents. This approach is both model-driven (informed by qualitative models of culture from the social sciences) and data-driven (able to capture and encode quantitative information from corpora of labeled culture-specific conversational behaviors). It is capable of modeling culture-specific variations in the selection of communication content, in the form of conversational behaviors and in the management of the conversation. Our approach goes beyond what many think of as traditional NLG by (1) considering both verbal and non-verbal behaviors, (2) focusing on social and personal behaviors, and (3) being adaptive to particular cultural backgrounds. In two evaluations, we demonstrated that observers of virtual agents are able to distinguish between culture-specific versions of both verbal and non-verbal conversational behaviors, and that they prefer those that reflect their own culture. This means that by incorporating culture-specific models into virtual agents, we can increase their believability and acceptability to users.

Although we demonstrated the feasibility and utility of modeling culture for several conversational behaviors, richer and more complete models may lead to further improvements in the generality and performance of culture-specific virtual agents. There is scope here for a variety of research projects including empirical ones that look at particular behaviors and/or particular cultures, and more engineering ones that concern themselves with statistical modeling and feature extraction for conversational behaviors. In particular, although our approach is partly data-driven, the collection and annotation of the corpus we used was time consuming and expensive. More research on the automatic analysis of conversational behaviors from video is also needed if we are to produce more human-like embodied conversational agents.

Acknowledgments

This work was funded by the European Commission under grant agreement eCute (FP7-ICT-2009-5).

References

Alexandris, C. and Fotinea, S.-E. (2004). Discourse particles: Indicators of positive and non-positive politeness in the discourse structure of dialog systems for modern Greek. *Sprache und Datenverarbeitung: International Journal for Language Data Processing*, **28**(1–2):19–29.

Aylett, R., Louchart, S., Dias, J., Paiva, A., and Vala, M. (2005). FearNot! – an experiment in emergent narrative. In *Proceedings of the International Conference on Intelligent Virtual Agents (IVA)*, pages 305–316, Berlin. Springer.

Brown, P. and Levinson, S. C. (1987). *Politeness: Some Universals in Language Usage.* Cambridge University Press, Cambridge, UK.

Bull, P. (1987). *Posture and Gesture.* Pergamon Press, Oxford, UK.

Byrne, D. E. (1971). *The Attraction Paradigm.* Academic Press, New York, NY.

Chartrand, T. L. and Bargh, J. A. (1999). The chameleon effect: The perception–behavior link and social interaction. *Journal of Personality and Social Psychology*, **76**(6):893–910.

Decision Systems Laboratory (2005–2007). GeNIe and SMILE. http://genie.sis.pitt.edu/. Accessed on 02/26/2013.

Ekman, P. (2001). *Telling Lies – Clues to Deceit in the Marketplace, Politics, and Marriage.* W. W. Norton & Co., New York, NY, 3rd edition.

Endraß, B., André, E., Rehm, M., Lipi, A. A., and Nakano, Y. (2011a). Culture-related differences in aspects of behavior for virtual characters across Germany and Japan. In *Proceedings of the International Joint Conference on Autonomous Agents and Multiagent Systems (AAMAS)*, pages 441–448, Taipei, Taiwan. International Foundation for Autonomous Agents and Multiagent Systems.

Endraß, B., Huang, L., Gratch, J., and André, E. (2010). A data-driven approach to model culture-specific communication management styles for virtual agents. In *Proceedings of the International Joint Conference on Autonomous Agents and Multiagent Systems (AAMAS)*, pages 99–108, Toronto, Canada. International Foundation for Autonomous Agents and Multiagent Systems.

Endraß, B., Nakano, Y., Lipi, A. A., Rehm, M., and André, E. (2011b). Culture-related topic selection in small talk conversations across Germany and Japan. In *Proceedings of the International Conference on Intelligent Virtual Agents (IVA)*, pages 1–13, Reykjavik, Iceland. Springer.

Endraß, B., Rehm, M., and André, E. (2009). Culture-specific communication management for virtual agents. In *Proceedings of the International Joint Conference on Autonomous Agents and Multiagent Systems (AAMAS)*, pages 281–287, Budapest, Hungary. International Foundation for Autonomous Agents and Multiagent Systems.

Gratch, J., Rickel, J., André, E., Badler, N., Cassell, J., and Petajan, E. (2002). Creating interactive virtual humans: Some assembly required. *IEEE Intelligent Systems*, **17**(4):54–63.

Hall, E. T. (1966). *The Hidden Dimension.* Doubleday, Garden City, NY.

Hall, E. T. and Hall, M. R. (1987). *Hidden Differences: Doing Business with the Japanese.* Anchor Books/Doubleday, New York. NY.

Hartmann, B., Mancini, M., Buisine, S., and Pelachaud, C. (2005). Design and evaluation of expressive gesture synthesis for embodied conversational agents. In *Proceedings of the International Joint Conference on Autonomous Agents and Multiagent Systems (AAMAS)*, pages 1095–1096, Utrecht, The Netherlands. International Foundation for Autonomous Agents and Multiagent Systems.

Hofstede, G. (2001). *Culture's Consequences – Comparing Values, Behaviours, Institutions, and Organizations Across Nations.* Sage Publications, Thousand Oaks, CA, 2nd edition.

Hofstede, G. (2011). http://www.geert-hofstede.com/. Accessed on 11/24/2013.

Hofstede, G., Hofstede, G.-J., and Minkov, M. (2010). *Cultures and Organisations. Software of the Mind. Intercultural Cooperation and its Importance for Survival.* McGraw-Hill USA, New York, NY, 3rd edition.

Hofstede, G. J., Pedersen, P. B., and Hofstede, G. (2002). *Exploring Culture – Exercises, Stories and Synthetic Cultures.* Intercultural Press, Yarmouth, ME.

House, J. (2006). Communicative styles in English and German. *European Journal of English Studies*, **10**(3):249–267.

Iacobelli, F. and Cassell, J. (2007). Ethnic identity and engagement in embodied conversational agents. In *Proceedings of the International Conference on Intelligent Virtual Agents (IVA)*, pages 57–63, Paris, France. Springer.

Isbister, K., Nakanishi, H., Ishida, T., and Nass, C. (2000). Helper agent: Designing an assistant for human–human interaction in a virtual meeting space. In *Proceedings of the ACM SIGCHI Conference on Human Factors in Computing Systems (CHI)*, pages 57–64, The Hague, The Netherlands. Association for Computing Machinery.

Johnson, W. L., Marsella, S., and Vilhjálmsson, H. (2004). The DARWARS Tactical Language Training System. In *Interservice/Industry Training, Simulation, and Education Conference*, Orlando, FL. National Training and Simulation Association.

Johnson, W. L., Mayer, R. E., André, E., and Rehm, M. (2005). Cross-cultural evaluation of politeness in tactics for pedagogical agents. In *Proceedings of the Conference on Artificial Intelligence in Education (AIED)*, volume 5, pages 298–305, Amsterdam, The Netherlands. International Artificial Intelligence in Education Society.

Johnson, W. L. and Valente, A. (2009). Tactical language and culture training systems: Using artificial intelligence to teach foreign languages and cultures. *AI Magazine*, **30**(2):72.

Jonsdottir, G. R., Thorisson, K. R., and Nivel, E. (2008). Learning smooth, human-like turntaking in realtime dialogue. In *Proceedings of the International Conference on Intelligent Virtual Agents (IVA)*, pages 162–175, Tokyo, Japan. Springer.

Kang, S.-H., Gratch, J., Wang, N., and Watt, J. H. (2008). Agreeable people like agreeable virtual humans. In *Proceedings of the International Conference on Intelligent Virtual Agents (IVA)*, pages 253–261, Tokyo, Japan. Springer.

Kipp, M. (2001). ANVIL – a generic annotation tool for multimodal dialogues. In *Proceedings of the European Conference on Speech Communication and Technology (EUROSPEECH)*, pages 1367–1370, Aalborg, Denmark. International Speech Communication Association.

Kipp, M., Neff, M., and Albrecht, I. (2007a). An annotation scheme for conversational gestures: How to economically capture timing and form. *Language Resources and Evaluation*, **41**(3–4):325–339.

Kipp, M., Neff, M., Kipp, K. H., and Albrecht, I. (2007b). Towards natural gesture synthesis: Evaluating gesture units in a data-driven approach to gesture synthesis. In *Proceedings of the International Conference on Intelligent Virtual Agents (IVA)*, pages 15–28, Paris, France. Springer.

Lipi, A., Nakano, Y., and Rehm, M. (2009). A parameter-based model for generating culturally adaptive nonverbal behaviors in embodied conversational agents. In *Proceedings of the International Conference on Universal Access in Human-Computer Interaction*, pages 631–640, San Diego, CA. Springer.

Mairesse, F. and Walker, M. A. (2008). Trainable generation of big-five personality styles through data-driven parameter estimation. In *Proceedings of the Annual Meeting of the Association for Computational Linguistics: Human Language Technologies (ACL-HLT)*, pages 165–173, Columbus, OH. Association for Computational Linguistics.

Marcus, A. and Alexander, C. (2007). User validation of cultural dimensions of a website design. In *Proceedings of the International Conference on Usability and Internationalization*, pages 160–167, Beijing, China. Springer.

Martin, J.-C., Abrilian, S., Devillers, L., Lamolle, M., Mancini, M., and Pelachaud, C. (2005). Levels of representation in the annotation of emotion for the specification of expressivity in ECAs. In *Proceedings of the International Conference on Intelligent Virtual Agents (IVA)*, pages 405–417, Berlin. Springer.

Mascarenhas, S., Dias, J., Afonso, N., Enz, S., and Paiva, A. (2009). Using rituals to express cultural differences in synthetic characters. In *Proceedings of the International Joint Conference on Autonomous Agents and Multiagent Systems (AAMAS)*, pages 305–312, Budapest, Hungary. International Foundation for Autonomous Agents and Multiagent Systems.

Pelachaud, C. (2005). Multimodal expressive embodied conversational agents. In *Proceedings of the ACM International Conference on Multimedia*, pages 683–689, Singapore. Association for Computing Machinery.

Rehm, M., André, E., Nakano, Y., Nishida, T., Bee, N., Endraß, B., Huan, H.-H., and Wissner, M. (2007). The CUBE-G approach – coaching culture-specific nonverbal behavior by virtual agents. In *Proceedings of the International Simulation and Gaming Association Conference*, page 313, Nijmegen, The Netherlands. International Simulation and Gaming Association.

Rist, T., André, E., and Baldes, S. (2003). A flexible platform for building applications with life-like characters. In *Proceedings of the International Conference on Intelligent User Interfaces (IUI)*, pages 158–165, Miami, FL. Association for Computing Machinery.

Sato, R., Higashinaka, R., Tamoto, M., Nakano, M., and Aikawa, K. (2002). Learning decision trees to determine turn-taking by spoken dialogue systems. In *Proceedings of the International Conference on Spoken Language Processing (INTERSPEECH)*, pages 861–864, Denver, CO. International Speech Communication Association.

Schneider, K. P. (1988). *Small Talk: Analysing Phatic Discourse*. Hitzeroth, Marburg, Germany.

Sidner, C. L. (2004). Building spoken-language collaborative interface agents. In Dahl, D., editor, *Practical Spoken Dialog Systems*, pages 197–226. Kluwer, Dordrecht, The Netherlands.

Ting-Toomey, S. (1999). *Communicating Across Cultures*. The Guilford Press, New York, NY.

Trompenaars, F. and Hampden-Turner, C. (1997). *Riding the Waves of Culture – Understanding Cultural Diversity in Business*. Nicholas Brealey Publishing, London, UK.

University of Maryland (2005). SHOP: Simple Hierarchical Ordered Planner. http://www.cs.umd.edu/projects/shop/. Accessed on 02/26/2012.

Wu, P. and Miller, C. (2010). Interactive Phrasebook^{TM} – conveying culture through etiquette. In *Proceedings of the International Workshop on Culturally-Aware Tutoring Systems*, pages 47–55, Pittsburgh, PA.

Yin, L., Bickmore, T., and Cortés, D.-E. (2010). The impact of linguistic and cultural congruity on persuasion by conversational agents. In *Proceedings of the International Conference on Intelligent Virtual Agents (IVA)*, pages 343–349, Philadelphia, PA. Springer.

11 Natural language generation for augmentative and assistive technologies

Nava Tintarev, Ehud Reiter, Rolf Black, and Annalu Waller

11.1 Introduction

Many people are not able to communicate easily, because of diversity in their physical and/or intellectual abilities. For example, someone with cerebral palsy may not have enough control over their vocal tract and mouth to speak (or enough manual dexterity control to use sign language); and someone with autism may be physically capable, but unable to speak because of their cognitive profile.

Augmentative and alternative communication (AAC) is a sub-field of assistive technology focusing on tools to help people with major motor and/or cognitive impairments communicate better. Traditionally, AAC has used technology to provide speech output for non-speaking users, either by converting text to speech or speaking out pre-stored phrases. The initial focus of computer-based AAC systems was on giving users access to a broad set of symbols/pictures (by using some kind of menu or search system), and on using speech synthesis to automatically speak out phrases corresponding to the selected symbols. However, computers can in principle do much more than this.

In particular, since many AAC systems are intended to help users produce utterances, there seems to be a natural role for Natural Language Generation (NLG) to play, as it is after all a technology for producing language. However, the use of NLG in AAC is somewhat different from most uses of NLG. Since the goal of AAC is to help the user communicate, the NLG system must be used interactively, under the user's control; we want to assist the user in communication, not replace the user with an automatic communicator. Also, since most human communication is social, NLG AAC systems often need to generate texts whose communicative goal is social interaction. Hence, NLG AAC systems are very different from systems that summarize information in a task-oriented context, which has been the focus of most NLG research for interactive systems.

In this chapter, we first give some basic background on AAC, and then survey some of the ways in which NLG can be used in AAC contexts. We then describe one particular NLG AAC system, "How was School Today...?", in detail, including the application context, how the system works from an NLG perspective, and the challenges faced in getting such a system used. We conclude this chapter with a summary of the novel challenges that AAC applications pose to NLG, such as the need to support social interaction.

11.2   Background on augmentative and alternative communication

Augmentative and alternative communication (AAC) is an area of clinical practice that attempts to compensate (either temporarily or permanently) for functional diversity with respect to expressive communication (i.e., ability patterns of individuals with severe speech-language and writing impairements; American Speech-Language-Hearing Association, 2002). Individuals using AAC are often said to have complex communication needs (CCN). These are people who, due to motor, language, cognitive, and/or sensory perceptual impairments (e.g., as a result of cerebral palsy), have not developed speech and language skills as expected. This heterogeneous group typically experiences restricted access to the environment, limited interactions with their communication partners, and few opportunities for communication (Light and Drager, 2007).

In other words, AAC is about developing and using strategies and techniques that support communication by individuals who have little or no functional speech due to a physical and/or intellectual disability. A key aspect of AAC is that the goal is to *complement* existing abilities; users should be encouraged to do as much as they can, and indeed to develop their communication abilities, and good AAC systems should be adaptive or adaptable so they are still appropriate to users as users' abilities change over time (Waller *et al.*, 1999). AAC users are very diverse; for example, someone with strong cognitive abilities and physical impairments (such as the physicist Stephen Hawking) needs very different assistance than someone with strong physical abilities and intellectual or cognitive impairment (such as people with severe autism or aphasia). This is further discussed in Section 11.2.3.

AAC can be used in many communicative contexts. Most current uses of AAC focus on needs-based communication (e.g., *I am thirsty*), but such communication is not very interesting from an NLG or dialogue perspective. Since that is the focus of this book, in this chapter we focus on AAC tools that support storytelling and personal narrative. A narrator combines many kinds of knowledge, such as knowledge of events and people, with knowledge about verb tense and linguistic connectives to generate a coherent and cohesive narrative. An AAC tool can be used to scaffold many of these narrative tasks. Personal narrative is a crucial part of social interaction and personal development. It is through narratives that we translate knowing into telling (McCabe and Peterson, 1991). Relating past experience provides a way of forming experience and is an essential part of one's make-up and sense of self (Quasthoff and Nikolaus, 1982; Polkinghorne, 1995).

11.2.1   State of the art

Table 11.1 gives an overview of commonly used AAC systems. In its simplest form, AAC can be based on communication boards or books that contain messages (single words or whole phrases) in the form of written text or symbols; the user communicates by pointing to the relevant entries in the board or book. A similar approach is used in many computer-based AAC systems, where the user selects a symbol or sequence of symbols on a computer screen and the computer system speaks out the underlying

Table 11.1. Overview of existing AAC devices. We define low-tech as systems that do not require any electrical power; high-tech as complex systems that have the capability to involve a language strategy or other computing power, e.g., for prediction; mid-tech as everything in between.

Type	Brand names (examples)	Level	Linguistic complexity	Scope
Photos, symbol cards, choice boards	PECS (Picture Exchange Communication System), PCS (Picture Communication Symbols)	Low-tech	Symbol cards for words, needs and wants	Limited vocabulary, mostly used for transactional communication
Communication books (symbol based)	PCS, BLISSYMBOLICS	Low-tech	Letters, words, sentences	Flexible, can be used for both transactional and social communication
Switch accessible audio recorder	BIGMack, Step-by-step	Mid-tech	Sounds, words, sentences	Recording and playback of one or more voice messages. Limited to content of previously made recordings.
Voice Output Communication Aid (VOCA)/ Speech Generating Device (SGD)	Maestro (DynaVox), Lightwriter (Toby Churchill Ltd), ECO2 (Prentke Romich Company)	High-tech	Letters, words, sentences	Highly flexible, for both transactional and social communication. Can be programmed according to linguistic level of user.

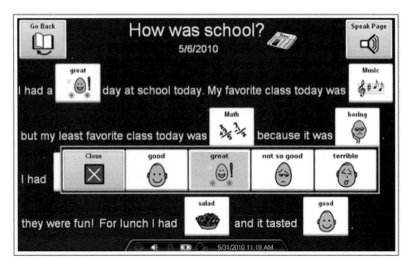

Figure 11.1   Example template for a child's story about their day at school used on a DynaVox system

message (word or phrase) using a speech synthesizer. Some systems, such as developed by AssistiveWare[1] or DynaVox,[2] use a hierarchical structure of pages, each of which typically focuses on a context (e.g., shopping) or a category (e.g., clothes, sports). Each page contains a number of symbols which, when pressed, communicate some stored word or phrase. Liberator[3] typically uses a fixed screen with a number of symbols that are combined to access a core vocabulary. Other facilities include moving to a different page, editing a page (less common), and creating sentences from multiple symbols and/or templates.

Figure 11.1 shows a filled template for a child's day at school. Such templates can be used to give users some flexibility and support. The layout is changeable by programming, but in reality this level of technical skill cannot be expected from most carers and family members.

Communication using existing AAC tools tends to be very slow. Compared to speaking rates of around 150 words per minute, word production rate using AAC is less than 15 words per minute (Beukelman and Mirenda, 1998; Nikolova *et al.*, 2010). Users with major physical impairments may use a scanning interface that is operated using a single switch (often a head-switch mounted to the wheelchair headrest). Such an interface, for example, iteratively highlights rows of icons until the row containing the right icon is reached, at which point the user presses the switch; the interface then iteratively highlights icons in the selected row until the user presses the switch again. Using this kind of interface, a user may only be able to speak 1–2 words per minute, which is very restrictive. Individuals who use these tend to be passive, resorting to yes/no responses to closed questions, and one-word communication of simple needs and wants

---

[1] www.assistiveware.com.
[2] www.dynavoxtech.com.
[3] www.liberator.co.uk.

to family members, carers, and other communicative partners. More complex types of communication, including in particular storytelling, seldom develop. The operational construction of narrative discourse is prohibitively slow and physically exhausting.

11.2.2   Related research

Current commercial AAC tools are poor at supporting personal narrative. Usually personal stories are told by recording a sequence of spoken narrative phrases on a simple AAC device; the AAC user "tells" the story by repeatedly pressing the "play" button (Waller, 2006). Solutions such as this rely on someone else preparing a narrative. In reality, recording a narrative for daily experiences is unsustainable. People who use more sophisticated AAC devices sometimes store extended texts for specific purposes such as personal introductions or lectures. These "story texts" are usually used in monologue form, offering little or no interaction with the listener/audience.

In research, utterance-based systems (see Todman *et al.* (2008) for an overview) are now starting to supplement natural language processing research that facilitates language production on lower levels such as improved scanning methods (Roark *et al.*, 2010) and predicting words (Nikolova *et al.*, 2010). There remain, however, several challenges to which users of utterance-based systems are subjected. These users must be able to (1) remember that they have messages pre-stored that are appropriate for a given situation; (2) remember where these messages are stored; (3) access a desired pre-stored message with only a few keystrokes; and (4) decide whether it is better to use the message as stored or to either edit or construct a new one (McCoy *et al.*, 2010).

Some research projects have developed prototype systems that attempt to address these challenges and facilitate personal narrative. At their most basic, these systems provide users with a library of fixed "conversational texts" that can be selected and uttered. The Talk system implements a retrieval system in which the user is supported to make conversational moves (Todman and Alm, 2003; Todman *et al.*, 2008). Based on a pragmatic conversational model, these moves allow the user to progress through a conversation via a series of gradual shifts of perspectives relating to the speaker, time (past, present, future), and event-related information (what, where, who, how, and why). By changing perspective, the system can predict possible next user utterances, e.g., *We went to Italy for a holiday. Where did you go?.* The system also provides access to phatic communication, the "glue" of conversation: standard conversational openings and closings, such as *Hello* and *How are you?*, and back-channel communication such as *Uh-huh, Great!*, and *Sorry, can you repeat that?.* McCoy *et al.* (2010) also developed a prototype utterance-based AAC system that uses a hierarchical script (akin to the frames described in Todman and Alm, 2003) to use in particular contexts, such as going to a restaurant.

Waller (2006) developed storytelling tools for AAC users that included ways to introduce a story, tell it at the pace required (with diversions) and give feedback to comments from listeners; but these tools were based on a library of fixed texts and templates. In the Prose system (Waller and Newell, 1997), the retrieval of story texts was based on relationships between events such as high-level conversation topics (e.g., pets) or the people involved in the story. A critical aspect of the subsequent Talk:About software, the commercial version of the Prose system, was the capability of modifying or adding to stories, so that stories could change each time they were told (Waller *et al.*, 2001). The TalksBac system also allowed the user to specify the categories of partners for whom an utterance would be relevant (Waller *et al.*, 1998a).

11.2.3 Diversity in users of AAC

A key issue in AAC is the diversity of users. People with complex communication needs differ enormously in terms of age, cognitive ability, linguistic skills, motor ability, and social ability as well as personal and environmental circumstances.

We have already mentioned the fundamental distinction between physical impairment and cognitive impairment. Someone with good cognitive skills but poor physical skills can internally create sentences and indeed stories; what he or she needs is help in actually communicating these stories to conversational partners. On the other hand, someone with cognitive impairments may need help in internally formulating utterances and stories in the first place. And someone with major social impairments may above all need help and encouragement in wanting to communicate at all to other people. Of course the above are very broad categories, and there is enormous variation even within a category. For example, cognitive impairments include impairments specifically in language skills, impairments in memory, impairments in vision and perception, and so on.

This wide variety of impairments is a major challenge when developing support for AAC. In principle, AAC systems should be tailored for individual users. Almost all commercial AAC systems support this to some degree, but in practice many users (carers, family members) lack the skills to do this. And the tailoring that is supported by commercial systems is usually at a fairly shallow level (e.g., which icons are on which page). Research AAC systems, on the other hand, sometimes seem to be tailored for a small number of specific individual users, with less thought being put into how the system could be adapted for users with different ability profiles.

One specific factor that has been investigated in prior research is the impact of the user's literacy skills (language comprehension, not language production) on the usability of AAC tools. For example, the BlissWord system looks at a symbolic representation of language that does not require the user to be literate to explore language and vocabulary (Andreasen *et al.*, 1998). The Friend system discusses three levels of interfaces, varying from iconic, to recognition of words, to scaffolding of grammatical constructs (Biswas, 2006; Biswas and Samanta, 2008). In the iconic interface the user can compose a sentence by selecting a sequence of icons. By contrast, the most complex interface requires some basic knowledge about parts of a sentence (e.g., who, action, what, from where, to where). Naturally, most users fall somewhere in the middle of the spectrum and can possibly read enough to recognize words. For example, with the TalksBac system aphasic users might recognize and select important topic words to help them retrieve more complete sentences and stories (Waller *et al.*, 1998b).

Last but not least, we need to keep in mind that people change. With children in particular, communication abilities are likely to progress with age and (hopefully) with the assistance of communication aids. On the other hand, some elderly people may lose communication skills over time. Hence, the ideal AAC system should not only be able to adapt to a user once, but should be able to keep on adapting to the user over time, as he or she changes.

11.2.4   Other AAC challenges

There are numerous difficulties in deploying AAC systems where they are most needed. One issue is size and battery life limitation in AAC hardware. These issues are now increasingly being resolved with the advent of smaller, more robust devices and mobile platforms. Cost and robustness of AAC hardware are also issues that can prevent adoption of technological solutions.

Another important consideration is the usability of the interface; interface usability is even more important for disabled users than for non-disabled users. One general guideline is to avoid overly dynamic interfaces – possibly the downside of many predictive systems (Newell *et al.*, 2008). If an interface is very dynamic, it may be too confusing to learn how to use (Biswas and Samanta, 2008). Dynamic systems that change the message selection interface and present content using prediction can be a challenge when, for example, trying to repeat previously uttered messages, because they might not be presented any more or are presented in a different place. Users who learn to use such a system are dependent on their environment for support, but carers and family members themselves might be overly challenged when supporting their usage.

As mentioned previously, the users of AAC tools are varied. Even for users with similar abilities, different AAC tools may have been used previously. Consequently, it is difficult to use evaluation methodologies requiring control groups, because the heterogenous population means large sample sizes are needed to reach statistical significance. But large sample sizes are difficult because experiments involving participants with impairments are very resource intensive (per participant). In addition, some participants may not be able to supply rich feedback about their impression of the system. For these reasons, it is often difficult to collect enough evaluation data to run the kinds of statistical hypothesis tests favored by the NLP community. Instead, many evaluations of AAC systems are done using appropriate qualitative techniques (such as analysis of recorded dialogue) that are well established in the AAC field, but unusual in NLP research (Dugard *et al.*, 2011).

For evaluation a number of approaches have been used, including conversational analysis of dialogues with and without the AAC tool (Waller and Newell, 1997), and simulated evaluation (Biswas, 2006). A simulated evaluation evaluates the coverage of the types of utterances that a system may produce, and potential production rates versus those seen in deployed systems.

11.3 Application areas of NLG in AAC

NLG techniques have been used in a variety of AAC applications. In very general terms, NLG technology is often used to "amplify" limited input from the user into full utterances. This amplification can include dealing with morphological, syntactic, and other linguistic details, and augmenting the user's input using information from external knowledge sources. This process reduces the amount of user input needed (hence helping people with physical impairments) and also automates much of the language production process (hence helping people with cognitive impairments).

11.3.1 Helping AAC users communicate

The earliest attempts to use NLG in AAC centered around generating grammatically correct sentences from incomplete user input (input that specified content but not linguistic details). For example, the Compansion system allowed users to specify content words (perhaps by selecting icons or words from a graphical display), and from these content words generated sentences (Demasco and McCoy, 1992). If the user specified APPLE EAT JOHN, Compansion would generate the sentence *The apple is eaten by John*; this involves syntactic operations such as adding function words and choosing appropriate morphological inflections. Compansion used a symbolic approach: it parsed the input into a semantic representation, and then generated a well-formed English sentence from this representation. The system only worked in a limited domain, because of the need to parse the user's input into a semantic representation. However, this NLG task could also be done using statistical language models; indeed, the process of converting abbreviated input into full English is in some ways analogous to machine translation from a minimal English used by AAC users to conventional English.

Dempster *et al.* (2010) extended the Compansion idea by using a knowledge base of information about the user's activities (holidays, sports, etc.). The knowledge base was partially entered by the user, and partially extracted from on-line sources. For example, if the user specified that he or she had attended a particular music concert, the system could download background information about the band, venue, etc. Once the knowledge base was established, the user could use it to generate English sentences, by selecting a fact from the knowledge base along with some guidance about how to express it.

The purpose of Dempster *et al.*'s system was to support social conversation ("pub talk"), not the needs-based communication (e.g., *I am hungry*) that most AAC focuses on. As mentioned above, current AAC techniques can work reasonably well for simple needs-based communication, but they provide little support for social interaction (Light and Drager, 2007).

A strength of Dempster *et al.*'s approach was that the system could suggest follow-up utterances to the user, by looking for knowledge base entries related to the current utterance. This meant that in a dialogue, the user could quickly respond to utterances from his or her conversational partner, by simply selecting one of the system's suggested follow-up utterances. In other words, while the Compansion system helped users create sentences, Dempster *et al.*'s system helped users participate in conversations. Evaluations of Dempster *et al.*'s system suggested that this ability to support dialogues was very valuable (e.g., increased participation by the AAC users), and indeed perhaps the largest benefit of using a user-specific knowledge base.

An extension to Dempster *et al.*'s approach would be to supplement the information in the knowledge base with sensor data from the user. For example, if the user had a GPS tracker and indicated that he or she attended a music concert recently, the system could automatically analyze the GPS data to determine which concert had been attended. Some of these ideas have been explored in the "How was School Today…?" project (which is further discussed in Section 11.4).

11.3.2   Teaching communication skills to AAC users

While most research in NLG and AAC has focused on communication aids, quite a bit of research has also been done on using NLG in teaching tools for AAC users. Of course, the line between communication aid and teaching tool is a fuzzy one, since a good communication aid will allow the user to communicate more, which enhances his or her communication skills. Perhaps even more importantly, a good communication aid can enhance the user's self-confidence and willingness to push the boundaries of his or her communication abilities.

One NLG system that was intended purely as an educational aid was STANDUP (Waller *et al.*, 2005). STANDUP allowed children to generate jokes (punning riddles) by controlling joke-generation software. Its goal was to allow children who use AAC to play with language, which is something that typically developing children do automatically, but children who use AAC rarely do.

11.3.3   Accessibility: Helping people with visual impairments access information

A related topic to using NLG to support AAC is using NLG to enhance accessibility of information, in particular for people with visual impairments. Such people typically use screen readers to access web pages; these devices use speech synthesis to verbalize web page content. Screen readers work well for textual web pages, but of course many of the most interesting web pages communicate visually as well as textually. For static images (such as photographs), the best solution is for the web page designer to manually add an ALT tag to describe the image; but this approach does not work well for graphs and diagrams that are automatically generated from data.

Some assistive technology tools support non-visual exploration of visual media. The iGraph system, for example, uses simple text generation techniques to allow visually impaired users to explore a graph (Ferres *et al.*, 2013). iGraph allows users to interactively explore a graph, getting descriptions of particular data points and also data summaries. A similar tool has looked at summarizing key points contained in information graphics (Demir *et al.*, 2011). Another example is Atlas.txt, which summarizes statistical data for visually impaired users; Atlas.txt focuses on summarizing how data varies across a spatial region (e.g., unemployment in different counties; Thomas and Sripada, 2008).

Using NLG for accessibility is quite different from the AAC uses of NLG mentioned above; essentially in accessibility the goal is to enhance the user's comprehension of data and information, while in AAC the goal is to enhance the user's ability to produce language and communicate. But the reality is that many people with communication impairments also have accessibility problems, and the ultimate solution is technology that takes a holistic approach.

11.3.4   Summary

A common theme behind much work in using NLG in AAC is converting limited and noisy user input into readable texts. Both motor and cognitive impairments limit the rate at which users can speak, sign, or enter information into a computer, and also increase the errors and noise in the language users can produce. The job of the NLG system is to take this limited and noisy input, and convert it into the equivalent of regular utterances. If the user is able to specify content, then the job of the NLG system is to ensure that the content is expressed correctly (as in Companion); if the user is unable to completely specify content, then the NLG system should also suggest content, based on domain knowledge, external data sources, and sensors. While early research in this area focused on generating individual utterances, current research looks into supporting dialogues between the AAC user and a conversational partner.

11.4   Example project: "How was School Today...?"

Many of the general issues that arise when using NLG and AAC are easier to understand in the context of a specific project. In this section we describe a specific example of an NLG AAC project that we are involved in: "How was School Today...?".

11.4.1   Use case

The "How was School Today...?" system is intended to be used by non-speaking children to create and tell a story about their day at school. The conversational partner of the child may be one of their parents or a member of the school staff they are meeting for the first time that day. The system generates a draft story automatically based on sensor data, and the child can then edit and narrate the story using an appropriate user interface.

The system addresses two of the aims of NLG in AAC presented in Section 11.2.3: helping AAC users communicate in fluent English, and teaching communication skills to AAC users. From a teaching perspective, the primary goal is to provide opportunities to learn dialogue skills (e.g., turn-taking) and storytelling skills in a social context; storytelling here means telling personal narratives, not reciting fictional stories.

The stories that are generated for each child are based on several different types of data. Figure 11.3 gives an example of the data that was collected and used in "How was School Today...?":[4]

- Location data each time the user enters a new room.
- Object interaction each time the user interacts with an object that has an RFID tag.
- Person interaction each time the user interacts with a person (e.g., a member of staff, a visitor).
- Voice messages – staff and teachers are encouraged to record voice messages, as if the user was speaking in the first person, that describe the user's recent activities.
- Photographs – staff and teachers can also link photos to voice recordings to aid recall of when and where something happened.
- Timetable used for inference together with location, e.g., Art in Classroom 3 between 10:15 and 11:20 a.m.
- Evaluations – positive and negative user sentiment annotations, used to comment on statements (based on the previous data types) that were previously generated.

11.4.2   Example interaction

Table 11.2 illustrates a dialogue extract as supported by the "How was School Today...?" system with Julie. Julie has little functional speech and severe motor impairment (no independent means of mobility), and interacts with the computer using a head-switch with row/column scanning: see Figure 11.2. Julie used the system on her DynaVox Vmax Voice Output Communication Aid (VOCA). Her cognitive skills were sufficient for her to master the interface on the second day. The dialogue extract starts with Julie telling her speech and language therapist about her morning.

In this example, Julie is able to quickly reply to context-related questions from her communication partner (*I wonder why that was? A visitor was there.*). In contrast to conversations usually observed between aided and unaided partners,[5] Julie is able to control the conversation, for example starting a new topic after talking about the morning break. When her communication partner prompts for more information about the new topic, Julie accesses the next generated phrase. When she is asked about the event she replies with the subjective evaluation *She was nice* (in this case, a positive one) of the message the system has previously generated *A visitor was there*.

11.4.3   NLG in "How was School Today...?"

The natural language generation in "How was School Today...?" follows the data-to-text architecture described in Reiter (2007). First the input data is analyzed and cleaned

---

[4] All examples and identifying names in this chapter have been anonymized.
[5] That is, the aided conversational partner is using an AAC device or requires other types of support such as conversational scaffolding, and the unaided partner can speak without additional support.

Table 11.2. A transcript of an extract of a conversation Julie had with her speech and language therapist (SLT) on day 3 about her experiences during day 2. The researcher (RA) had been present all day for technical support.

1	Julie	{next} [I had break.]
2	Julie	{next} [Lesley was there.]
3	SLT	Lesley was there?
4	Julie	((Opens mouth in agreement, then turns back to screen))
5	SLT	Ok mhmh. So what happened?
6	Julie	{positive evaluation} [It was fun.]
7	SLT	Oh good! ((laughs)) I'm glad to hear it!
8	RA	We like Lesley.
9	SLT	((nods in direction of RA))
10	Julie	((smiles))
11	Julie	{next} [Then I went to junior primary instead of reading class.]
12	SLT	Right, you went to junior primary? I wonder why that was?
13	Julie	{next} [A visitor was there.]
14	SLT	Oh, a visitor, right. Wonder what the visitor was doing?
15	Julie	{next} [*The dental hygienist came to give a talk.*]
16	SLT	Oh, dental hygienist.
17	Julie	{previous} [A visitor was there.]
18	SLT	That was the visitor, okay. Thats why you went to junior primary, uhm, what did you think of the talk?
19	Julie	{positive evaluation} [She was nice.]
20	SLT	She was nice, that was good! ((laughs))

Notation:
- Switch selected button by Julie: {curly brackets}
- Natural speech: standard text.
- Computer generated language accessed using one button: [standard text in square brackets].
- Recorded messages accessed using one button: [*standard text in square brackets*].
- Paralinguistic behaviors: ((standard text in double brackets)).

Figure 11.2   Participating pupil: the prototype "How was School Today…?" system is mounted on the wheelchair, and the pupil has access to the system via head-switch controlled row/column scanning

09:34:00, Voice Recording, *A man came to talk to me in gym. I signed an important document.*
09:36:00, Voice Recording, *My dad was in school today.*
09:40:00, Location, Gym.
09:40:00, Object, Skittles
09:40:00, Object, Baton
09:40:00, Person, Mrs Roberts
10:48:00, Voice Recording, *I was doing relay racing at school and I joined in really well.*
10:48:00, Location, Changing
10:50:00, Voice Recording, *I did some high jump and every time I jumped I had a lie down on the mat.*
11:08:00, Location, Classroom
11:12:00, Object, Blackboard
11:31:00, Person, Mary
11:36:00, Location, Tutorial Room
11:36:00, Object, Money
11:39:00, Object, Monkey Game
11:58:00, Location, Classroom
12:00:00, Location, Outside
12:08:00, Object, Bike
12:18:00, Location, Classroom
12:29:00, Voice Recording, *I asked Jim this morning what he did for his weekend.*
12:30:00, Location, Dining Room.

Figure 11.3   Example data from the "How was School Today…?" project

up; then the data is abstracted and interpreted using a simple knowledge base; then a document/narrative structure is created; and finally correctly formed sentences are produced.

Data analysis and interpretation

As mentioned above, the input data to "How was School Today…?" (see example in Figure 11.3) includes information about location, interactions with objects and people, and voice messages recorded by school staff. The first step is to clean up the data. For example, depending on the location sensor used, location information may be replicated (e.g., we could be told several times that the child is in the classroom), and there may also be spurious location readings (for example, an RFID location sensor triggered when the child went by but did not enter a room). This is done using simple heuristics that were tuned using actual data. The second step, which is probably of more interest to an NLG audience, is to supplement the sensor readings using a simple knowledge base, which in particular records the child's timetable. The timetable information specifies where the child is supposed to be at a particular time, with what teacher and for what subject. The system infers that if the child is in the timetabled location with the timetabled teacher, then he or she is doing the timetabled subject; hence it can generate *I had art with Mrs. Smith* instead of *I was in the art room. Mrs. Smith was there.* The timetable also allows the system to identify and describe some unusual events, such as a substitute or replacement teacher.

The output of data analysis and interpretation is a set of messages, each of which more or less corresponds to a single sentence.

Document structuring

The most interesting and innovative aspect of "How was School Today…?" from an NLG perspective is probably its document structuring, which selects which messages to include in the story, and orders and groups them into a document. This is essential to convert what has so far been a data-mining and analysis exercise into a narrative-generation system, which produces a dialogue with proper narrative structure and a consistent style. The importance of narrative in exchanging information is well-researched; for an example see Reiter *et al.* (2008).

In other words, the goal is to arrange the sensor-based input into a solid narrative structure, which also accurately relates events that happened, similar to the type of story that equivalent typically developing adults and children might tell. This problem can be split into two separate phases: "clustering" the data into events; and deciding which events would be of interest to the user and to his or her conversation partner. We have started looking at the first problem (Reddington and Tintarev, 2011). We tested three different approaches: location-based clustering, time-based clustering, and recording-based clustering. Each of these approaches is motivated by a different model of how children structure a narrative about an important event in their day. Our preliminary evaluations suggest that most subjects prefer narratives generated using recording-based clustering.

Location-based clustering Location-based clustering assumes that a change of location is likely to denote a change of context, and can thus be used as an event barrier. A potential disadvantage with this method is that some target users conduct much of their activities in the same location, e.g., in a home classroom. To compensate, this approach also considers an upper bound for time, creating a new event once a certain amount of time has passed. Once a new event with location is identified, the additional sensor data (who was there, what objects were interacted with, any voice recordings) is tied to the event.

For example, a cluster generated from Figure 11.3 would be:

11:36, Location, Tutorial Room
11:36, Object, Money
11:39, Object, Monkey Game

When converted into English text, the above cluster gives the story: *I played with Money and Monkey Game. This happened at a Tutorial Room.*

Time-based clustering The time-based clustering approach is a hierarchical clustering based on temporal proximity. The motivation behind this approach is to create a "clean" approach for clustering that prioritizes temporal closeness over the supremacy of any given data type. This approach groups data elements strictly according to when things happened. That is, things that happened around the same time are likely to belong to the same event. Time-based clustering first considers each element to be its own unique cluster, and then seeks the two elements that are closest together in time without being in the same cluster. Once found, the two clusters are merged in classic greedy processing fashion. Clustering continues until the desired number of clusters are obtained. The number of events that are produced must be specified in advance; no data is left out of the clusters.

Recording-based clustering   Clustering based on voice recordings was designed to capitalize on the richness of the data in voice recordings, and replicate the sentence structure used by the target age group. This approach takes the set of recordings as the set of events, and uses the other data to establish where the recording took place, and what people and objects were involved. This approach works on the premise that if a voice recording has been made, this should be a reportable event. During preprocessing, voice recordings within a certain time window (e.g., 15 minutes) of each other are viewed as being part of the same event. This potentially removes "uninteresting" stories early on, but may remove too much information.

Sentence planning and surface realization

The final NLG task is to convert the messages into sentences and utterances. This is done in a fairly straightforward fashion in "How was School Today…?". Sentence planning performs basic referring expression generation (e.g., use of pronouns), aggregation (e.g., *Mary, Tom and Bob were there*), and lexicalization. The main lexicalization challenge is expressing postive/negative evaluations entered by the child. For example, a positive evaluation of a teacher would be expressed as *nice* (e.g., *She's nice*), but a positive evaluation of a food object would be expressed as *Yum*.

One reason that sentence planning is simple is that we are trying to produce simple language, because we are trying to give a voice to children who would use simple language if they were able to speak. For example, in one version of "How was School Today…?" the microplanner dynamically determined the tense of sentences using a Reichenbach-like model (Reichenbach, 1947). But the resulting texts were judged to be too syntactically sophisticated (especially when sentences appeared using perfect constructions), so we reverted to using simple past tense throughout. Realization is done using the SimpleNLG software package (Gatt and Reiter, 2009).

Interaction

The "How was School Today…?" system supports the children that use it by scaffolding the communication. The idea is to provide pragmatic support for personal narrative, by recalling relevant data and providing well-formed English sentences from this data, thereby reducing the cognitive load on the child. Therefore, once the text has been generated, a second function of the system is to support the telling of stories. The system supports simple forms of editing and narration. In particular, the following functions are supported: *hiding* and *showing* events, and *selection* of what to say next. The flow of storytelling in turn can be controlled on a number of levels: the child can choose what day to talk about, what event during that day to tell, and what pieces of information (or event segments) to share about that event. In addition, the child can

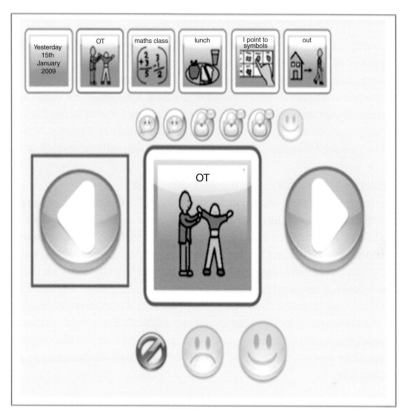

Figure 11.4   Example screenshot from the first version of the "How was School Today…?" interface. Buttons from top to bottom, left to right. Navigation: day and date of story, maximum of five story events (here: "occupational therapy (OT)," "maths class," "lunch," "I point to symbols," "exit"). Event messages, numbers vary for each event. (The selected event, occupational therapy (OT), has: two computer-generated messages, three recorded messages, one user-added positive evaluation. The user told the first three items, and the remaining three remain grayed out until they are told.) Sequential message navigation: previous, repeat, next. Evaluation: delete event, negative evaluation, positive evaluation.

select recently told or favorite (by frequency of telling) events. The number of functionalities available to the child relative to the support worker/carer will differ. The idea is that the amount of control changes according to the abilities of the child (see Section 11.4.4).

The piece of information that can be relayed at any point in a conversation supported by the "How was School Today…?" system is either a generated phrase or a voice recording (see Figures 11.4 and 11.6). For children with several motor and cognitive impairments, event segments may be communicated in a sequential manner by repeatedly pressing a switch, while children with less complex impairments may select event segments in the order they see fit.

11.4.4   Current work on "How was School Today...?"

As a result of evaluating our initial "How was School Today…?" prototype we identified a number of areas for potential improvement, which are likely to apply to other NLG systems designed for assistive technology. The second iteration of the system is being developed to accommodate a number of different functionalities: varied stages of users, increasing user mobility, and giving more autonomy for staff to run the system, as well as story generation from limited data. Figures 11.5 and 11.6 present screenshots of the "How was School Today…? in the Wild" system currently in development.

"How was School Today...? in the Wild" architecture

The prototype system ("How was School Today…?") required participants to transport a heavy radio frequency identification (RFID) reader. In addition, the main system ran on a tablet PC (8 to 12-inch touch screen, generic or VOCA hardware). Julie, for example, used the system on her VOCA via a head-switch using row/column scanning. Essentially, this meant that the system could only be used by individuals who use a wheelchair and would not be useful for more mobile children. For "How was School Today…? in the Wild", we switched to using a mobile phone as the primary data collection device.

The prototype system also ran entirely on a tablet PC; this did not allow for a separate mobile phone for data collection, and also meant that stories could only be taken home if the tablet was physically taken home. For this reason we switched to a client–server architecture.

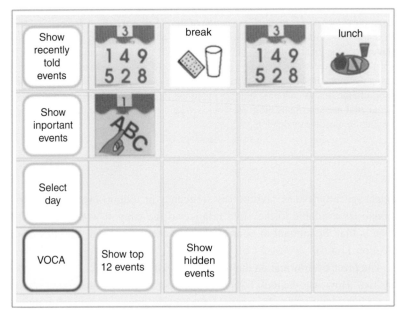

Figure 11.5   The "How was School Today…? in the Wild" interface for children at "Stage 3," who are able to select a day they want to talk about. This day has no data, but shows the first five events that are on the timetable for the day.

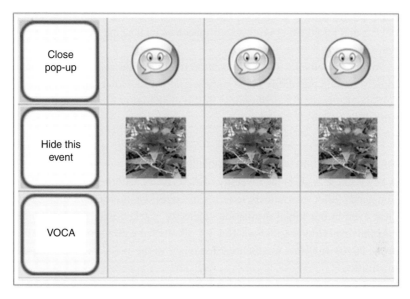

Figure 11.6 Screenshot from "How was School Today…? in the Wild" for a single event with three generated utterances and three audio recordings. The three recordings are actually part of a single larger message and were taken at the same time as a single photograph (a plant that was drawn during art class). Or to be more precise, each recording is a single phrase, but all relate to the same event and are therefore tied to the same photograph.

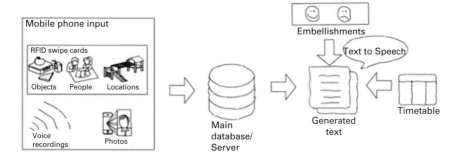

Figure 11.7 Types of input that can be collected by a mobile device: voice recording, RFID scans (of objects, people, and locations), and photos. This is supplemented by a timetable that teachers can modify. This data is then sent to a central database. A server-side process generates the texts and allows the children to present photographs, synthesized utterances, and voice recordings as well as add their own evaluations.

Figure 11.7 displays the architecture of the new system. A mobile phone serves as a data collection device for RFID tags (locations, people, and objects) as well as photos and voice recordings. A tailored mobile phone interface was created in Java (J2ME) for the purpose of collecting this data (see Figure 11.8). The data is encrypted and sent over a 3G network to a Java servlet on a main server where this data is written to a main database. Then, when a child wants to use the data to tell a story, the relevant data for that child is retrieved. The generated stories are shown on a PC or DynaVox device, and retold using synthesized speech as well as the stored photographs and voice recordings.

We experimented with Wi-Fi location tracking (Liu *et al.*, 2007), which seems to be rapidly gaining popularity in the commercial world. However, this technology is not yet sufficiently robust for deployment in the challenging environment of a special-needs school.

Supporting children with different abilities

A key issue in AAC is the diversity of AAC users. Children with complex communication needs differ enormously in terms of cognitive, motor, and social ability. This was clear even in our initial evaluation where we worked with just two children, and discovered that "How was School Today…?" worked well for one but not the other (Black *et al.*, 2010). In "How was School Today…? in the Wild," we will support three stages of abilities:

- **Stage 1**: Children with very limited skills, who in particular may not be able to do much more then press a "Next" button on an interface. For such children, we would like to support the kind of storytelling suggested by Grove (2010). Teachers will edit content and set up a narrative, which the child can sequentially step through. Since stage 1 children are likely to have memory problems, the system will generate a summary of events over short periods of time as well as over an entire day.
- **Stage 2**: This corresponds to users, such as Julie, who would be comfortable with the initial version of "How was School Today…?".
- **Stage 3**: This is for more able children, and brings in the ability to tell stories about previous days (as well as the current day), including favorites and frequently told events.

In the long term, as we broaden the range of children we work with, there may be overlaps between our work and research on tools to help typically developing children create stories (such as Robertson and Good, 2005), and also between our work and research on tools to help adults with complex communication needs tell personal narratives (such as Dempster, 2008). Ideally we could combine these efforts and create a storytelling tool that could be used across the age and impairment spectrum.

Reducing staff load

"How was School Today…? in the Wild" is more likely to be used if it does not impose a large burden on staff at the child's school. We have made a number of changes to help staff easily enter and update information about children (e.g., their timetables) and sensor tags (e.g., if a new tag is given to a visitor), using a web-based interface. We are

Figure 11.8   Data collection interface on a Nokia 6212 for RFID (by swiping RFID stickers), voice
recordings, and photos

also experimenting with using 2D bar codes (QRCodes[6]), which would allow encoding
of alphanumeric input data without reference to a central database.

    After "How was School Today…?" we wanted to make it easier for staff to listen to
and change previously recorded messages. We also want to allow parents to record messages about events at home. Both of these are supported in "How was School Today…?
in the Wild" via the tailor-made mobile phone application displayed in Figure 11.8.

Story generation

The main story generation and NLG challenges in "How was School Today…? in the
Wild" center on making it more robust. For example, in order to support the different
stages of children described in Section 11.4.4, we need to be able to generate stories
about parts of a day, and also about previous days. We are also putting considerable
effort into creating a system that can generate stories based on incomplete data; the goal
is to do the best we can with whatever data we have, instead of giving up and not having
a story to tell.

---

[6] http://www.denso-wave.com/qrcode/index-e.html.

Stage 3 children will be able to tell stories from previous days, which means they can tell multiple stories in one dialogue. We are experimenting with changing how a story is told based on previously told stories in a dialogue; it is certainly technically feasible to adjust referring expressions, time phrases, etc. based on the discourse context, but we need to establish whether this is in fact appropriate, or whether it is better (from the child's perspective) for a story to be told in exactly the same way regardless of discourse context.

Future direction: Analysis of voice messages

The voice recordings coming in to the system contain information that is potentially very useful. We are currently investigating whether the voice recordings can be translated into text using speech recognition technology, and then semantically analyzed (probably in a fairly simple way, e.g., identifying people who are mentioned in the recording). This would support research in topic progression in terms of person, aspect, and time as described in Todman *et al.* (2008): for example, the user could shift the topic to other events where a particular person was present. In principle this should be feasible using standard speech recognition technology, but a special-needs school is not always an ideal environment for recording from an acoustic perspective. If a message is unclear, or likely to be incorrect, the system would ideally have mechanisms for clarification and correction.

11.5   Challenges for NLG and AAC

From an NLG perspective, AAC applications such as "How was School Today...?" offer a number of interesting challenges, which we discuss below.

11.5.1   Supporting social interaction

Most NLG research focuses on generating texts (or dialogue contributions) for task-oriented information provision. That is, the assumption is that the system is helping the user carry out a task, such as booking an airline ticket or analyzing a medical data set. However, most human communication is social rather than task-oriented. "How was School Today...?" stories do not help the child or conversational partner perform a task; their purpose is to encourage social interaction. From an AAC perspective, existing technical approaches work reasonably well for communicating needs and supporting tasks; what these approaches cannot do well is social communication.

We are just beginning to explore the challenges in generating texts whose purpose is social dialogue. Probably the biggest differences are in content selection and document structuring. In a task-oriented dialogue we need to discover what information is useful and structure it in a way that is easy to scan and comprehend, whereas in a social dialogue we need to discover what information is interesting and supports social bonding, and structure it as an enjoyable narrative. The implications extend all the way to lexical choice and prosody – the system should be able to generate utterances that have affect (cf. de Rosis and Grasso, 2000). Last but not least, sensitivity and adaptation to the conversational partner is essential. For example, it may be useful to note whether a particular conversational partner has heard this story before.

11.5.2 Narrative

Existing NLG systems do not do a good job of generating factual narratives, and this has been found to cause problems in task-oriented contexts (Portet *et al.*, 2009). We need better models of narrative structure at the document planning level; existing narrative models mostly come from the computational creativity community and assume that the system has the freedom to create new content, which is true for fictional stories but not for personal narratives. We also need better ways of linking messages together for maximum coherence, and of introducing and concluding a narrative.

At the sentence planning and lexicalization levels, we need better ways of communicating time (e.g., *This happened* **this morning** *at art*). Perhaps the biggest challenge is deciding when time needs to be communicated in the first place, and when omitting it does no damage to the narrative flow.

11.5.3 User personalization

AAC tools give support to individual users for communication, and ideally a user's AAC tool should have a personalized style which is their own "voice." This may include creating a synthetic voice for an individual user (Yarrington *et al.*, 2008) and/or creating sentences in a particular style associated with an individual user. To take a simple example from "How was School Today…?," one child said she would like to talk like a teenager, i.e., he or she wanted "How was School Today…?" texts to be similar to utterances used by normally developing teenagers in social contexts.

11.5.4 System evaluation

Evaluation of NLG systems is different from evaluations in related areas such as natural language understanding (Mellish and Dale, 1998; Reiter and Belz, 2009). However, the techniques commonly used in NLG were not suitable for "How was School Today…?," and may perhaps not be suitable for AAC NLG systems in general. One standard approach is corpus evaluation, but often there is little to no baseline communication to compare with, i.e., no corpus available. Arguably, a corpus of communications of similarly aged able-bodied peers could be elicited, but this may not be a suitable comparison. An evaluation of the system by the AAC users themselves is not always possible, and is likely to be influenced by the users wanting to "do well" or please the evaluators. The number of subjects is also likely to be small and have highly varied profiles, making task-based evaluation difficult.

11.5.5   Interaction and dialogue

Last but not least (especially in the context of this book), NLG AAC systems pose challenges from a dialogue and interaction perspective. The usual dialogue model is that a person is talking to a computer, but in AAC the model is that one person is using a computer to help him or her have a dialogue with another person. So, the computer is not a full dialogue partner, it is an assistant to one of the dialogue partners. What implications does this have for dialogue technology?

Another particular challenge concerns the creation of an AAC device for users with social impairments. For example, people with autism may not be able to gauge the uptake or agreement of their conversational partner. An AAC system may be able to help with this, but this is not something we have addressed in "How was School Today…?".

The other aspect of this is that the user must be able to control the NLG system, so that it does what the user intends it to do. From a human–computer interaction perspective there are still open questions in terms of the best way for a user to exercise such control; little research has been done on this topic.

11.6   Conclusions

In this chapter we have introduced augmentative and alternative communication as a very relevant application area for NLG, particularly NLG for interactive systems. We presented a research AAC tool, "How was School Today…?," that has been implemented and evaluated in a special needs school. Using our experiences in developing this tool, we outlined several potentially fruitful areas for future research on NLG for interactive systems.

Acknowledgments

We would like to express our thanks to the children, their parents, and the staff and the special schools where this project had its base. Without their valuable contributions and feedback this research would not have been possible. We would also like to thank DynaVox Inc. for supplying the communication devices to run our system on and Sensory Software for supplying their AAC software. This research was supported by the UK Engineering and Physical Sciences Research Council (EPSRC) and the Digital Economy Programme under grants EP/F067151/1, EP/F066880/1, EP/E011764/1, EP/H022376/1, and EP/H022570/1.

References

American Speech-Language-Hearing Association (2002). Augmentative and alternative communication: Knowledge and skills for service delivery [knowledge and skills]. Available from http://www.asha.org/policy. Accessed on 11/24/2013.

Andreasen, P., Waller, A., and Gregor, P. (1998). BlissWord – full access to Blissymbols for all users. In *Proceedings of the Biennial Conference of the International Society for Augmentative and Alternative Communication*, pages 167–168, Dublin, Ireland. International Society for Augmentative and Alternative Communication.

Beukelman, D. R. and Mirenda, P. (1998). *Augmentative and Alternative Communication: Management of Severe Communication Disorders in Children and Adults*. Brookes Publishing Company, Baltimore, MD, 2nd edition.

Biswas, P. (2006). A flexible approach to natural language generation for disabled children. In *Proceedings of the International Conference on Computational Linguistics and the Annual Meeting of the Association for Computational Linguistics (COLING-ACL)*, pages 1–6, Sydney, Australia. Association for Computational Linguistics.

Biswas, P. and Samanta, D. (2008). Friend: A communication aid for persons with disabilities. *IEEE Transactions on Neural Systems and Rehabilitation Engineering*, 16(2):205–209.

Black, R., Reddington, J., Reiter, E., Tintarev, N., and Waller, A. (2010). Using NLG and sensors to support personal narrative for children with complex communication needs. In *Proceedings of the NAACL HLT Workshop on Speech and Language Processing for Assistive Technologies*, pages 1–9, Los Angeles, CA. Association for Computational Linguistics.

de Rosis, F. and Grasso, F. (2000). Affective natural language generation. In Paiva, A., editor, *Affective Interactions*, pages 204–218. Springer LNCS, Berlin, Germany.

Demasco, P. W. and McCoy, K. F. (1992). Generating text from compressed input: An intelligent interface for people with severe motor impairments. *Communications of the ACM*, 35(5): 68–78.

Demir, S., Carberry, S., and McCoy, K. F. (2011). Summarizing information graphics textually. *Computational Linguistics*, 38(3):527–574.

Dempster, M. (2008). Using natural language generation to encourage effective communication in non-speaking people. In *Proceedings of the Young Researchers Consortium at the International Conference on Computers Helping People with Special Needs*, Linz, Austria.

Dempster, M., Alm, N., and Reiter, E. (2010). Automatic generation of conversational utterances and narrative for augmentative and alternative communication: A prototype system. In *Proceedings of the NAACL HLT Workshop on Speech and Language Processing for Assistive Technologies*, pages 10–18, Los Angeles, CA. Association for Computational Linguistics.

Dugard, P., File, P., and Todman, J. (2011). *Single-case and Small-n Experimental Designs: A Practical Guide To Randomization Tests*. Routledge, New York, NY, USA, 2nd edition.

Ferres, L., Lindgaard, G., and Sumegi, L. (2013). Evaluating a tool for improving accessibility to charts and graphs. *ACM Transactions on Computer-Human Interaction (TOCHI)*, 20(5).

Gatt, A. and Reiter, E. (2009). SimpleNLG: A realisation engine for practical applications. In *Proceedings of the European Workshop on Natural Language Generation (ENLG)*, pages 90–93, Athens, Greece. Association for Computational Linguistics.

Grove, N. (2010). *The Big Book of Storysharing*. Special Educational Needs Joint Initiative for Training (SENJIT), Institute of Education, University of London, London, UK.

Light, J. and Drager, K. (2007). AAC technologies for young children with complex communication needs: State of the science and future research directions. *Augmentative and Alternative Communication*, 23(3):204–216.

Liu, X., Sen, A., Bauer, J., and Zitzmann, C. (2007). A software client for wi-fi based real-time location tracking of patients. In *Proceedings of the International Conference on Medical Imaging and Informatics*, pages 141–150, Beijing, China. Springer.

McCabe, A. and Peterson, C. (1991). Getting the story: A longitudinal study of parental styles in eliciting narratives and developing narrative skill. In McCabe, A. and Peterson, C., editors, *Developing Narrative Structure*, pages 217–253. Lawrence Erlbaum Associates, Hillsdale, NJ.

McCoy, K. F., Bedrosian, J., and Hoag, L. (2010). Implications of pragmatic and cognitive theories on the design of utterance-based AAC systems. In *Proceedings of the NAACL HLT Workshop on Speech and Language Processing for Assistive Technologies*, pages 19–27, Los Angeles, CA. Association for Computational Linguistics.

Mellish, C. and Dale, R. (1998). Evaluation in the context of natural language generation. *Computer Speech & Language*, **12**(4):349–373.

Newell, A. F., Carmichael, A., Gregor, P., Alm, N., and Waller, A. (2008). Information technology for cognitive support. In Sears, A. and Jacko, J. A., editors, *The Human-Computer Interaction Handbook: Fundamentals, Evolving Technologies and Emerging Applications*, pages 811–828. Lawrence Erlbaum Associates, Mahwah, NJ, 2nd edition.

Nikolova, S., Tremaine, M., and Cook, P. (2010). Click on bake to get cookies: Guiding word-finding with semantic associations. In *Proceedings of the ACM SIGACCESS Conference on Computers and Accessibility (ASSETS)*, pages 155–162, Orlando, FL. Association for Computing Machinery.

Polkinghorne, D. (1995). Narrative configuration in qualitative analysis. In Hatch, J. A. and Wisniewski, R., editors, *Life History and Narrative*, pages 5–24. Taylor & Francis, Bristol, PA.

Portet, F., Reiter, E., Gatt, A., Hunter, J., Sripada, S., Freer, Y., and Sykes, C. (2009). Automatic generation of textual summaries from neonatal intensive care data. *Artificial Intelligence*, **173**(7–8):789–816.

Quasthoff, U. M. and Nikolaus, K. (1982). What makes a good story? Towards the production of conversational narratives. In Flammer, A. and Kintsch, W., editors, *Discourse Processing*. North-Holland Publishing Company, Amsterdam.

Reddington, J. and Tintarev, N. (2011). Automatically generating stories from sensor data. In *Proceedings of the International Conference on Intelligent User Interfaces (IUI)*, pages 407–410, Palo Alto, CA. Association for Computing Machinery.

Reichenbach, H. (1947). *Elements of Symbolic Logic*. Macmillan Co., New York, NY.

Reiter, E. (2007). An architecture for data-to-text systems. In *Proceedings of the European Workshop on Natural Language Generation (ENLG)*, pages 97–104, Saarbrücken, Germany. Association for Computational Linguistics.

Reiter, E. and Belz, A. (2009). An investigation into the validity of some metrics for automatically evaluating NLG systems. *Computational Linguistics*, **35**(4):529–558.

Reiter, E., Gatt, A., Portet, F., and van der Meulen, M. (2008). The importance of narrative and other lessons from an evaluation of an NLG system that summarises clinical data. In *Proceedings of the International Conference on Natural Language Generation (INLG)*, pages 147–156, Salt Fork, OH. Association for Computational Linguistics.

Roark, B., de Villiers, J., Gibbons, C., and Fried-Oken, M. (2010). Scanning methods and language modeling for binary switch typing. In *Proceedings of the NAACL HLT Workshop on Speech and Language Processing for Assistive Technologies*, pages 28–36, Los Angeles, CA. Association for Computational Linguistics.

Robertson, J. and Good, J. (2005). Story creation in virtual game worlds. *Commununications of the ACM*, **48**(1):61–65.

Thomas, K. E. and Sripada, S. (2008). What's in a message? Interpreting geo-referenced data for the visually-impaired. In *Proceedings of the International Conference on Natural Language Generation (INLG)*, pages 113–120, Salt Fork, OH. Association for Computational Linguistics.

Todman, J. and Alm, N. (2003). Modelling conversational pragmatics in communication aids. *Journal of Pragmatics*, **35**(4):523–538.

Todman, J., Alm, N., Higginbotham, J., and File, P. (2008). Whole utterance approaches in AAC. *Augmentative and Alternative Communication*, **24**(3):235–254.

Waller, A. (2006). Personal narrative and AAC: From transactional to interactional conversation. In *Proceedings of the International Society for Augmentative and Alternative Communication (ISAAC) Research Symposium*, Düsseldorf, Germany. International Society for Augmentative and Alternative Communication.

Waller, A., Dennis, F., Brodie, J., and Cairns, A. Y. (1998a). Evaluating the use of TalksBac, a predictive communication device for nonfluent adults with aphasia. *International Journal of Language & Communication Disorders*, **33**(1):45–70.

Waller, A., Dennis, F., Cairns, A., Whitehead, N., Brodie, J., Newell, A. F., and Morrison, K. (1998b). The future development of TalksBac: A predictive augmentative communication system for adults with nonfluent aphasia. *International Journal of Language & Communication Disorders*, **33**(1):45–70.

Waller, A., Francis, J., Tait, L., Booth, L., and Hood, H. (1999). The WriteTalk project: Story-based interactive communication. In Bühler, C. and Knops, H., editors, *Assistive Technology on the Threshold of the New Millennium*, pages 180–184. IOS Press, Amsterdam, The Netherlands.

Waller, A. and Newell, A. (1997). Towards a narrative-based communication systems. *European Journal of Disorders of Communication*, **32**(S3):289–306.

Waller, A., O'Mara, D., Manurung, R., Pain, H., and Ritchie, G. (2005). Facilitating user feedback in the design of a novel joke generation system for people with severe communication impairment. In *Proceedings of the International Conference on Human-Computer Interaction (HCII)*, Las Vegas, NV. Lawrence Erlbaum Associates.

Waller, A., O'Mara, D. A., Tait, L., Booth, L., Brophy-Arnott, B., and Hood, H. E. (2001). Using written stories to support the use of narrative in conversational interactions: Case study. *Augmentative and Alternative Communication*, **17**(4):221–232.

Yarrington, D., Pennington, C., Bunnell, H. T., Gray, J., Lilley, J., Nagao, K., and Polikoff, J. (2008). ModelTalker Voice Recorder (MTVR) – a system for capturing individual voices for synthetic speech. In *Proceedings of the ISAAC Biennial Conference*, Montreal, Canada. International Society for Augmentative and Alternative Communication.

Part V

Evaluation and shared tasks

12 Eye tracking for the online evaluation of prosody in speech synthesis

Michael White, Rajakrishnan Rajkumar, Kiwako Ito, and Shari R. Speer

12.1 Introduction

The past decade has witnessed remarkable progress in speech synthesis research, to the point where synthetic voices can be hard to distinguish from natural ones, at least for utterances with neutral, declarative prosody. Neutral intonation often does not suffice, however, in interactive systems: instead it can sound disengaged or "dead," and can be misleading as to the intended meaning.

For concept-to-speech systems, especially interactive ones, natural language generation researchers have developed a variety of methods for making contextually appropriate prosodic choices, depending on discourse-related factors such as givenness, parallelism, or theme/rheme alternative sets, as well as information-theoretic considerations (Prevost, 1995; Hitzeman *et al.*, 1998; Pan *et al.*, 2002; Bulyko and Ostendorf, 2002; Theune, 2002; Kruijff-Korbayová *et al.*, 2003; Nakatsu and White, 2006; Brenier *et al.*, 2006; White *et al.*, 2010). In this setting, it is possible to adapt limited-domain synthesis techniques to produce utterances with perceptually distinguishable, contextually varied intonation (see Black and Lenzo, 2000; Baker, 2003; van Santen *et al.*, 2005; Clark *et al.*, 2007, for example). To evaluate these utterances, listening tests have typically been employed, sometimes augmented with expert evaluations. For example, evaluating the limited domain voice used in the FLIGHTS concept-to-speech system (Moore *et al.*, 2004; White *et al.*, 2010) demonstrated that the prosodic specifications produced by the natural language generation component of the system yielded significantly more natural synthetic speech in listening tests and, in an expert evaluation, compared to two baseline voices. As Swift *et al.* (2002) and van Hooijdonk *et al.* (2007) have noted, however, offline evaluation methods such as these do not offer insight into how listeners actually process synthetic speech. Thus, as an alternative, online (and objective) method of evaluation, we present in this chapter a case study that – for the first time to our knowledge – investigates the potential of using eye tracking, a method frequently used to investigate prosodic effects in spoken language processing (see Speer, 2011, for review), for the online evaluation of varied intonation in synthetic speech.[1]

In the speech synthesis community, offline evaluation methods have remained predominant, since they are relatively simple to administer. For example, in the most recent

---

[1] The chapter integrates, revises, and extends two earlier conference papers (White *et al.*, 2009; Rajkumar *et al.*, 2010).

editions of the Blizzard Challenge (Clark *et al.*, 2007; Karaiskos *et al.*, 2008; King and Karaiskos, 2009), speech synthesizers have been evaluated via listening tests involving mean opinion scores of how natural or human a synthetic voice sounds, along with word error rates in transcribing semantically unpredictable sentences. As an alternative method of evaluation, Swift *et al.* (2002) were the first to propose using eye tracking to investigate how synthetic speech is processed incrementally in comparison to human speech. They showed that human listeners process segmental information in synthetic speech incrementally at both the lexical and discourse levels, though with processing delays in comparison to human speech. They also found subtle differences in the online processing of two synthetic voices, demonstrating the potential of eye tracking to serve as a fine-grained evaluation measure.

Subsequently, van Hooijdonk *et al.* (2007) used eye tracking to investigate the impact of both segmental and supersegmental information on how human listeners process synthetic speech, comparing both a diphone voice and a unit selection voice to human speech. In their experiment, participants followed two consecutive instructions (in Dutch) to click on an object within a visual display. The first instruction mentioned an initial referent (e.g., *roze vork / pink fork*) with neutral intonation. The second instruction mentioned a target referent using a double accent pattern, the choice of which was forced by the unit selection synthesizer, as it typically produced these patterns and did not allow the accent pattern to be controlled. The target could either be of the same type but a different color (e.g., *blauwe vork / blue fork*), or a different type and color (e.g., *blauwe vos / blue fox*). In both cases, the other possible referent served as a competitor (there was also an unrelated distractor). When the target was of a different type, the double accent pattern was considered felicitous; in contrast, when the target was of the same type, the double accent pattern was considered infelicitous. The results of the experiment showed that the diphone voice induced significantly more fixations to the competitor than the unit selection voice or human speech, which could be explained by the relatively poor segmental intelligibility of diphone synthesis. Perhaps surprisingly, though, in all three voice conditions, there were significantly more anticipatory looks to the competitor when the noun was of a different type, despite the expected felicity of the double accent pattern in this context. As the disambiguating segmental information in the noun arrived, looks to the competitor subsided more quickly with the human speech, echoing the findings of Swift *et al.* (2002) of processing delays with synthetic speech.

It is not entirely clear how unexpected the anticipatory looks to the competitor in the different type condition should be taken to be, as van Hooijdonk *et al.* (2007) did not report whether listeners perceived the accents as contrastive, and did not provide an acoustic analysis. Moreover, as they did not compare different accent patterns, there was no way to observe the impact of felicitous and infelicitous prosody in the same context.

Going beyond the study of van Hooijdonk *et al.* (2007), we present an experiment that investigates whether different accent patterns in synthetic speech yield significant differences in anticipatory eye movements. The experiment replicates with synthetic speech Ito and Speer's (2008; 2011) eye-tracking experiment, where participants followed recorded instructions to decorate holiday trees with ornaments laid out on a grid. The decoration sequences were carefully constructed to include contrasts between

consecutively mentioned ornaments (e.g., *Hang a red star. Next, hang a yellow star.*),
as well as locally non-contrastive sequences (e.g., *Hang a yellow tree. Next, hang
a green ball.*) The noun phrases in these critical utterances had one of two pitch accent
patterns: (1) a contrastive L+H* accent on the adjective, and no accent on the noun,
e.g., *hang a YELLOW$_{L+H*}$ star$_{\emptyset}$*; (2) H* on the adjective and !H* on the noun, e.g.,
hang a yellow$_{H*}$ star$_{!H*}$. The results demonstrated a robust effect of the contrastive
L+H* accent together with the prosodically attenuated noun, which produced very early
looks to the target cell in contrastive sequences, significantly faster than with the double
accent pattern. In addition to this facilitative effect of L+H*, the study showed an intonational "garden-path" effect in non-contrastive sequences, with increased looks to the
contrastive competitor and delayed looks to the target. A subsequent experiment with
size instead of color adjectives established the statistical significance of this effect.

Our experiment used a custom Festival Multisyn (Clark *et al.*, 2007) unit selection voice with prosodic specifications given in APML (De Carolis *et al.*, 2004), the
same kind of voice used in our offline evaluation of the synthesized utterances in the
FLIGHTS system (White *et al.*, 2010). Since the quality of speech synthesized with a
unit selection voice depends in large part on how well the speech database covers the target utterances, we were immediately confronted with the question of how to design the
speech database for the voice used in our experiment. Since we did not know whether
to expect to find similar facilitative and garden-path effects with varied intonation in
synthetic speech as Ito and Speer found with human speech, we decided to construct a
database that would enable high-quality (though still non-trivial) synthesis.

The plan of the rest of the chapter is as follows. In Section 12.2 we give a detailed
description of our materials and experimental procedure. In Section 12.3 we present
the results of our eye-tracking experiment, where we found a lack of facilitation overall
with synthetic speech, though even stronger garden-path effects. We provide an interim
discussion of these results in Section 12.4, where we suggest possible reasons for the
discrepancy between the results with natural and synthetic speech. To help ascertain
whether a context-independent measure of quality can predict online effects, we present
the results of an offline rating task in Section 12.5; then, in Section 12.6, we present an
acoustic analysis of the stimuli used in these eye-tracking experiments, with the aim of
identifying specific acoustic factors influencing the processing of synthetic speech. In
Section 12.7 we discuss the implications of our findings, and in particular why the most
complete picture of the quality of the synthetic speech emerges from examining the eye-
tracking experiment, the offline ratings, and the acoustic analysis together. In Section 12.8
we summarize our results and discussion and conclude with a discussion of future work.

12.2   Experiment

12.2.1   Design and materials

Participants decorated holiday trees following pre-synthesized auditory instructions.
Each participant decorated three trees using the ornaments laid out on three separate

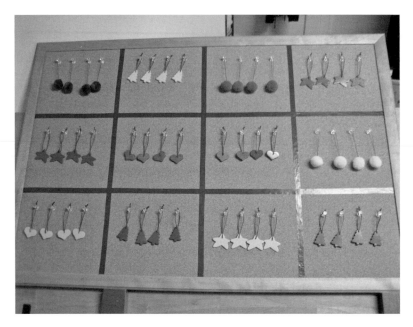

Figure 12.1   Example ornament layout on a grid

grids. Four types of ornaments (3 targets: star, tree, ball, and 1 filler: heart) were painted in three colors (red, yellow, and green), yielding 12 ornament sets that occupied 12 cells on each grid. (The size of ornaments was unified within a grid but altered across the three grids to distract participants from the experimental manipulation.) Each cell contained four identical ornaments. The three target ornaments in three colors were distributed to occupy nine out of ten peripheral cells surrounding the two central cells. The two central cells and the remaining one peripheral cell were occupied by the filler ornaments (i.e., hearts). The locations of ornaments were altered across the three boards. An example ornament grid is shown in Figure 12.1. Each participant decorated the tree with 26 ornaments; as mentioned earlier, the orders of decoration were constructed to include locally contrastive and non-contrastive sequences.

In the original Ito and Speer (2008; 2011) experiment, the auditory instructions were recorded by a trained female phonetician who maintained her overall pitch range and speech rate within and across conditions. All the instruction utterances were ToBI transcribed[2] by an annotator blind to the experimental design. Example F0 traces and the ToBI transcriptions for the natural speech are given in Figure 12.2 (from Ito and Speer, 2008); the F0 traces for the synthesized speech are very similar. Table 12.1 shows the mean durations and F0 values for the adjectives and nouns across conditions, for both the synthesized and natural speech.[3]

---

[2] ToBI (Tones and Break Indices) is a set of conventions for annotating the prosodic structure of spoken language (Beckman *et al.*, 2005).

[3] Samples of the synthetic stimuli are available at http://www.ling.ohio-state.edu/~mwhite/tree-stimuli/.

Table 12.1. Mean duration and F0 of target NPs across conditions; corresponding natural speech values are in italic type

Contrast? / Tune	Adj dur (ms)	Adj F0 (Hz)	N dur (ms)	N F0 (Hz)
Y / L+H* Ø	356 *330*	332 *299*	458 *489*	148 *148*
Y / H* !H*	366 *332*	223 *207*	524 *549*	192 *164*
N / L+H* Ø	343 *320*	332 *300*	462 *491*	152 *150*
N / H* !H*	368 *316*	223 *208*	516 *558*	197 *163*

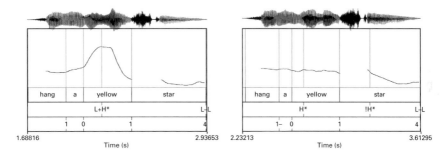

Figure 12.2   Example F0 traces and ToBI annotations for [L+H* no-accent] (left) and [H* !H*] (right) (from Ito and Speer, 2008)

To produce the synthetic stimuli, 192 pseudo-instructions (with ToBI tune annotations) were recorded by the same speaker as in the original experiment, and used to construct a Festival unit selection speech database. The pseudo-instructions – e.g., *hang a greedy$_{H*}$ ball$_{!H*}$* – were designed to ensure that the stimuli would require at least two joins, while otherwise providing excellent coverage of diphones in context. Festival was then used to generate critical phrases like *hang the green$_{L+H*}$ ball$_Ø$*. Another trained ToBI annotator then marked F0, adjective, and noun durations and certified that the tunes were clear in all the items. Synthesized critical phrases were spliced in at the end of the natural speech stimuli of the original experiments. Volume levels across the segments were normalized according to the default settings of Adobe Audition.

12.2.2   Participants and eye-tracking procedure

Thirty-three undergraduate students at the Ohio State University participated in partial fulfilment of a course requirement. Data from 29 native speakers of American English are analyzed below. Participants sat in front of a drafting table with the top tilted at 35 degrees to support the ornament display board. They wore lightweight headgear fitted with an eye-camera and a magnetic transmitter that functioned to correct measured eye positions for head movement. Participants followed the synthesized instructions to choose an ornament from the grid and place it on a small tree located to their right. The *x* and *y* coordinates of eye-fixations on the board were recorded at 60 Hz using an ASL Eye-Trac 6 data-collection system. As illustrated in Figure 12.3, the experimenter monitored the participant's eye locations and body orientations via a ceiling-mounted

Figure 12.3   Experimental setup

camera, and pressed a key to play each instruction when the participant had finished hanging an ornament and had faced back to the board.

12.3   Results

Participants had nine trials in each of the four critical conditions. The dependent variables were the mean proportion of fixations to the target and to the competitor. The fixation proportion was calculated for each time point by dividing the total number of actual fixations to the target/contrastive competitor by the total number of possible fixations. Trials in which a participant was looking at the intended target at the onset of the adjective were eliminated. After this correction, we use repeated measures analysis of variance (ANOVA) for calculating the significance of the various effects we describe in this section. For eleven 100 ms windows starting from the onset of the adjective in each instruction, we did ANOVA calculations for both subjects and items.

As Figure 12.4 shows, when synthetic speech stimuli are used in a contrastive discourse sequence, [L+H* no-accent] does not have any facilitative effect as compared to [H* !H*] in searching for a contrastive target. This is unlike the response of subjects to natural speech. For natural speech, the two lines diverged at the onset of the noun, and until about 300 ms into the noun, fixation proportions to the target were significantly higher for [L+H* no-accent] than for [H* !H*] trials. But for synthetic speech, the lines are almost together throughout the entire length of the noun. In addition, the synthesized speech is processed more slowly than the natural speech, as Figure 12.5 shows.

The left panel of Figure 12.6 shows that in non-contrastive sequences using synthetic speech, [L+H* no-accent] does evoke contrast, as looks to the contrastive competitor

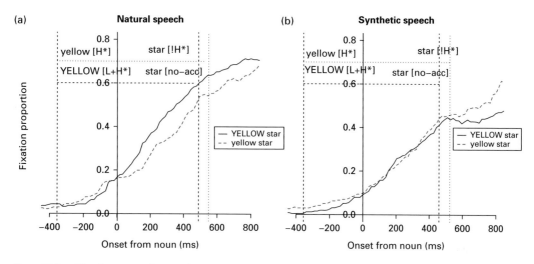

Figure 12.4 Fixation proportions to the target in two contrastive sequences, e.g., *red star → YELLOW/yellow star*

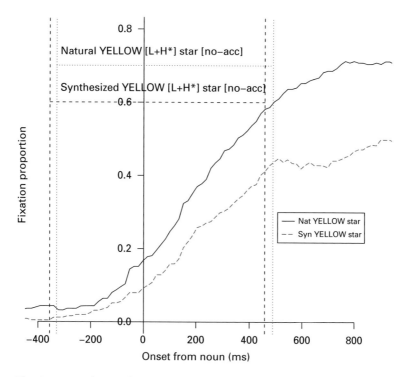

Figure 12.5 Fixation proportions to the target due to contrastive accent in contrastive sequences with natural and synthetic speech

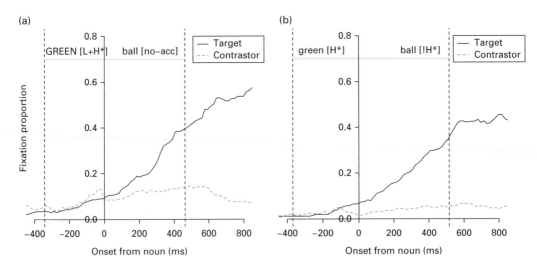

Figure 12.6   Fixation proportions to the target and contrastive competitor in two non-contrastive sequences with synthetic speech, e.g., *yellow tree → GREEN/green ball*

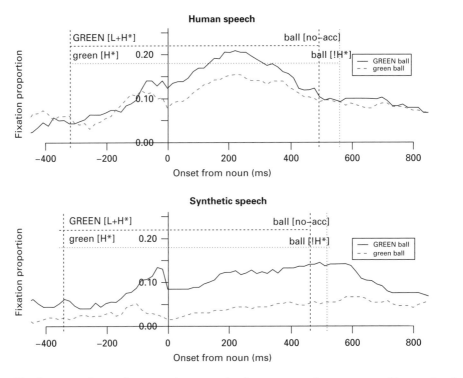

Figure 12.7   Fixation proportions to the contrastive competitor in non-contrastive sequences with natural and synthetic speech

keep increasing past the end of the noun. By contrast, the right panel shows that [H* !H*] in non-contrastive sequences does not result in more looks to the contrastive competitor. This is similar to the behavior using natural speech (not shown here for space reasons). For both kinds of speech, a direct comparison of fixations to the contrastive competitor in the two non-contrastive sequences is shown in Figure 12.7. For natural speech, the two lines diverge at around 50 ms before the noun onset, with fixations to the contrastive competitor rising after that point. Ito and Speer reported the relative increase in looks to the competitor with [L+H* no-accent] as a trend in their experiment using color adjectives (the one we replicated), and a subsequent experiment using size adjectives – which involved a more difficult visual search task – established the statistical significance of the effect. In the current experiment with synthetic speech, from just before the noun onset there was a clear separation between the fixation proportions to the contrastive competitor in the two conditions. Until 100 ms into the noun, this effect is statistically significant for both subjects as well as items (at $p < 0.05$). The effects for subjects are in fact very prominent: for all time windows up to 300 ms into the noun, significance with $p < 0.0001$ is observed, and the remaining two windows of the noun show significant results with $p < 0.001$. But for items (of which there are fewer), this effect is significant at $p < 0.05$ for just the two windows surrounding the onset of the noun. Thus the experiment confirms the effect of synthesized L+H* on the adjective in evoking a contrast set for the upcoming noun, in contrast to H* accents; indeed, the results show a relative garden-path effect that is more robust than the one found with natural speech.

12.4   Interim discussion

It is remarkable that despite our best attempts to produce high-quality synthetic stimuli, eye movement monitoring was able to clearly distinguish how listeners process these stimuli in comparison to human speech. Our informal listening tests suggested that the stimuli were of generally excellent quality, with only quite subtle audible joins or other artifacts, and these impressions were largely confirmed by our offline ratings study described in the next section. Additionally, our ToBI annotator had no trouble identifying the intended prosody in a blind test, and few of the participants suspected that the experiment involved synthetic speech prior to the debriefing.

As our results failed to replicate the facilitative effect of contextually appropriate prosody with human speech, while producing a relatively stronger garden-path effect, one might be tempted to conclude that contextually appropriate prosody in synthetic speech does not help listeners, though contextually inappropriate prosody can certainly hamper their processing. However, since our experiment confirms the presence of processing delays with synthetic speech that have been observed in earlier studies (Swift et al., 2002; van Hooijdonk et al., 2007), we consider it likely that a processing delay in interpreting the segmental and supersegmental information in the adjective means that the disambiguating segmental information in the noun in a sense arrives too soon – that is, before there is time to see any anticipatory facilitative effect of a contrastive L+H* accent on the adjective. This suggests that designers of future eye-tracking experiments for online synthesis evaluation should take into account possible delays in processing, and accordingly lengthen the time before the disambiguating segmental information arrives, for example by using longer or extra adjectives in the stimuli.

Another possibility is to use a more difficult visual search, which can lengthen the time window in which facilitation plays out, as in the related experiment of Ito and Speer (2011) using size rather than color adjectives, where the task of identifying relative size was found to be more difficult than identifying color.

With the relatively stronger garden-path effect seen with the synthetic stimuli, we might also seek an explanation in terms of processing delays. If it takes listeners longer to process the contrastive adjective information, then listeners might be updating their referential domain for the target at the same time as the conflicting information from the noun is arriving, causing additional delays in identifying the correct referent. Another possibility is that with somewhat less intelligible segmental information, due to some imperfect joins, listeners are relying more heavily on prosody to guide their interpretation.

Returning to the somewhat surprising results of van Hooijdonk *et al.* (2007), it is worth noting that they likewise demonstrated a garden-path, rather than facilitative, effect, using a much simpler visual layout. In their case, the increased looks to the contrastive competitor could have been due to the first accent (on the adjective) receiving a contrastive interpretation by listeners. That this possibility was not investigated shows the importance of comparing the effects of different accent patterns, not just different contexts, in evaluating speech synthesis with eye-tracking. Finally, we note that their finding that the diphone voice had more looks to the competitor than either the unit selection voice or natural speech is consistent with the hypothesis that listeners pay more attention to the prosody when the segmental intelligibility is lower.

In the next two sections, we present an offline ratings study and an acoustic analysis of our experimental stimuli and examine how these interact with the results of the eye-tracking experiment. As we shall see, the emergence of adjective duration as a significant factor in predicting facilitation lends credence to the idea that processing delays inhibit the observation of anticipatory effects of contextually appropriate prosody. In addition, the extent of F0 drop from the adjective to the noun emerges as a specific acoustic factor that may hinder processing. Interestingly – given the somewhat unexpected difficulty subjects had in rating utterances with marked intonation out of context – it will turn out that a full understanding of the roles that duration and F0 drop play in synthesis quality can only be gained by examining the online and offline results together.

12.5   Offline ratings

12.5.1   Design and materials

Quality ratings were collected by means of an online interface. For each item, two types of responses were elicited from the participants: (1) a naturalness rating on a scale ranging from 1 (bad) to 7 (excellent); and (2) a forced-choice judgment as to whether the item was natural or synthetic speech. The stimuli for the rating experiment consisted of 144 items, presented as two lists. Each list contained 18 synthetic items from the eye-tracking experiment, 18 natural items from the experiment reported in Ito and Speer (2008, 2011), and 36 filler items (18 synthetic and 18 natural items) from a direction-giving domain.[4] Each list was rated by 10 native speakers of American English between the ages of 18 and 40. The filler items were introduced to ensure that each item was presented without any notion of discourse context. We made a conscious decision to collect context-independent naturalness ratings in order to see whether some notion of absolute or context-independent quality had any relationship with the eye-tracking effects observed. Inside each list, domains were strictly interleaved (items from the directions domain alternated with tree decorating domain items) so that subjects did not infer (even if inadvertently) contextual dependencies between successive items presented. Inside each list, sound files were arranged in a pseudo-random order. To this end we ensured that there were no cases where two consecutive files in the same domain had the same word sequence repeated. We also ensured that in a single domain, the same tune did not appear more than three times in a row.

12.5.2   Results

Table 12.2 summarizes the results of the offline ratings experiment for each domain. In both the tree and directions domains, the synthetic items received a lower average rating compared to the natural items in that domain. We also calculated identification accuracy (percentage of times that the subjects were actually right in their classification decision) and the percentage of times they classified items as synthetic (irrespective of whether the outcome was right or wrong). There was a positive correlation between ratings and identification accuracy of natural speech items and a negative correlation in the case of synthetic speech items (the overall correlation adjusts for this inversion). In the tree decorating domain, identification accuracy was lower for both natural and synthetic speech items in comparison to the directions domain, which we can attribute to the perceivable quality difference in the case of items in the directions domain as compared to the tree decorating domain. The ratings in Table 12.2 also directly illustrate that the synthetic speech in the tree decorating domain was of higher quality than typical unit selection speech, of which the speech from the directions domain is fairly representative; this was an expected result, given that the tree decorating speech corpus was carefully constructed for excellent coverage and that the phones were hand-aligned.

Table 12.3 illustrates the results of the experiment for each tune. Here we observe that [L+H* no-accent] items were rated much lower than the [H* !H*] items even for natural speech, where subjects were essentially at chance in guessing whether the [L+H* no-accent] items were synthetic. We suggest that this discrepancy between the tunes, which was much stronger than anticipated, can be explained by the fact that a contrastive [L+H* no-accent] tune presented out-of-context can sound odd for listeners.

---

[4] The directions domain data had L*-L*-L*-L*-L* and H*-!H*-!H*-!H*-!H* tunes. Utterances from this domain formed the basis of the text selection experiments reported in Espinosa *et al.* (2010).

Table 12.2. Domain-wise split-up of item ratings

Domain	Mean rating	%Acc	Correlation	%Syn guesses
Overall	4.92	72.00	0.90	48.00
Synthetic tree	4.64	65.83	−0.83	65.80
Natural tree	5.47	66.95	0.80	33.08
Synthetic directions	3.82	74.44	−0.90	74.40
Natural directions	5.74	80.83	0.86	19.16

Table 12.3. Tone-wise split-up of item ratings

Tone	Type	Mean rating	%Acc	Correlation	%Syn guesses
L+H* Ø	Overall	4.34	63.61	0.68	62.00
H* !H*		5.77	69.17	0.92	40.00
L+H* Ø	Synthetic	3.96	75.56	−0.65	75.50
H* !H*	Tree	5.32	56.11	−0.88	56.00
L+H* Ø	Natural	4.72	51.67	0.089	48.3
H* !H*	Tree	6.22	82.22	0.83	17.8

12.6   Acoustic analysis using Generalized Linear Mixed Models (GLMMs)

As the results described in Section 12.2 show, L+H*-accented adjectives in natural speech induce both facilitation and garden-pathing effects, while synthetic speech causes only the latter effect. To isolate specific acoustic properties that are potentially influencing listeners' processing of synthetic speech, we first examined the extent of item-wise variability in our experimental stimuli. For this analysis (and for all the statistical analyses reported from here on), Generalized Linear Mixed Models (GLMMs) were used. For these analyses, we adopted a binary choice dependent variable encoding the presence or absence of looks to the desired area of interest within the first 300 ms of the noun. This particular time window was postulated on the basis of the ANOVA analyses, which demonstrated significant subjects and item effects until this time. Once listeners encountered adjectival information, after a gap of 100–150 ms (time for planning and executing saccades), they started fixating on a particular nominal referent. The eye-tracker provides eye location data every 17 ms and thus, in this case, this involved considering 17 data points. Table 12.4 presents a comparison between two models – with and without item random effects – for both types of speech using a log-likelihood test, where the two models are fitted to the data and their log-likelihoods are compared. The models are presented in R GLMM format.[5] The results indicate that a GLMM with an item intercept was significantly different from a model without an item intercept

---

[5] The dependent variable occurs to the left of "~" and independent variables occur to the right; "*" denotes an interaction between two fixed effects; random effects are represented after the "|" symbol.

Table 12.4. Model comparison using the log-likelihood test (df is 1)

Speech/model	Condition	Chi-sq
Synthetic	Contrastive sequence	$7.168, p < 0.01$
	Non-contrastive sequence	0.1539
Natural	Contrastive sequence	0.1194
	Non-contrastive sequence	4.687×10^{-8}
Model 1 (D.f 3)	looks $\sim$ cond $+$ (1 | subj)	
Model 2 (D.f 4)	looks $\sim$ cond $+$ (1 | subj) $+$ (1 | item)	

only in the facilitation case with synthetic speech, the only experimental condition that did not induce an effect akin to natural speech. Thus there is item-wise variation in the eye-tracking results with the synthetic speech stimuli not seen with the natural speech stimuli. So for all the subsequent analyses, item effects were considered only in the case of those involving synthetic speech facilitation.

12.6.1 Acoustic factors and looks to the area of interest

As predictors of looks to the desired area of interest, we examined several acoustic factors including **duration of the adjective and noun**, **F0 drop** (difference in F0 between the adjective and noun), **adjective F0 latency** (time from the onset of the adjective to the adjective F0 peak time-point), and Festival's **join cost** for the synthetic items. Various acoustic factors were treated as the fixed effects of the model, and subjects and items were considered to be the random effects. The basic regression equations (in R GLMM format) for predicting looks to the desired area of interest are:

Synthetic speech looks $\sim$ acoustic-factor*cond $+$ (1|subj) $+$ (1|item)
Natural speech looks $\sim$ acoustic-factor*cond $+$ (1|subj)

These equations were used to answer the question: *Is a given acoustic factor a significant predictor of looks to the desired area of interest?* We built models where each factor was considered separately, instead of a single model with all these factors considered together, since the individual factors were not strongly correlated with each other. All the analyses were done by considering the interaction between acoustic factor and condition as predictors of looks to the target. Such an interaction term was considered because many of the acoustic factors were very related to the experimental condition in which they were used (see Table 12.1); for example, L+H* items have consistently higher adjective F0 peaks compared to the H* items.

Tables 12.5 and 12.6 show the results of GLMM-based acoustic analyses for synthetic speech facilitation and natural speech garden-pathing, respectively. For synthetic speech contrastive sequences, longer L+H*-accented adjectives induced more looks to the area of interest. This lends credence to the hypothesis that given more processing time (as in longer adjectives), listeners would be able to integrate adjectival

Table 12.5. GLMMs testing the effect of acoustic factors on looks to the target in the synthetic speech facilitation case

Factor	Coeff	z-value
adjDur	0.001633	0.5702, $p < 0.05$
condL+H*	−3.591332	−2.3068, $p < 0.05$
adjDur:condL+H*	**0.010052**	2.4045, $p < 0.05$
F0Drop	−0.006239	−0.4679
condL+H*	10.170012	2.7537, $p < 0.01$
F0Drop:condL+H*	**−0.050244**	−2.1107, $p < 0.05$

Table 12.6. GLMMs testing the effect of acoustic factors on looks to the competitor in the natural speech garden-pathing case

Factor	Coeff	z-value
adjDur	−0.013232	−3.392, $p < 0.001$
condL+H*	−4.778460	−2.963, $p < 0.01$
adjDur:condL+H*	**0.017399**	3.348, $p < 0.001$
F0Drop	−0.05591	−2.291, $p < 0.05$
condL+H*	0.41848	0.231
F0Drop:condL+H*	**0.05436**	2.089, $p < 0.05$

information more effectively to constrain upcoming nominal referent choices, and early looks to the target would then be visible at the onset of the noun itself. A preliminary item analysis revealed that duration was related to adjective identity: synthetic speech items with longer adjectives *yellow* (411 ms) and *green* (368 ms) did exhibit a facilitation trend, as opposed to *red* (287 ms) which did not exhibit this trend (Figure 12.8).

The analysis also shows that for synthetic speech, larger F0 drops induced fewer looks to the target in contrastive sequences. This leads to the possibility that items with large F0 drop values were considered to be unnatural and distracted listeners from focusing on the desired nominal referent. For natural speech garden-pathing, Table 12.6 also shows that longer adjectives induced more looks to the contrastive competitor, but here greater F0 drops induced more looks to this area of interest. Thus, as might be expected for natural speech, L+H* items with longer adjectives and large F0 drop values – and thus more readily perceived contrastive tunes – induced more looks to the competitor in infelicitous sequences. Similar analyses revealed that none of the other acoustic factors we looked at were significant in predicting natural speech facilitation or synthetic speech garden-pathing. In particular, for synthetic speech, join cost (the one automatic measure of acoustic fit we considered) was not a significant predictor of any of the effects discussed above.

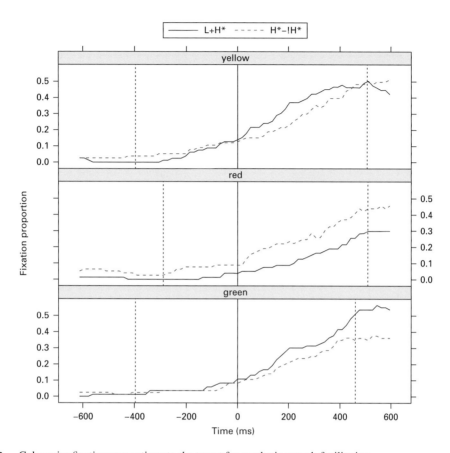

Figure 12.8   Color-wise fixation proportions to the target for synthetic speech facilitation

12.6.2   Relationship between ratings and looks

To investigate whether a context-independent measure of quality could predict processing-based effects, we analyzed the relationship between the quality ratings from the offline rating task and looks to the relevant areas of interest in the eye-tracking experiment. Table 12.7 shows the results of the GLMM analyses for items in a contrastive sequence. Note that for both kinds of speech, the first row shows the analysis of rating by itself, while the second row shows the analysis of rating in its interaction with the tune condition. For synthetic speech, rating by itself was not a significant predictor of looks to the target. This is probably because of the high correlation between rating and condition (condition does not predict looks to the target for synthetic speech facilitation), as seen in Table 12.3, where L+H* items are rated lower across the board. But the rating–condition interaction actually predicts looks to the target. The positive coefficient estimated by the regression model indicates that higher-rated items induced more looks to the target in the L+H* condition. Since the independent quality ratings for the class of items rated as least natural predicted facilitatory looks to the target

Table 12.7. GLMMs testing the effect of rating on looks to the target in the facilitation case

Speech	Factor	Coeff	z-value
Synthetic	rating	−0.21359	−1.30
	rating	−1.3554	−2.806, $p < 0.001$
	condL+H*	−7.2991	−2.288, $p < 0.05$
	rating:condL+H*	**1.4334**	2.109, $p < 0.05$
Natural	rating	−0.1783	−1.6023
	rating	−0.04947	−0.1798
	condL+H*	−1.09159	−0.4253
	rating:condL+H*	0.29348	0.6000

Table 12.8. GLMMs testing the effect of rating on looks to the competitor for garden-pathing

Speech	Factor	Coeff	z-value
Synthetic	**rating**	**−0.6577**	−3.639, $p < 0.001$
	rating	−0.5529	−1.5310
	condL+H*	−0.9457	−0.3996
	rating:condL+H*	0.3700	0.6972
Natural	**rating**	**−0.4779**	−3.441, $p < 0.001$
	rating	−0.7097	−1.5487
	condL+H*	1.0430	0.2908
	rating:condL+H*	−0.3258	−0.4959

during the processing of synthetic speech, we might have expected the ratings to predict facilitation in the case of the better-rated natural speech items as well. However, for natural speech neither the rating nor the rating–condition interaction was a significant predictor of looks to the target, suggesting that the natural speech items all sound natural in context. In the case of garden-pathing, Table 12.8 shows the corresponding GLMM analyses. Here, although rating by itself is a significant predictor of looks to the competitor – which is not surprising, given that lower ratings are indicative of the L+H* condition – the analysis that considers the interaction between rating and condition does not significantly predict looks to the target for either type of speech.

12.6.3   Correlation between rating and acoustic factors

In light of our findings about F0 drop and adjective duration influencing processing of synthetic speech (described in Section 12.6.1), we examined whether specific acoustic factors also influenced the offline ratings, in order to see whether the same factors were at work in both situations. The quality ratings were highly correlated with F0 drop ($r = −0.76$ for synthetic speech and $r = −0.87$ for natural speech; both at $p < 0.001$), while adjective duration was not correlated with the ratings ($r = 0.07$ for synthetic speech and $r = 0.01$ for natural speech; both at $p < 0.001$). Thus F0 drop affected both online processing and the offline naturalness ratings, while adjective duration was a factor specific to online processing. Now, with F0 drop, since larger values are indicative of the [L+H* no-accent] tune, the negative correlation with the offline ratings may simply have been the result of the absence of context making it difficult to judge the naturalness of utterances with marked prosody, as mentioned before. But, as the results of the eye-tracking experiment show, in the facilitation case of synthetic speech, L+H*-accented adjectives with larger F0 drops hindered looks to the target. Thus, to confirm that F0 drop is a relevant factor influencing the quality of items as revealed by the offline rating task, we also need the results from the eye-tracking experiment. On the other hand, adjective duration was not correlated with the ratings in any of the conditions, while it is a factor that is relevant to the processing of synthetic speech. Thus, while adjective duration has an important effect on the time course of processing, it does not appear to be a factor related to the quality of the speech items.

12.7 Discussion

A close look at the results of the eye-tracking experiment and the rating study suggests that online measures of unconscious processing and offline measures of conscious judgments together contribute to a more comprehensive evaluation of synthetic speech than is possible using either method alone. The acoustic analysis of the eye-tracking stimuli revealed that larger values of F0 drop in a [L+H* no-accent] contrastive discourse sequence hinder processing of synthetic speech. The ratings study confirms the effect of F0 drop, since synthetic items with larger F0 drop values were given lower ratings (as were natural items with larger F0 drop). The fact that both studies demonstrate the influence of F0 drop points to the conclusion that this is a factor meriting attention in future speech synthesis work.

The analysis also brings to light the fact that longer adjectives facilitate more looks to the target for synthetic speech in [L+H* no-accent] contrastive sequences. Consistent with this finding, we noted in an item-specific analysis of the synthetic speech stimuli that longer color adjectives tended to facilitate more looks to the target. However, adjective duration was not seen to be a factor having any bearing on the offline ratings, suggesting that longer adjectives are simply providing more time for listeners to exhibit anticipatory effects of contrastive prosody.

The ratings study indicates that the synthetic stimuli used in the eye-tracking experiment are of higher quality than typical unit selection speech, as is evident from the difference in ratings between the tree decorating data and the directions domain data. The analysis that looks at both types of measure reveals that higher synthetic speech ratings do indicate increased looks to the target in the facilitation case. It is interesting to note, however, that the ratings are not otherwise predictive of eye-tracking effects, at least when their interaction with condition is taken into account. This could be because context-independent ratings do not provide an accurate picture of the quality of items, especially in the case of contrastive prosody presented out of context. Thus the studies demonstrate that to make a statement about the quality of synthetic speech, one can obtain a more complete picture by relying on the results of the eye-tracking experiment as well as the offline rating study.

12.8   Conclusions

In this chapter, we have presented an experiment that indicates that eye tracking has the potential to be a highly sensitive and objective method for the online evaluation of prosody in synthetic speech. Even with high-quality unit selection synthesis, the results failed to replicate the facilitative effect of contextually appropriate accent patterns found with human speech, while producing a more robust intonational garden-path effect than natural speech with contextually inappropriate patterns. Since we observed processing delays with the synthetic speech that could explain the absence of facilitation, we suggest that experimental designs for eye-tracking evaluations should make allowances for processing delays, thereby providing sufficient time for any facilitative effects to arise.

An acoustic analysis revealed that, for synthetic speech, adjectives with longer durations were associated with an increase in looks to the target in felicitous trials. This lends credence to our conjecture that more processing time may simply be needed to see facilitative effects with synthetic speech, in order to offset the observed delays in processing synthetic speech compared to natural speech. Using an offline rating task, we found that synthetic items that were rated as more natural also produced more facilitation in looks to the target in the online eye-movement monitoring experiment, suggesting that the less natural items were hindering the effect of condition observed with the natural speech data. We also found that larger drops in F0 between the accented adjective and following noun in the case of synthetic speech resulted in reduced looks to the target. This suggests that unnaturally large F0 drops may be a specific acoustic factor that hinders online processing, and thus merits specific attention in future work on improving synthesis quality.

Somewhat to our surprise, we found that subjects had difficulty in rating utterances with marked intonation out of context, as items with the [L+H* no-accent] tune were rated much lower than the [H* !H*] items even for natural speech, where subjects were essentially at chance in guessing whether the [L+H* no-accent] items were synthetic. For this reason especially, we found that a full understanding of the role that F0 drop played in synthesis quality could only be gained by examining the online and offline results together, as the eye-tracking results with natural speech gave no indication that the [L+H* no-accent] items were unnatural in context, and there were no negative effects of F0 drop on processing (unlike with the synthetic items).

Beyond unnaturally large F0 drops, we suspect that the presence of even barely noticeable joins in the synthesized speech were responsible for the observed processing delays in the eye-tracking experiment, given the multi-dimensional scaling results of Mayo *et al.* (2011) that listeners place a great deal of perceptual importance on the presence of artifacts and discontinuities in the speech. For this reason, it would be interesting in future work to carry out eye-tracking experiments using statistical parametric synthesis techniques (Zen *et al.*, 2009), which are much less prone to introducing such artifacts in comparison to unit selection synthesis. In addition, it would be interesting to design an eye-tracking experiment with more items and specifically controlling for F0 drop and adjective duration, the two acoustic factors that the present study brought to light, and examine the eye-tracking effects observed.

Acknowledgments

We thank Cynthia Clopper and the OSU Clippers and Speerlab discussion groups for providing feedback, Ping Bai for assistance with data analysis, Dominic Espinosa for providing the filler stimuli of the offline rating task, Laurie Maynell for serving as our voice talent, Ross Metusalem for assistance with ToBI-annotating our spoken stimuli, and Rob Clark for help with Festival. This work was supported in part by an OSU Arts & Humanities Innovation Grant.

References

Baker, R. E. (2003). *Using Unit Selection to Synthesise Contextually Appropriate Intonation in Limited Domain Synthesis.* Master's thesis, Department of Linguistics, University of Edinburgh.

Beckman, M. E., Hirshberg, J., and Shattuck-Hufnagel, S. (2005). The original ToBI system and the evolution of the ToBI framework. In Jun, S.-A., editor, *Prosodic Typology: The Phonology of Intonation and Phrasing.* Oxford University Press, Oxford, UK.

Black, A. and Lenzo, K. (2000). Limited domain synthesis. In *Proceedings of the International Conference on Spoken Language Processing (INTERSPEECH)*, pages 411–414, Beijing, China. International Speech Communication Association.

Brenier, J., Nenkova, A., Kothari, A., Whitton, L., Beaver, D., and Jurafsky, D. (2006). The (non)utility of linguistic features for predicting prominence in spontaneous speech. In *Proceedings of the IEEE Spoken Language Technology Workshop*, pages 54–57, Palm Beach, FL. Institute of Electrical and Electronics Engineers.

Bulyko, I. and Ostendorf, M. (2002). Efficient integrated response generation from multiple targets using weighted finite state transducers. *Computer Speech and Language*, **16**(3–4): 533–550.

Clark, R. A. J., Richmond, K., and King, S. (2007). Multisyn: Open-domain unit selection for the Festival speech synthesis system. *Speech Communication*, **49**(4):317–330.

De Carolis, B., Pelachaud, C., Poggi, I., and Steedman, M. (2004). APML, a markup language for believable behavior generation. In Prendinger, H. and Ishizuka, M., editors, *Life-like Characters. Tools, Affective Functions and Applications*, pages 65–85. Springer, Berlin, Germany.

Espinosa, D., White, M., Fosler-Lussier, E., and Brew, C. (2010). Machine learning for text selection with expressive unit-selection voices. In *Proceedings of the International Conference on Spoken Language Processing (INTERSPEECH)*, pages 1125–1128, Makuhari, Chiba, Japan. International Speech Communication Association.

Hitzeman, J., Black, A. W., Mellish, C., Oberlander, J., and Taylor, P. (1998). On the use of automatically generated discourse-level information in a concept-to-speech synthesis system. In *Proceedings of the International Conference on Spoken Language Processing (ICSLP)*, Sydney, Australia. International Speech Communication Association.

Ito, K. and Speer, S. R. (2008). Use of L+H* for immediate contrast resolution. In *Proceedings of Speech Prosody*, Campinas, Brazil. Speech Prosody Special Interest Group.

Ito, K. and Speer, S. R. (2011). Semantically-independent but contextually-dependent interpretation of contrastive accent. In Frota, S., Elordieta, G., and Prieto, P., editors, *Prosodic Categories: Production, Perception and Comprehension*, pages 69–92. Springer, Dordrecht, The Netherlands.

Karaiskos, V., King, S., Clark, R. A. J., and Mayo, C. (2008). The Blizzard Challenge 2008. In *Proceedings of the Blizzard Challenge (in conjunction with the ISCA Workshop on Speech Synthesis)*, Brisbane, Australia. Blizzard.

King, S. and Karaiskos, V. (2009). The Blizzard Challenge 2009. In *Proceedings of the Blizzard Challenge*, Edinburgh, Scotland. University of Edinburgh.

Kruijff-Korbayová, I., Ericsson, S., Rodríguez, K. J., and Karagjosova, E. (2003). Producing contextually appropriate intonation in an information-state based dialogue system. In *Proceedings of the Conference of the European Chapter of the Association for Computational Linguistics (EACL)*, pages 227–234, Budapest, Hungary. Association for Computational Linguistics.

Mayo, C., Clark, R. A. J., and King, S. (2011). Listeners' weighting of acoustic cues to synthetic speech naturalness: A multidimensional scaling analysis. *Speech Communication*, **53**(3): 311–326.

Moore, J. D., Foster, M. E., Lemon, O., and White, M. (2004). Generating tailored, comparative descriptions in spoken dialogue. In *Proceedings of the Florida Artificial Intelligence Research Society Conference (FLAIRS)*, pages 917–922, Miami Beach, FL. The Florida Artificial Intelligence Research Society.

Nakatsu, C. and White, M. (2006). Learning to say it well: Reranking realizations by predicted synthesis quality. In *Proceedings of the Annual Meeting of the Association for Computational Linguistics (ACL)*, pages 1113–1120, Sydney, Australia. Association for Computational Linguistics.

Pan, S., McKeown, K., and Hirschberg, J. (2002). Exploring features from natural language generation for prosody modeling. *Computer Speech and Language*, **16**:457–490.

Prevost, S. (1995). *A Semantics of Contrast and Information Structure for Specifying Intonation in Spoken Language Generation*. PhD thesis, University of Pennsylvania.

Rajkumar, R., White, M., Ito, K., and Speer, S. R. (2010). Evaluating prosody in synthetic speech with online (eye-tracking) and offline (rating) methods. In *Proceedings of the ISCA Workshop on Speech Synthesis (SSW)*, pages 276–281, Kyoto, Japan. International Speech Communication Association.

Speer, S. R. (2011). Eye movements as a measure of spoken language processing. In Cohn, A. C., Fougeron, C., and Huffman, M. K., editors, *The Oxford Handbook of Laboratory Phonology*, pages 580–592. Oxford University Press, Oxford, UK.

Swift, M. D., Campana, E., Allen, J. F., and Tanenhaus, M. K. (2002). Monitoring eye movements as an evaluation of synthesized speech. In *Proceedings of the IEEE Workshop on Speech Synthesis*, pages 19–22, Santa Monica, CA. Institute of Electrical and Electronics Engineers.

Theune, M. (2002). Contrast in concept-to-speech generation. *Computer Speech and Language*, **16**(3–4):491–531.

van Hooijdonk, C., Commandeur, E., Cozijn, R., Krahmer, E., and Marsi, E. (2007). The online evaluation of speech synthesis using eye movements. In *Proceedings of the ISCA Workshop on Speech Synthesis*, pages 385–390, Bonn, Germany. International Speech Communication Association.

van Santen, J., Kain, A., Klabbers, E., and Mishra, T. (2005). Synthesis of prosody using multi-level unit sequences. *Speech Communication*, **46**(3–4):365–375.

White, M., Clark, R. A. J., and Moore, J. (2010). Generating tailored, comparative descriptions with contextually appropriate intonation. *Computational Linguistics*, **36**(2):159–201.

White, M., Rajkumar, R., Ito, K., and Speer, S. R. (2009). Eye tracking for the online evaluation of prosody in speech synthesis: Not so fast! In *Proceedings of the International Conference on Spoken Language Processing (INTERSPEECH)*, pages 2523–2526, Brighton, UK. International Speech Communication Association.

Zen, H., Tokuda, K., and Black, A. W. (2009). Statistical parametric speech synthesis. *Speech Communication*, **51**(11):1039–1064.

13 Comparative evaluation and shared tasks for NLG in interactive systems

Anja Belz and Helen Hastie

13.1 Introduction

Natural Language Generation (NLG) has strong evaluation traditions, in particular in the area of user evaluation of NLG-based application systems, as conducted for example in the M-PIRO (Isard *et al.*, 2003), COMIC (Foster and White, 2005), and SumTime (Reiter and Belz, 2009) projects. There are also examples of embedded evaluation of NLG components compared to non-NLG baselines, including, e.g., the DIAG (Di Eugenio *et al.*, 2002), STOP (Reiter *et al.*, 2003b), and SkillSum (Williams and Reiter, 2008) evaluations, and of different versions of the same component, e.g., in the ILEX (Cox *et al.*, 1999), SPoT (Rambow *et al.*, 2001), and CLASSiC (Janarthanam *et al.*, 2011) projects. Starting with Langkilde and Knight's work (Knight and Langkilde, 2000), automatic evaluation against reference texts also began to be used, especially in surface realization. What was missing, until 2006, were comparative evaluation results for directly comparable, but independently developed, NLG systems.[1]

In 1981, Spärck Jones wrote that information retrieval (IR) lacked consolidation and the ability to progress collectively, and that this was substantially because there was no commonly agreed framework for describing and evaluating systems (Sparck Jones, 1981, p. 245). Since then, various sub-disciplines of natural language processing (NLP) and speech technology have consolidated results and progressed collectively through developing common task definitions and evaluation frameworks, in particular in the context of shared-task evaluation campaigns (STECs), and have achieved successful commercial deployment of a range of technologies (e.g. speech recognition software, document retrieval, and dialogue systems).

However, as recently as 2005, Spärck Jones's 1981 analysis could still be said to hold of NLG. Pre-2006, there was virtually no comparative evaluation in NLG; there were no commonly agreed task definitions (with the possible exception of referring expression generation); there were no commonly agreed frameworks for evaluating systems; there had been little consolidation of results or collective progress; and there still was virtually no commercial deployment of NLG systems or components.

The past six years have seen big changes in the NLG field. In 2005, a discussion began about sharing data and system components, and how to make it possible to directly

---

[1] Where we use the term comparative evaluation in this chapter we have this sense in mind, i.e., evaluating directly comparable, but independently developed, systems.

compare independently developed systems and components. Informal discussions at the 2005 NLG workshops, The European Workshop on Natural Language Generation (ENLG) and the Workshop on Using Corpora in Natural Language Generation (UCNLG), were followed by a Special Session on Sharing Data and Comparative Evaluation at the International Natural Language Generation Conference (INLG) in 2006. The papers at this event[2] showed clearly that while some NLG researchers were enthusiastic about the introduction of shared tasks in NLG (Reiter and Belz, 2006), there was also deep unease in parts of the NLG community, about competitive evaluation, in particular from Di Eugenio (2007) and Scott and Moore (2007). One concern was that shared-task competitions would introduce de facto standards for task definitions, inputs, and evaluation methods, and would end up marginalizing and stifling research that did not conform to the standards. Both sides of the discussion came together at the 2007 NSF Workshop on Shared Tasks and Comparative Evaluation in NLG,[3] where some of the break-out working groups started hammering out the details of potential shared tasks, while another working group began drafting guidelines for NLG evaluation that would avoid the adverse side effects people were worried about.

All four specific task proposals that were first presented at the NSF workshop have since come to fruition (TUNA-REG, GREC, GIVE, and QG). A further two shared tasks have been run (SR and HOO), and two more have been proposed and are likely to run in the near future (SR-Spanish and GRUVE).[4] That makes a total of eight NLG STECs in five years, ranging from referring expression generation to instruction generation in virtual environments. Many STECs evaluate systems implementing NLG subtasks (TUNA, several of the GREC tasks, SR), but end-to-end systems have also been covered (GREC-Full, HOO), and one STEC centered around evaluating NLG components embedded in an interactive system (GIVE); we will come back to these three different evaluation contexts in the following section. The variety of tasks has been mirrored by variety in evaluation methods: NLG STECs have employed a wide range of extrinsic and intrinsic, automatic, and human-assessed evaluation methods.

The overall benefits from the shared tasks have been substantial: legacy task definitions, data, and evaluation resources have been made available to the wider community[5] and have led to a great deal of follow-on research and associated publications. Many researchers from outside traditional NLG have been drawn in as a result of the shared tasks (Basile and Bos, 2011; Boyd and Meurers, 2011), comparative evaluation is increasingly expected, and the Generation Challenges sessions have been part and parcel of the alternating INLG and ENLG events since 2008.

Outside of the context of the Generation Challenges STECs and of related research that uses legacy tasks and data from these STECs, comparative evaluation continues to be less prevalent. To date, the only other examples of evaluations involving directly comparable, independently developed end-to-end generation systems all use

---

[2] http://www.itri.brighton.ac.uk/~Anja.Belz/inlg06-specsess.html.

[3] All position papers presented at the workshop can be found at: http://www.ling.ohio-state.edu/nlgeval07/accepted.html.

[4] All eight tasks are surveyed in later sections of this chapter.

[5] See https://sites.google.com/site/genchalrepository/.

the SumTime data (Belz and Reiter, 2006; Belz and Kow, 2009; Angeli *et al.*, 2010; Langner, 2010). As for evaluation of subtasks, several groups of researchers have reported results for regenerating the Penn Treebank prior to the SR Task (Langkilde-Geary, 2002; Callaway, 2003; Nakanishi *et al.*, 2005; Zhong and Stent, 2005; Cahill and van Genabith, 2006; White *et al.*, 2007; White and Rajkumar, 2009), although the systems were not directly comparable as they used inputs of different levels of specificity.

NLG can sometimes look like the poor cousin of NLA (Natural Language Analysis), especially because two sizable NLP subfields that used to comprise NLG as a subtask (MT and summarization) for the most part no longer do, and are no longer even seen as overlapping with NLG. The advent of comparative and competitive evaluation has the potential to change the status of NLG, and the next five years will be crucial in this respect. Important trends are likely to be toward reusable plug-and-play system components, and toward applications. Next to pure data-to-text applications, NLG components embedded in interactive systems are likely to provide some of the most important opportunities for technological advances. Comparative evaluations and shared-task competitions can help to further increase the critical mass of the research field and speed up progress in developing NLG for applications.

In this chapter, our aim is to take a look at existing evaluation methods and shared tasks in NLG and interactive systems, and to use this as the basis for a discussion of how to evaluate NLG within interactive systems, outlining requirements and sketching possible future shared tasks. The structure of the chapter is as follows. In Section 13.2, we present an overview of different types of evaluation in a categorization framework that is particularly suited to NLG in the context of interactive systems and that will provide us with a standard terminology to describe specific examples of evaluations in the rest of this chapter. We survey evaluation in NLG in Section 13.3, and evaluation in interactive systems in Section 13.4, before bringing the two themes together in Section 13.5, where we discuss requirements and methods for evaluating NLG embedded in interactive systems, and describe two possible future shared tasks.

13.2   A categorization framework for evaluations of automatically generated language

13.2.1   Evaluation measures

In Table 13.1, we present an overview of the categorization framework of evaluation types we discuss in this section, populated with the systems and shared tasks we survey in Sections 13.3 and 13.4. The categories reflect *what* is being evaluated (indexing the rows), and *how* it is being evaluated (indexing the columns).

Along the *what*-dimension, we distinguish (a) **evaluation of components in isolation**, i.e., evaluations that do not involve the embedding systems or modules (examples of NLG components that have been evaluated in isolation include referring expression generation (REG) modules and surface realizers); (b) **evaluation of components**

Table 13.1. Overview of the system evaluations (represented by name of project or shared task) that we discuss in Sections 13.3 and 13.4, categorized according to *what* is being evaluated (components in isolation, embedded components, end-to-end systems), and *how* it is evaluated (evaluation mode and type of measure)

		HOW			
		Intrinsic		Extrinsic	
Evaluation mode:					
Measure type:		Output quality	User like	User task success	System purpose success
WHAT	**Components in isolation**	TUNA, GREC, SR'11, Prodigy-METEO	?	GREC-MSR, TUNA-AS, TUNA-REG	?
	Embedded components	SkillSum, GIVE, SPOT, M-PIRO, COMIC, GRUVE	SkillSum, GIVE, ILEX, M-PIRO, CLASSiC, GRUVE	SkillSum, GIVE, ILEX, M-PIRO, CLASSiC, GRUVE	SkillSum, ILEX, GIVE
	End-to-end systems	GREC-Full, SumTime-METEO, QG'10, BT-Nurse, HOO	BT-Nurse	SumTime-METEO	STOP

embedded in systems, where evaluation is at the system level (e.g., assessing the effect alternative NLG front-ends have on the quality/performance of the same embedding interactive system); and (c) **evaluation of stand-alone, end-to-end systems**, which are treated as black boxes for the purposes of the evaluation (e.g., evaluation of weather forecast generation systems).

Along the *how*-dimension we distinguish two evaluation modes: **intrinsic** measures assess properties of systems or components in their own right, for example comparing their outputs to model outputs in a corpus, whereas **extrinsic** measures assess the effect of a system on something that is external to it, for example the effect on human performance for a given task or the value added to an application (Sparck Jones, 1994).

We further distinguish the following two subcategories of intrinsic types of evaluation measures:

1. **Output quality measures:** Measures of this type do not involve users, but assess the outputs produced by systems/components directly, measuring some aspect of the outputs (similarity to human-produced model outputs, brevity, grammaticality, etc.) either automatically or by trained human assessors.

a. *Automatic:* Automatic output quality measures either compare word strings to model outputs (e.g., BLEU [Papineni *et al.*, 2002], NIST [Doddington, 2002], string edit distance), or compute a property of the generated text directly (text length, parsability, Flesch reading ease score, etc.).

b. *Human-assessed:* Intrinsic human output quality evaluation usually takes the form of running an experiment in which trained evaluators[6] are asked to assess the quality of output texts in terms of a given set of quality criteria, again either comparing to model outputs (meaning similarity, Human TER) or to properties of the generated text (fluency, clarity, appropriateness).

2. **User like measures:** Measures in this category cannot by definition be assessed automatically; collecting user like measures involves asking users (as opposed to trained evaluators) for their (necessarily subjective) assessment of given aspects of a system or component. This can take the form of questions such as "How helpful did you find the reports?" (SkillSum[7]), or "Would you recommend this game to a friend?" (GIVE), as well as the more direct "Did you like the system?" or "Which of these two systems did you like better?".

Note that we do not mean to imply that output quality measures assess the actual goodness of outputs in some objective way. Rather, deciding what constitutes "quality" is an integral part of designing systems and evaluations. Depending on the context, quality may appropriately be taken to mean brevity, similarity to a given set of human-produced model outputs, or greater use of high-frequency words (among many other possibilities).

Among the extrinsic types of evaluation measures we distinguish the following two subcategories:

1. **User task success measures:** Measures of this type assess some measurable aspect of user actions and behaviors; this can include reading speed (TUNA, GREC), comprehension accuracy (GREC), or completion of an application-specific task such as finding a treasure (GIVE). User task success measures by definition involve a human in the loop, although they are typically computed unobtrusively, in the course of monitoring user behavior, or from logs recorded during user–system interaction.

2. **System purpose success measures:** Measures that assess system purpose success take as a starting point the explicitly specified intended purpose of a given system or module and seek to measure the extent to which the system or component fulfills it. A classic example in NLG is the STOP system (Reiter *et al.*, 2003a) whose purpose was to generate personalized smoking cessation letters. The project team was able to collect statistics regarding how many people actually did stop smoking after receiving STOP-generated letters, compared to a non-personalized baseline (see also below). Carrying out such controlled experiments is not always possible. For example, it is hard to imagine how it might be done in the case of the SumTime-METEO system, whose purpose is to generate maritime weather forecasts that enable companies

---

[6] Training can simply take the form of a detailed explanation of a quality criterion and how to apply it.

[7] Summaries of SkillSum and the other projects mentioned in this section can be found in Section 13.3 below.

running drilling platforms in the North Sea to adapt activity according to the expected weather.[8]

System purpose success measures have a slightly different status from the other measure categories, in that it is possible for a system purpose success measure to coincide with a measure of another type. For example, a system's purpose can be simply to entertain, and a simple way of assessing this is to ask the user directly whether they are feeling entertained (as in the GIVE Shared Task, see below), which is a user like measure.

Clearly, it is important to consider precisely what it is that is being measured in an evaluation experiment. If users are asked in an exit questionnaire how clear/readable/reliable they found a system's instructions (and especially if the same questionnaire also asks for explicitly subjective judgments such as "would you recommend the system to a friend"), then one cannot conclude from high ratings that the system's instructions *are* indeed clear and readable, but only that users *felt* that they were (so this would count as user like measures). Providing evaluators with detailed definitions of clarity/readability/reliability and instructions about how to apply them (resulting in a human-assessed intrinsic output quality measure) might yield very different results. User judgments are always likely to be colored by personality and personal experience with the system, both of which one would aim to eliminate (to the extent that this is possible) in an assessment of system quality.

In any of the evaluation types, a system-level score can be obtained either by directly assigning scores to systems, or indirectly, by assessing (the effect of) outputs for a given set of inputs and then deriving a system-level score from the output scores (e.g., by computing the mean). In principle, any of the evaluations can involve single systems or sets of systems, but, for example, intrinsic assessments of output quality are most appropriately regarded as producing a system ranking (thus requiring multiple systems), rather than individually meaningful scores.

13.2.2   Higher-level quality criteria

The taxonomy of evaluation measures represented by the columns in Table 13.1 is a generic one, and the idea is that any given evaluation method can, independently of the context in which it is developed/applied, be assigned to one of the columns. For example, a measure that computes the percentage of times users play a game to completion is simply an extrinsic measure – more specifically it is simply a task-success measure – regardless of the type of system that is being evaluated or the purpose of the evaluation.

Designers of evaluation experiments often talk about assessing higher-level **quality criteria** such as dialogue performance, dialogue quality, system usability, user satisfaction, efficiency, etc. It does not make sense to pair up individual evaluation measures with single abstract, higher-level quality criteria. Rather, the decision about which measure(s) to use to assess a given quality criterion is part and parcel of the design of

---

[8] Platform managers are unlikely to agree to testing (potentially imperfect) automatic weather forecasting systems if it means running the risk of being ill-prepared for bad weather, and consequently of accidents.

evaluations and is very much context-dependent. For example, it would not make sense to say that the percentage of times users play a game to completion is, generically, a user satisfaction measure. However, in the context of evaluating a specific type of system, a system or evaluation designer may well decide to use it in this capacity.

What we wish to highlight here is the distinction between generic types of evaluation measures (columns in Table 13.1) and evaluation criteria (what the evaluation measures are used as a surrogate measure for). In Section 13.5, we describe a set of evaluation criteria for embedded evaluation of NLG in interactive systems and suggest specific evaluation measures for assessing them.

13.2.3   Evaluation frameworks

For comprehensive comparative evaluation of systems or components, multiple evaluation methods, moreover in more than one of the categories of evaluation methods discussed above, tend to be applied. This has been the case for all NLG Shared Task Evaluations to date, and also for the NIST-run MT and Summarization evaluations. This approach yields a set of evaluation scores and corresponding ranking of systems for each evaluation measure that is applied, and it frequently happens that different systems come top of the ranking table for different measures. This means that there tends not to be an overall winning system (that has been found acceptable, or even beneficial, in most of the shared-task evaluations mentioned above), unless one of the measures is identified as the "lead" measure (e.g., BLEU in MT-Eval).

In some contexts it is found desirable that an overall winning system should be identified. Deciding on an overall winner on the basis of a single measure is not entirely satisfactory because it is rare that a single measure can be found that encapsulates all aspects of system quality. This is particularly true of complex systems comprising multiple components, and of evaluation situations where stakeholders other than system developers (end users, funding bodies seeking to accelerate technological progress, companies that will turn systems into marketable products, etc.) play a role. Here, **evaluation frameworks** offer a solution: evaluation frameworks compute a single overall score on the basis of the separate scores produced by multiple individual measures. The chief purpose of designing such frameworks is to decide on a set of quality criteria that collectively describe all aspects of system quality identified as important in a given evaluation situation, to then select a set of evaluation measures that capture the criteria, and also to quantify the relative importance of each measure, for example by assigning weights to them.

For example, the overall goal of a dialogue system evaluation may be to assess "Dialogue Performance." The evaluation designers may decide that Dialogue Performance is comprised of the quality criteria Dialogue Quality, User Satisfaction, and Usability. They select several intrinsic output quality measures to capture Dialogue Quality, several user like measures to put into exit questionnaires to capture User Satisfaction, and several task-success measures to capture Usability. Finally, the designers select a method for combining the selected measures into a single score; this method could be as simple as computing the (weighted) average of the measures, or more complex, involving, for example, methods from decision theory. We discuss several existing examples of evaluation frameworks in Section 13.4 below.

In Section 13.5, we use the term **evaluation model** to refer to a complete specification of the goals and implementation of a specific evaluation. We intend evaluation models to subsume evaluation frameworks as described above and in Section 13.5, but would recommend that all evaluations are based on explicit evaluation models, not just large-scale shared-task evaluations such as those for which the evaluation frameworks in Section 13.4 have been developed. In Section 13.5, we also propose a specific evaluation model for embedded evaluation of NLG in interactive systems.

13.2.4   Concluding comments

In this section, we presented an overview of different types of evaluation in a categorization framework that is particularly suited to NLG in the context of interactive systems, and we will use this standard terminology to describe specific examples of evaluations in the rest of this chapter. In the following section, we describe existing approaches to evaluation in NLG and summarize the shared tasks that have so far been run.

13.3   An overview of evaluation and shared tasks in NLG

NLG in 2011 is a very different field from what it was in 2005. One of the most important changes has been the introduction of systematic comparative evaluation methodologies. In this section, we provide a broad survey of evaluation methodologies in NLG. As outlined in the previous section, a primary dimension along which evaluation methods differ is whether the object of evaluation is (a) an NLG system module (which implements an NLG subtask) that is evaluated in its own right, (b) an end-to-end NLG system evaluated as a black box, or (c) an NLG system that is a module in an embedding interactive system and where evaluation is at the level of the embedding system. We use this subcategorization to structure this section, looking at evaluation methods and shared tasks in NLG component evaluation in Sections 13.3.1 and 13.3.2; in end-to-end NLG system evaluation in Sections 13.3.3 and 13.3.4; and in embedded NLG module evaluation in Sections 13.3.5 and 13.3.6.

13.3.1   Component evaluation: Referring Expression Generation

Referring Expression Generation (REG) is one of the most lively and thriving subfields of NLG. Traditionally, it addressed the following question:

[G]iven a symbol corresponding to an intended referent, how do we work out the semantic content of a referring expression that uniquely identifies the entity in question? (Bohnet and Dale, 2005, p. 1004)

The first REG competitions were organized against a background of growing interest in empirically evaluating REG algorithms, which had hitherto often been justified on the basis of theoretical and psycholinguistic principles but lacked a sound empirical grounding. Early empirical evaluations of REG algorithms either used existing corpora such as COCONUT (Gupta and Stent, 2005; Jordan and Walker, 2005), or constructed new datasets based on human experiments (Viethen and Dale, 2006; Gatt *et al.*, 2007; van der Sluis *et al.*, 2007). These evaluations focused on classic approaches to the REG problem, such as Dale's Full Brevity and Greedy algorithms (Dale, 1989), and Dale and Reiter's Incremental Algorithm (Dale and Reiter, 1995). In studies by Gupta and Stent (2005) and Jordan and Walker (2005), these algorithms were evaluated in a dialogue context and were extended with novel features to handle dialogue and/or compared to new frameworks such as Jordan's Intentional Influences model (Jordan, 2000).

The TUNA shared tasks

The TUNA shared tasks were run between 2007 and 2009 and focused on Referring Expression Generation (REG) in the definition given above. They were based on the TUNA Corpus (van Deemter *et al.*, 2012), and focused on the generation of full, identifying, definite noun phrases in visual domains where the entities are either furniture items or people.

The XML format used in the shared tasks was adapted slightly from the original corpus. Each corpus instance (an example is shown in Figure 13.1) in TUNA-STEC data is

```
<TRIAL ID="t101">
  <DOMAIN>
    <ENTITY ID="e1" TYPE="target">
      <ATTRIBUTE NAME="type" VALUE="sofa" />
      <ATTRIBUTE NAME="colour" VALUE="blue" />
      <ATTRIBUTE NAME="size" VALUE="large" /> ...
    </ENTITY>
    <ENTITY ID="e2" TYPE="distractor">
      <ATTRIBUTE NAME="..." VALUE="..." />
      ...
    </ENTITY> ...
  </DOMAIN>
  <WORD-STRING>
    the blue sofa
  </WORD-STRING>
  <ANNOTATED-WORD-STRING>
    the
    <ATTRIBUTE NAME="colour">blue</ATTRIBUTE>
    <ATTRIBUTE NAME="type">sofa</ATTRIBUTE>
  </ANNOTATED-WORD-STRING>
  <ATTRIBUTE-SET>
    <ATTRIBUTE NAME="colour" value="blue"/>
    <ATTRIBUTE NAME="type" value="sofa"/>
  </ATTRIBUTE-SET>
</TRIAL>
```

Figure 13.1   A TUNA corpus instance

an XML file that pairs a DOMAIN node with a human-written description identifying one of the entities (the *target*) from the remaining entities (the *distractors*). There are three representations for the human description (WORD-STRING, ANNOTATED-WORD-STRING, ATTRIBUTE-SET).

There were three TUNA shared tasks: TUNA-AS (focusing on attribute selection), TUNA-R (focusing on referring expression realization), and TUNA-REG (combining attribute selection and realization).

TUNA-AS

Input and output data: The TUNA-STEC data, made up of only the singular descriptions from the original TUNA corpus, was randomly divided into 60% training set, 20% development set, and 20% test sets, each set containing items from both the furniture and people domains.

Task definition: The TUNA-AS Task, organized in 2007 and 2008, focused on selecting the content for the description of a target entity; it required peers to develop a method that takes as input a DOMAIN, in which one ENTITY was the target, and return an ATTRIBUTE-SET consisting of a subset of the target attributes that distinguishes it from the other objects in the domain.

Evaluation: The *intrinsic, automatic output-quality measures* applied were (a) Dice and MASI (Passonneau, 2006), assessing the degree of overlap between ATTRIBUTE-SETs generated by a peer system and those produced by a human; (b) *accuracy*, the proportion of peer outputs that are identical to human outputs; and (c) *minimality*, the proportion of peer outputs that contain no more attributes than necessary to distinguish the target (Dale and Reiter, 1995).

Two *extrinsic user-task-success measures* were obtained in laboratory-based experiments in which both system and human-written ATTRIBUTE-SETs (realized as strings using purpose-built software) were compared in terms of (a) the time it took human subjects to read the descriptions and identify the entity being described, and (b) the proportion of identification errors they made.

TUNA-R

Input and output data: The training and development data consisted of the same TUNA corpus instances as used for TUNA-AS. A new test set, consisting of 112 corpus instances, was developed for this task using the same methodology as that used for the original TUNA corpus collection.

Task definition: The TUNA-R Task, organized in 2008 and 2009, required participants to develop a method that, given an ATTRIBUTE-SET representing the semantic content of an identifying description for a target entity, outputs a WORD-STRING, that is, a realization in English of the semantic representation.

Evaluation: Systems were evaluated using four *intrinsic, automatic output-quality measures*: accuracy (see above), string edit distance, BLEU, and NIST.

TUNA-REG

Input and output data: The training, development, and test data consisted of the same TUNA corpus instances as used for TUNA-AS.

Task definition: TUNA-REG, which was also organized in 2008 and 2009, can be seen as combining TUNA-AS and TUNA-R, in that it required participants to develop a method that, given an input DOMAIN, outputs an identifying description for the target referent, that is, a WORD-STRING (thus, such systems would need to select the ATTRIBUTEs and realize them, though not necessarily in two separate steps).

Evaluation: As in TUNA-R, peer and human outputs were compared by computing BLEU and NIST scores (*intrinsic, automatic output-quality measures*). *Intrinsic, human-assessed output-quality measures* were obtained in an experiment in which system-generated and human-produced outputs were judged by trained evaluators for their fluency and adequacy (Gatt and Belz, 2010). In addition, three *extrinsic user-task-success measures* were obtained in a lab-based experiment where human subjects read the descriptions and identified the entity being described: (a) the time it took to read the description; (b) the time it took subjects to identify the target referent given the description; and (c) the identification error rate.

The GREC shared tasks
GREC-MSR

Input and output data: The GREC Corpus (version 2.0) consists of roughly 2000 introductory sections in Wikipedia articles. In each text, three broad categories of Main Subject Reference (MSR) have been annotated (13 000 referring expressions in total). The GREC-MSR shared task version of the corpus was randomly divided into 90% training data (of which 10% was randomly selected as development data) and 10% test data.

Figure 13.2 shows one of the texts in the GREC-MSR training/development data set. REFs indicate an instance of referring, REFEX is the selected referring expression, and ALT-REFEX is a list of alternative REs for the referent. ALT-REFEX lists were generated for each text by an automatic method that collects all the (manually annotated) REs for the referent in the text and adds several defaults: pronouns and reflexive pronouns in all subdomains; and category nouns (e.g., *the river*), in all subdomains except people. Outputs generated by GREC-MSR systems are in the same format as the inputs, except that there are no ALT-REFEX lists and there is exactly one REFEX for each REF.

Task definition: The task in GREC-MSR is to develop a method for selecting one of the REFEXs in the ALT-REFEX list, for each REF in each TEXT in the test sets. The test data inputs are identical to the training/development data, except that REF elements contain only an ALT-REFEX list, not the preceding selected REFEX. The main objective in the 2009 GREC-MSR Task was to get the word strings contained in REFEXs right (whereas in REG'08 it was the REG08-TYPE attributes).

Evaluation: A wide range of *intrinsic, automatic output-quality measures* were applied: (a) accuracy of REFEX word strings, i.e., the proportion of REFEX word

```
<TEXT ID="36">
<TITLE>Jean Baudrillard</TITLE>
<PARAGRAPH>
  <REF ID="36.1" SEMCAT="person" SYNCAT="np-subj">
    <REFEX REG08-TYPE="name" EMPHATIC="no" HEAD="nominal" CASE="plain">
      Jean Baudrillard
    </REFEX>
    <ALT-REFEX>
      <REFEX REG08-TYPE="name" EMPHATIC="no" HEAD="nominal" CASE="plain">
        Jean Baudrillard
      </REFEX>
      <REFEX REG08-TYPE="name" EMPHATIC="yes" HEAD="nominal" CASE="plain">
        Jean Baudrillard himself
      </REFEX>
      <REFEX REG08-TYPE="empty">_</REFEX>
      <REFEX REG08-TYPE="pronoun" EMPHATIC="no" HEAD="pronoun" CASE="nominative">
        he
      </REFEX>
      <REFEX REG08-TYPE="pronoun" EMPHATIC="yes" HEAD="pronoun" CASE="nominative">
        he himself
      </REFEX>
      <REFEX REG08-TYPE="pronoun" EMPHATIC="no" HEAD="rel-pron" CASE="nominative">
        who
      </REFEX>
      <REFEX REG08-TYPE="pronoun" EMPHATIC="yes" HEAD="rel-pron" CASE="nominative">
        who himself
      </REFEX>
    </ALT-REFEX>
  </REF>
  (born June 20, 1929) is a cultural theorist, philosopher, political commentator,
  sociologist, and photographer.
  ...
</PARAGRAPH>
</TEXT>
```

Figure 13.2   Example text from the GREC-MSR training data

strings selected by a participating system that are identical to the one in the corpus; (b) accuracy of REG08-Type, i.e., the proportion of REFEXs selected by a participating system that have a REG08–TYPE value identical to the one in the corpus; (c) average accuracy of three automatic coreference resolution tools when applied to texts containing system-generated referring expressions; (d) string edit distance; (e) BLEU-3; and (f) NIST-5. The latter three string-comparison metrics assessed just the referring expressions selected by peer systems (ignoring the surrounding text).

Three *intrinsic, human-assessed output-quality measures* were assessed: Fluency, Clarity, and Coherence of referring expressions within the textual context, as described in the STEC results report (Belz *et al.*, 2009). In addition, *extrinsic user-task-success measures* were obtained in a reading/comprehension experiment in which the task for subjects was to read texts one sentence at a time and then to answer three brief multiple-choice comprehension questions after reading each text. The measures recorded were reading time, question-answering time, and question-answering accuracy.

GREC-NEG

Input and output data: The GREC'10 data is derived from the GREC-People corpus which (in its 2010 version) consists of 1100 annotated introduction sections from

Wikipedia articles in the category People, divided into training, development, and test data.

People mentions in the GREC-People texts were annotated by marking up the referring expressions word strings and annotating them with coreference information, semantic category, syntactic category and function, and various supplements and dependents. Annotations included nested references, plurals and coordinated referring expressions, certain unnamed references, and indefinites. For full details see the GREC'10 documentation (Belz, 2010).

The manual annotations were automatically checked and converted to XML format. The REF, REFEX, and ALT-REFEX elements were the same as in the GREC-MSR annotations described above, except that here all alternative referring expressions are collected in a single list, appended at the end of the text, rather than to each reference. Also, arbitrary-depth embedding of references is allowed.

The training, development, and test data for the GREC-NEG task is exactly as described above. The test data inputs are the same, except that REF elements in the test data do not contain a selected REFEX element.

Task definition: The task in GREC-NEG is to select one REFEX from the ALT-REFEX list for each REF in each TEXT in the test set, including any embedded REFs. The aim is to select referring expressions that make the text fluent, clear, and coherent.

Evaluation: The *intrinsic, automatic output-quality measures* applied were: (a) REG08-Type Precision, i.e., the proportion of REFEXs selected by a participating system which match the reference REFEXs; (b) REG08-Type Recall, i.e., the proportion of target REFEXs for which a participating system has produced a match; and (c) String Accuracy, i.e., the proportion of word strings selected by a participating system that match those in the reference texts.

GREC-NER

Input and output data: The GREC-NER training and development data come in two versions. The first is identical to the format described above (containing information about correct system outputs). The second is the test data input format, where texts do not have REFEXs, REFs, or ALT-REFEXs. Moreover, a proportion of REFEXs have been replaced with standardized named references. System outputs have the same format as test data inputs, plus ALT-REFEX and REFEX tags inserted around recognized people references.

Task definition: The task in GREC-NER is a straightforward combined named-entity recognition and coreference resolution task, restricted to people entities. Systems insert REF and REFEX tags with coreference IDs around recognized mentions. The aim is to match the "gold-standard" tags in the GREC-People data.

Evaluation: The evaluation measures applied were three *intrinsic, automatic output-quality measures*: MUC-6 (Vilain *et al.*, 1995), CEAF (Luo, 2005), and b-cubed (Bagga and Baldwin, 1998), which are three performance measures commonly used in coreference resolution.

13.3.2   Component evaluation: Surface Realization

Surface Realization is the NLG subtask of mapping representations of the meaning, and in some cases the structure, of a sentence to a fully realized sentence (word string). Many different surface realizers have been developed over the past three decades or so. While symbolic realizers dominated for much of this period, the past decade saw the development of a variety of statistical surface realizers. A significant subset of statistical realization work (Langkilde-Geary, 2002; Callaway, 2003; Nakanishi *et al.*, 2005; Zhong and Stent, 2005; Cahill and van Genabith, 2006; White and Rajkumar, 2009) produced results for regenerating the Penn Treebank (Marcus *et al.*, 1994). The basic approach in all this work was to remove information from the Penn Treebank parses (the word strings themselves as well as some of the parse information), and then convert and use these underspecified representations as inputs to the surface realizer, whose task is to reproduce the original Treebank sentence.

While publications reporting this type of work referred to each other and (tentatively) compared BLEU scores, the results were not in fact directly comparable, because of the differences in the input representations automatically derived from Penn Treebank annotations. In particular, the extent to which they were underspecified varied from one system to the next.

The SR'11 shared task

The aim in developing the Surface Realization (SR) Task was to make it possible, for the first time, to directly compare different, independently developed surface realizers by developing a "common-ground" input representation, which could be used to generate realizations by all participating systems. In fact, the organizers created two different input representations, one shallow, one deep, in order to enable more teams to participate. While the SR'11 Task was for English, there are plans to develop a Spanish version. For full details, see the results report (Belz *et al.*, 2011).

Input and output data: The SR Task data has two input representations, one for each track, shallow and deep. In both, sentences are represented as sets of unordered labeled dependencies (with the exception of named entities that are ordered). The input representations were created by post-processing the CoNLL 2008 Shared Task data (Surdeanu *et al.*, 2008). For the preparation of the CoNLL'08 Shared Task data, selected sections of the Penn Treebank were converted to syntactic dependencies via the LTH Constituent-to-Dependency Conversion Tool for Penn-style Treebanks, or Pennconverter for short (Johansson and Nugues, 2007). The resulting dependency bank was then merged with the Nombank (Meyers *et al.*, 2004) and Propbank (Palmer *et al.*, 2005) corpora. Named entity information from the BBN Entity Type corpus was also incorporated. The SR'11 shallow representation is based on the Pennconverter dependencies. The deep representation is derived from the merged Nombank, Propbank, and syntactic dependencies in a process similar to the graph completion algorithm outlined in Bohnet *et al.* (2010).

Task definition: The task for SR'11 systems is to map from the shallow (or deep) inputs to word strings that are readable, clear, and as similar as possible in meaning to the original corpus sentences.

Evaluation: System outputs were evaluated with a range of *intrinsic, automatic output-quality measures* which compare system outputs to the original corpus sentences: BLEU, NIST, TER (Snover *et al.*, 2006), and METEOR (Lavie and Agarwal, 2007). In addition there were *intrinsic, human-assessed output-quality measures*: Meaning Similarity (compared to the original corpus sentences), Readability, and Clarity evaluated in human evaluation experiments where the evaluators assessed system outputs in terms of the three criteria, using sliding scales as rating tools.

13.3.3   End-to-end NLG systems: data-to-text generation

Many well-known NLG systems are data-to-text systems, i.e., they map some non-language representation of information (typically consisting of data structures of numbers and symbols) to word strings conveying the information: from recipe generation in EPICURE (Dale, 1990) and weather forecast generation in FOG (Goldberg *et al.*, 1994) to museum exhibit descriptions in M-PIRO (Androutsopoulos *et al.*, 2005) and nursing reports in BT-Nurse (Sambaraju *et al.*, 2011). Data-to-text systems are often devised for real-life tasks currently carried out by humans, meaning that (unlike the isolated module evaluations of the preceding section) evaluations can involve real users and real-world situations. For example, Reiter *et al.* (2005) evaluated the SumTime-METEO weather forecast generation system in terms of how much post-editing meteorologists did before deeming the forecasts ready for release to clients. Reiter *et al.* (2003a) also evaluated the STOP system that generated personalized stop-smoking letters to smokers from health-care providers; they looked at the proportion of smokers who managed to stop after receiving a personalized letter, compared to a control group who received non-personalized letters.

Prodigy-METEO data and task

The METEO data is, as far as we aware, the only resource that has been used by multiple research groups to develop directly comparable systems outside the context of NLG shared tasks (Reiter *et al.*, 2005; Belz, 2007; Reiter and Belz, 2009; Angeli *et al.*, 2010; Langner, 2010). The Prodigy-METEO corpus (Belz, 2009) contains just a.m. wind forecasts and corresponding input data vectors, whereas the original SumTime-METEO corpus (Sripada *et al.*, 2002) contains complete weather forecasts and the complete original weather data files.

Input and output data:[9]   The Prodigy-METEO inputs are data vectors of meteorological data about predicted wind characteristics; the outputs are corresponding wind statements that form part of weather forecasts written by meteorologists for offshore oil platforms. The outputs were extracted from the SumTime-METEO corpus, and the inputs were created by extracting information from the outputs and from the weather

---

[9] Available from https://sites.google.com/site/genchalrepository/.

data files corresponding to outputs. For example, the following is the target output for input 5Oct2000_03.num.1:

```
SSW 16-20 GRADUALLY BACKING SSE THEN FALLING VARIABLE 04-08 BY LATE
EVENING
```

The wind data inputs are vectors of time stamps and wind characteristics (speed, direction, gusts, etc.); e.g., the following is the input vector for output 5Oct2000_03.prn.1:

```
[[1, SSW,16,20,-,-,0600],[2, SSE,-,-,-,-,-1],[3, VAR,04, 08,-,-,0000]]
```

Task definition: In order to be directly comparable with existing results using the Prodigy-METEO data, systems must map from the inputs described above (which may be augmented by supplementary information not obtained by copying or converting other SumTime-METEO data) to wind statements. The aim for outputs is to be as clear and fluent as weather forecast texts, not as ordinary English texts.

Evaluation: Prodigy-METEO work has been evaluated by *intrinsic, automatic output-quality measures* using the BLEU metric (NIST scores are also sometimes reported), and by *intrinsic, human-assessed output-quality measures* using the criteria of fluency and clarity assessed with discrete rating scales and absolute quality judgments (rather than preference judgments).

13.3.4   End-to-end NLG systems: text-to-text generation

There are many different text-to-text applications, including machine translation, summarization, paraphrasing and text improvement. All involve (possibly shallow) analysis of the input text followed by (possibly shallow) synthesis of the output text. Many approaches do not comprise a clearly identifiable NLG subtask. Below we review the three text-to-text shared tasks that have so far been organized by the NLG community.

The GREC-Full shared task
The immediate motivating application context for the GREC-Full Task is the improvement of referential clarity and coherence in extractive summaries and multiply edited texts (such as Wikipedia articles) by regeneration of referring expressions contained in them.

Input and output data: Inputs are as described for GREC-NER, and outputs as for GREC-NEG (see Section 13.3.1).

Task definition: The overall aim for the GREC-Full Task is to improve the referential clarity and fluency of input texts. Systems replace referring expressions as and where necessary to produce as clear, fluent, and coherent a text as possible. This task can be viewed as composed of three sub-tasks: (1) named entity recognition (as in GREC-NER, see Section 13.3.1); (2) generation of lists of possible referring expressions for each entity; and (3) named entity generation by selection of one of the possible referring expressions (as in GREC-NEG, see Section 13.3.1).

Evaluation: GREC-Full systems were evaluated by four *intrinsic, automatic output-quality measures*: (a) BLEU-3, (b) NIST-5, (c) string edit distance, and (d) length-normalized string edit distance; and by two *intrinsic, human-assessed output-quality measures*: preference-strength judgments using sliders for assessing Fluency and Referential Clarity.

The HOO shared task

The Helping Our Own (HOO) Task is an automated text correction task focused on correcting mistakes by non-native speakers of English in English academic papers in the NLP domain. A pilot version of this task was run in 2011 (Dale and Kilgarriff, 2011).

Input and output data: The HOO data is a set of fragments of text, averaging 940 words in length, derived from a set of 19 ACL Anthology papers donated to the task by their authors. The papers were copy-edited as if for publication by two professional copy-editors and corrections were converted into a list of edits in standardized notation in a separate file where each edit looked as follows:

```
<edit type="MY" index="0001-0004" start="631" end="631">
  <original><empty/></original>
  <corrections>
    <correction>both </correction>
  </corrections>
</edit>
```

Outputs produced by participating systems were either (a) a set of plain text files that contain corrected text in situ (the organizers provided a tool that extracts the changes made to produce a set of XML edit structures for evaluation), or (b) a set of edit structures as above.

Task definition: HOO systems have to (a) identify stretches of text to be edited, and (b) provide the replacement text to be inserted instead. The aim is to edit not just spelling and grammatical errors, but any incorrect usage.

Evaluation: In the pilot run, HOO systems were automatically evaluated in terms of the following *intrinsic, automatic output-quality measures*: Recall and Precision scores for Detection (approximate identification of stretches of text to be edited), Recognition (precise identification of stretches of text to be edited), and Correction (identification of corrections that were also provided by the human annotators).

The QG shared tasks

The Question Generation (QG) Task was run in 2010 and comprised two tracks (see below). As no report is available detailing the evaluation methods that were finally applied to system outputs, we restrict the following overview to task definitions and data. Preliminary reports state that the automatically generated questions have been rated by several raters according to five criteria (relevance, question type, syntactic correctness and fluency, ambiguity, and variety).

QG-A

Input and output data: QG-A inputs are self-contained paragraphs covering a diverse set of topics selected from Wikipedia, OpenLearn, and Yahoo!Answers, 65 for the development data and 60 for the test data. Also provided were discourse relations based on the discourse parser HILDA. For the development data set six questions per paragraph were manually created and scored.

Task definition: In the Question Generation from Paragraphs Task, systems generate a list of six questions from a given input paragraph at three levels of granularity: broad (one question based on the entire input paragraph), medium (two questions each based on multiple sentences), and specific (three questions each based on a single sentence or less).

QG-B

Input and output data: Sixty input sentences were selected from OpenLearn, Wikipedia, and Yahoo!Answers (20 from each source), avoiding overly long or short sentences. The development data set consisted of sentences paired with questions, and one or more target question types.

Task definition: System inputs consisted of (a) a single sentence, and (b) a question type (who, why, how, when). For each input, the task was to generate two questions of the specified type.

13.3.5 Embedded NLG components

In this section, we survey evaluations from several different projects (part of) whose aim was to develop NLG components embedded in application systems. All of the applications are interactive to some degree, and some of the NLG components take aspects of the interaction into account when generating outputs. For example, the ILEX system varies descriptions depending on browsing history and descriptions can refer back to previously viewed items (Cox *et al.*, 1999). In Section 13.3.6, we describe the GIVE Task, in slightly more detail than the other shared tasks we surveyed in this section, as it is the only NLG shared task so far that involves user–system interaction (competing NLG components being embedded in the same interactive system). Finally, in Section 13.5.9, we discuss a related new embedded task, GRUVE (Janarthanam and Lemon, 2011).

COMIC

The COMIC system is a multimodal interactive environment for designing bathrooms. One of the project evaluations focused on NLG during one specific phase of the interaction where the user does guided browsing of available tiles, and aimed to answer the specific question whether adaptive features that take into account dialogue history and user preferences made a big enough difference to system turns for users to notice (Foster and White, 2005). This was assessed in two separate studies, both with an overhearer evaluation design, in which subjects judged the quality of recorded interactions between the system and a user. The measures used were *intrinsic, automatic output-quality measures* implemented as preference judgments between the adaptive and the non-adaptive outputs in answer to specific questions. Outputs were presented first in spoken, then in written form, and judgments were collected for both forms. In both studies, the generator with adaptive features (dialogue history and user modeling) was selected significantly more often than the non-adaptive version, in both the speech and the text modality.

ILEX

ILEX is a browser for electronic catalogues of museum artifacts (jewelry) and has an NLG module that dynamically generates text descriptions of the exhibits (Cox *et al.*, 1999). The ILEX team carried out an evaluation of the NLG module in which learning outcomes of users interacting with a dynamic version of the system (which was tailored to user type and browsing history) were compared to those for users who interacted with a static version (which always generated the same texts regardless of user type and browsing history).

The two versions were compared in terms of the following extrinsic and intrinsic measures: (a) a *user-task-success measure*, comprehension accuracy (multiple-choice questions about style, materials, etc.); (b) a *system-purpose-success measure*, learning gain (the ability to classify previously unseen objects); and (c) several *intrinsic user like measures* (10 questions, each answered on a 5-point rating scale, e.g., *I think I would like to use ILEX frequently.*). Results for the latter measure type were reported in terms of mode: the dynamic version had a better mode for 3 out of 10 questions, and there was no difference for 5 questions. The other two measures also failed to yield statistically significant differences between the dynamic and non-dynamic versions of the NLG module.

It may be the case that the differences between the two versions were too subtle for the measures applied, and that differences would have been detected by intrinsic measures of text quality. This was the case in the GREC-MSR evaluations (Belz *et al.*, 2009), where comprehension accuracy and reading speed measures yielded no differences, but text quality measures did.

M-PIRO

The M-PIRO system builds on ILEX, and also has an NLG module that generates descriptions of museum items. The text structuring module was tested against a base-line version, with the specific objective of testing the added benefit from the modules implementing aggregation and comparison to previously viewed items (Karasimos and Isard, 2004). The researchers evaluated the M-PIRO system against a baseline that did not have these functionalities. They used an *extrinsic user-task-success measure*, comprehension accuracy (multiple-choice questions); an *intrinsic, human-assessed output-quality measure*, intrinsic text quality (fluency) implemented as preference judg-ments; and several *user like measures*, including numerical ratings of "interestingness," enjoyment, and learning gain. M-PIRO was shown to be better on comprehension accuracy and user-assessed learning gain, but not on the other measures.

SPOT

SPOT is a trainable sentence planner developed for the NLG module of a mixed-initiative travel planning system (AMELIA). The SPOT developers carried out an evaluation of SPOT, comparing it to AMELIA's native hand-crafted template generator, as well as two rule-based and two baseline sentence planners (Rambow *et al.*, 2001). Subjects read five transcripts of real interactions with AMELIA. At 20 points, AMELIA's actual utterance was displayed alongside alternative outputs from all systems. Subjects then rated each of the outputs on a 5-point Likert scale expressing their degree of agreement with the statement *The system's utterance is easy to understand, well-formed, and appropriate to the dialogue context* (an *intrinsic, human-assessed output-quality measure*). SPOT's and AMELIA's template-based generators were both judged significantly better than the remaining variants, but neither was better than the other.

One potential issue here is that three different quality criteria are conflated into a single rating, which is hard to interpret. For example, an utterance can be perfectly clear and well-formed, yet entirely inappropriate in a given context, and it is hard to see what rating such an utterance should be given on the rating scale described above.

SkillSum

SkillSum is an NLG system that generates a personalized feedback report for low-skilled readers who have just completed a screening assessment of their basic literacy and numeracy skills (Williams and Reiter, 2008). It is deployed as a module in a literacy screening application and was developed as an improved replacement for the screening application's original report generator, which glued together pieces of canned text. The system's purpose is to determine whether students at further education colleges have the appropriate literacy skill levels for their chosen course modules. A secondary system purpose is to improve students' ability to accurately assess their own skill levels.

Following several smaller pilot evaluations during system development, the SkillSum team conducted a final large-scale evaluation with 230 further education students aiming to assess the readability and usefulness of the SkillSum module that used hand-crafted rules compared to (a) a version of SkillSum based on corpus frequency, and (b) the original canned-text report generator. To assess readability, they automatically computed several *intrinsic, automatic output-quality measures* (mean sentence length, mean word length, Flesch reading ease, Flesch–Kincaid grade); several *extrinsic user-task-success measures* (reading-aloud speed, reading-aloud errors, silent-reading speed, comprehension accuracy); and some *intrinsic user like measures* (preference judgments between alternative versions of reports). The team found no statistically significant differences for any of these measures.

The SkillSum team also measured how the baseline and SkillSum report generators affect students' ability to accurately assess their own skill levels, asking them once before and once after they used the screening application whether they thought their literacy skills were good enough for a chosen course. The idea was that if the reports were useful, reading them should improve self-assessment. Here, the team found that fewer students changed their self-assessment with the SkillSum version than the baseline, but significantly more students correctly changed their self-assessment with the SkillSum version, and significantly fewer students changed it incorrectly.

As in the case of ILEX, differences between report versions were fairly subtle. Important conclusions that the project team drew (Williams and Reiter, 2008) are that (a) the variation between types of low literacy skills, and therefore between the needs of students with low literacy skills, is probably too great for a single user-tailored approach to report generation, and (b) modeling multiple different user types, and evaluating these different models, might have worked better. Both of these could have contributed to the sparsity of significant results.

CLASSiC

As part of the CLASSiC EC FP7 project,[10] a scheduling system was built for making appointments for a France Telecom Engineer to come to the caller's home. Two NLG components were created; one was a statistical adaptive module trained using reinforcement learning and the other was a rule-based adaptive baseline component.

The baseline NLG module was swapped out half way through data collection, and the resulting two system variants were compared in terms of dialogue system performance and usability (Janarthanam *et al.*, 2011). In a PARADISE-style evaluation (see Section 13.4.3), users were asked to fill out a questionnaire that included *intrinsic user like measures*, e.g., *During your call, did you understand what the system said?*, and a self-assessment question (perceived task success, or the user's perception of whether they completed the task successfully or not), as well as *intrinsic, automatic output-quality measures* such as dialogue length, and *extrinsic user-task-success measures* (e.g., actual task success). The statistical NLG module showed significant improvement (+23.7%) in terms of perceived task success compared to the baseline, although no significant difference was observed in terms of actual task success. The statistical version of the NLG module also had significantly higher (+5%) overall user satisfaction (the average of all user like scores), and dialogues were significantly shorter, both in terms of time (call duration: −15.7%) and in terms of number of words (average words per system turn: −23.93%).

13.3.6   Embedded NLG components: the GIVE shared task

GIVE participants implement NLG modules that generate instructions to guide a human user in solving a treasure-hunt task in a virtual 3D world in real time. In evaluations, the NLG modules are plugged into the same interactive virtual environment, and the performance and experience of users is evaluated.

The first edition of the GIVE Task was limited in that only discrete steps of the same size, and rotation in one of four directions, were possible in the virtual environments. As a result, the NLG task was easier than intended (as noted in the results report [Koller *et al.*, 2010], one of the best participating NLG modules generated instructions of the type *Move three steps forward*). In the second edition of the task, users were able to

---

[10] http://www.classic-project.org.

navigate the virtual environment freely, turning at any angle and moving any distance at a time. In this section, we provide an overview of this second version of GIVE.

Input and output data and task definition: The task for GIVE-2 NLG modules is to generate, in real time, natural-language instructions that enable users to successfully complete the treasure hunt. The virtual environment consists of multiple rooms containing objects (as landmarks), various buttons (with one of several functions, including opening the treasure safe, triggering an alarm, and no function), and floor tiles that trigger alarms. The possible outcomes of a single game are "success" (treasure found), "alarm," and "user leaves."

From the point of view of the NLG module, the interaction between it and the embedding GIVE-2 system looks as follows: The NLG module receives a message every 200 milliseconds containing the position and orientation of the user and the list of currently visible objects. Furthermore, the NLG module receives a message every time the user manipulates an object. The NLG module can, at any point, request from the GIVE-2 system a plan of action to get the user from his/her current position to the treasure. It is up to the NLG module to decide when to request plans. So for any single instance of generating an instruction, the NLG module has the following available to it as input:

- A representation of the world the user is in
- Object manipulations by the user so far
- Current position and orientation of the user
- Current list of visible objects
- Current plan of actions to reach treasure.

Certain properties of the embedding GIVE-2 system affect the kind of instructions that the NLG module should generate. For example, the GIVE-2 system displays the instructions for a fixed length of time, so if an instruction is too long, the user cannot read all of it. Furthermore, as the user is free to move at any point (including while the next instruction is being generated), it is possible for the next instruction to be out of date by the time it reaches the user, and they may therefore not be able to understand it (e.g., if objects that the NLG module assumes are visible are no longer visible from the user's new position).

Non-NLG tasks that the NLG module needs to perform include deciding when to request a new plan of action and what to do with the information about objects manipulated by the user. Content-selection tasks performed by the module include deciding which subset of actions to select to be covered by the next instruction, and selecting objects from the list of visible objects (or even the world representation) to refer to as landmarks. Most of the NLG modules created for the GIVE-2 Task perform discourse-planning tasks such as aggregating actions and objects. Other typical NLG tasks that participating modules could (but do not have to) perform include referring expression generation, lexicalization, and surface realization.

Evaluation: Participating NLG modules were evaluated in terms of the following *extrinsic user-task success measures* and *intrinsic, automatic output-quality measures* that were computed by the GIVE-2 organizers (apart from the first measure, these were computed on successfully completed games only):

- Task success rate: percentage of games in which user found trophy
- Task duration: mean number of seconds from start of game to trophy retrieval
- Distance covered by user: mean number of distance units of the virtual environment covered between start of game and retrieval of trophy
- Actions required: mean number of object manipulation actions in games
- Instructions required: mean number of instructions produced by the NLG module in games
- Instruction length: mean number of words in instructions.

The GIVE-2 organizers also obtained ratings for a further 22 *intrinsic user like measures* from players via an exit questionnaire filled in at the end of each game. Questions ranged from the perceived clarity and readability of instructions to how enjoyable the game was. If one deemed entertainment the purpose of the GIVE-2 environment, some of these measures could be classified as *system-purpose-success measures*. Some measures aim to assess the quality of part of the module output (the instructions), while most of the measures are more clearly user like measures.

Discussion

While GIVE-2 is an interactive system, it is not linguistically interactive like the dialogue systems we focus on in the next section. The system outputs (a) continuously updated visualizations of the world the user is moving through, and (b) verbal instructions to the user, at points determined by the NLG module. The user does not directly input any information, but can move through a given virtual world and manipulate some of the objects in it, both of which can affect system behavior. The input can be assumed to be reliable and the NLG task can be viewed as data-to-text generation, except that strategies have to be developed for what to do if the user strays from the intended path. All in all, this is a relatively simple interactive setup, making the GIVE tasks ideal first shared tasks on NLG embedded within an interactive system.

The evaluations that have been carried out are at the system level but compare systems that differ only in the NLG module, so they count as embedded module evaluations in the terminology of Table 13.1.

One point that we return to in Section 13.5 is that while NLG module developers receive feedback (in the form of exit questionnaires filled in after test games) on how highly users rate the performance of their module, they do not receive any feedback about how to improve their instructions. Such feedback could be provided in the form of model instructions. Human-authored model instructions could conceivably be created, perhaps for a set of specific individual game situations, against which modules could be automatically measured, or which could be used to train modules.

Finally, the status of some of the automatic measures is unclear. To count as output quality measures, it would have to be clear which end of a scale corresponds to higher quality, which is not the case for instruction length, and possibly some of the other measures.

13.3.7   Concluding comments

In this section, our aim was to provide a high-level overview of system evaluations in NLG, with a focus on the recent move toward comparative and competitive forms of evaluation. In the following section, we look at evaluation of spoken dialogue systems, an interactive system type that is likely to play an increasingly important role in NLG research in the future.

13.4   An overview of evaluation for spoken dialogue systems

13.4.1   Introduction

For the last fifteen years, the field of spoken dialogue system (SDS) evaluation has strived to find a common evaluation framework to enable systems to be compared and ranked using single system-level scores derived from multiple evaluation measures that in combination capture all relevant aspects of system quality. Evaluation frameworks are particularly important for SDSs due to the number and complexity of components that make up SDSs. In the past, evaluation of SDSs has typically been in the form of task-based experiments where paid participants perform a task in the laboratory and then fill out a questionnaire, yielding a range of *intrinsic user like measures*.

However, it can be argued that lab-based evaluations are inefficient, and involve unrealistic scenarios and user goals. Studies have shown that dialogue quality measures vary significantly, depending on whether the pool of callers includes real users or paid subjects. For example, using the same system, Ai *et al.* (2007) showed that paid subjects talk more with the system and speak faster, while real users barge in more frequently, use touchtone input more, and ask for help more. Evaluation of academic systems is starting to move more toward evaluations "in the wild," for example, in the Let's Go! shared task (Black *et al.*, 2011). With evaluations in the wild, collecting *intrinsic user like measures* (for example, to establish user satisfaction) is challenging, because real, busy users are likely to hang up once they have the information they need, rather than staying on the line to fill out a questionnaire. User like measures that one can realistically collect in the wild are limited to single ratings such as *Please rate your experience from 0 to 9 using the keypad*, or perceived task success questions such as *Did you find what you were looking for?*. Such single scores may not necessarily be reliable measures of quality criteria such as dialogue quality. Furthermore, asking users to explicitly rank different SDSs is simply not practical. Evaluations in the wild therefore tend to rely on unobtrusively obtained task-success measures such as task time and task success.

In this section, we start by discussing several dimensions along which an evaluation can differ in terms of realism and control (including paid vs. real users, simulated vs. actual system turns, simulated vs. human users). Next we describe a number of existing evaluation frameworks and shared tasks for SDSs.

13.4.2   Realism and control

There are different degrees of realism and "controlledness" in experiments involving users, with a corresponding associated increase/decrease in cost. Below, we look at a variety of experimental settings where the system is running either in ordinary mode or with standardized user–system exchanges that are easier to control but are potentially less realistic.

User experiments with system in ordinary running mode

The least expensive but most difficult to control are experiments "in the wild," where real users with real goals interact with the system in any way they choose; for example, residents of Pittsburgh in the middle of the night trying to get a bus home by calling the Let's Go! system (Black *et al.*, 2011). Obtaining *user like measures* in the wild is particularly tricky. While it may be possible to ask one or two questions at the end of the dialogue such as *Did you find what you needed?*, users frequently exit the interaction once they have the information they need.

A step up from experiments "in the wild" in terms of control and cost is task-based evaluation. Here, the systems can be run in a realistic setting, but subjects are typically paid and given specific tasks to accomplish, e.g., calling a system on a mobile phone to find a restaurant with certain criteria.

Moving the same type of experiment out of the real world and into the lab comes with a further decrease in realism but an increase in control. Crowdsourcing platforms such as Amazon Mechanical Turk (AMT) provide an alternative environment for traditional lab-based evaluation. With AMT, participants are paid a small amount of money to perform short tasks. The key advantage of crowdsourcing is the ease of recruitment. With regard to evaluating NLG systems that are embedded, task-based experiments have been shown to be effectively evaluated using AMT. For example, Jurčiček *et al.* (2011) describe a telephone infrastructure for interacting with the system and a web interface for feedback collection. Crowdsourcing platforms provide a simple and cost effective way of evaluating interactive systems and in particular the output of NLG systems. The main drawback is that it is hard to monitor the cooperativeness and accuracy of the participants, although initial results from evaluations of participant accuracy are encouraging (Yang *et al.*, 2010; Jurčiček *et al.*, 2011).

Changing the experiment from one in which the user interacts directly with the interactive system to an "overhearer-style" method in which the user observes (part of) an interaction (Stent *et al.*, 2004; Dethlefs and Cuayáhuitl, 2011) gives even greater control, but less realism and higher costs.

Wizard of Oz experiments

Wizard of Oz (WOZ) experiments, where a human generates system responses, provide a powerful mechanism for testing certain NLG strategies, such as different referring expression generation (Janarthanam and Lemon, 2010) and information presentation strategies (Liu *et al.*, 2009). The major drawback of WOZ experiments is that WOZ environments are costly to build, and running experiments and training the wizards is very time-consuming. In addition, there may be discrepancies between the different styles of wizards; this is particularly prevalent in NLG WOZ studies. Despite these drawbacks, an important benefit of WOZ experiments is that the human-generated system turns can be treated as model outputs, and having model outputs that facilitate NLG strategy optimization (Liu *et al.*, 2009) is very valuable.

User simulations

Given the increasing use of statistical methods within various components of SDSs such as NLG and dialogue management components (Williams and Young, 2007; Henderson *et al.*, 2008; Janarthanam and Lemon, 2010), and given the huge cost and overhead of evaluations, many researchers are opting for evaluation with simulated users (Eckert *et al.*, 1997; Schatzmann *et al.*, 2005; Janarthanam and Lemon, 2009). The general goal of a user simulation is to produce as many natural, varied, and consistent interactions as necessary, trained on as little data as possible. The quality of the user simulation can be of crucial importance because it can dramatically influence the results in terms of SDS performance analysis and strategies learned (Schatzmann *et al.*, 2005).

Running evaluations with simulated users has a number of benefits. First, each system can be tested on the same user simulation. Second, one has control over the amount of variability encoded into the user simulation; and third, one can train user simulations to reflect different types of users, such as novice and expert (Janarthanam and Lemon, 2010).

With user simulations one can explore a large variety of user responses to different system-generated utterances (Rieser and Lemon, 2010). In fact, simulated environments can evaluate the entire range of utterances that a system may produce rather than just those seen in deployed systems. In addition, one can evaluate various NLG strategies with user simulations by comparing a cumulative reward measure to determine the best performance. For example, in Dethlefs and Cuayáhuitl (2011), user simulations are used to compare various learned policies with a baseline in terms of cumulative reward. These are also followed up with human evaluations, which is recommended. In Chapter 11 by Waller and colleagues in this volume, user simulations are described in the context of evaluating Assistive and Augmentative Communication (AAC) systems.

The main drawback is that user simulation evaluation is not generally viewed as sufficient on its own: user simulation evaluation may be fine for optimizing and testing strategies, but for final system evaluation, additional evaluations with human users are seen as necessary.

13.4.3   Evaluation frameworks

In this section, we give an overview of a number of evaluation frameworks for SDSs that have influenced the field in recent years.

PARADISE

One of the first complete evaluation frameworks to emerge was PARADISE (PARAdigm for DIaLogue System Evaluation; Walker *et al.*, 2000). PARADISE aimed to create a useful single evaluation score for SDSs computed from a variety of evaluation measures. It has been widely adopted in SDS evaluations (Bonneau-Maynard *et al.*, 2000; Lamel *et al.*, 2000; Rahim *et al.*, 2001).

The PARADISE model posits that the overall goal of a spoken dialogue agent is to maximize usability. Usability is expressed in the overall evaluative goal of user satisfaction and is broken down into two contributing quality criteria: task success and dialogue cost. Two further quality criteria are identified as relevant contributors to dialogue cost: dialogue efficiency and dialogue quality, both assessed by *intrinsic output quality measures*. Examples of dialogue efficiency measures are elapsed time and number of user turns; dialogue quality measures include mean recognition rate and number of barge-ins (Walker *et al.*, 2000, p. 367). Using methods from decision theory (Keeney and Raiffa, 1976), the PARADISE model assigns a positive or negative weight to each of these measures using multiple linear regression. This model can help developers understand the contribution made by different measures in the framework; for example, number of turns may have a large negative weight, meaning that users generally prefer shorter dialogues.

The PARADISE model may be over-ambitious, attempting to equate dialogue quality with the single overall evaluative goal of user satisfaction, and perhaps a goal based on perceptual quality judgments would be more appropriate. Furthermore, linear combination of factors may not be a good fit for all features. For example, a very low number of system turns may be as much an indication of problematic dialogues as a very large number of system turns.

Möller and Ward

Möller and Ward (2008) argue that both perception and user-judged quality are "events" that happen in a particular personal, spatial, temporal, and functional context. They propose a tripartite framework with one part modeling the behavior of the user and system during the interaction (broadly corresponding to *extrinsic task-success measures* and *automatically computed intrinsic output quality metrics*), the second part the perception and judgment processes taking place inside the user (broadly corresponding to *user like measures*), while the third part models what matters to system designers and service providers (weights assigned to different measures to reflect their relative importance). We describe each of the three parts in more detail below.

The **behavior model** translates the characteristics of the system and the user into predicted interaction behavior. One way to do this is to collect "interaction parameters" (including system errors, user errors, points of confusion, dead time, etc.) that quantify the behavior of the interaction participants in a similar way that PARADISE's dialogue efficiency measures do. Möller and Ward discuss running user simulations to obtain values for the behavior model, thus avoiding using real participants (see also Section 13.4.2 for user simulations).

The second part of this framework, the **perception and judgment model**, captures the interaction quality perceived by the user and is analogous to the set of user satisfaction measures used in the PARADISE model. Specifically, the user is asked to describe each perceptual event and/or quality event either qualitatively through open questions, or quantitatively on rating scales.

Möller and Ward argue that different stakeholders may focus on different quality dimensions and weigh them differently based on the current task, context of use, price, etc. Their **value model** assigns weights to the different evaluation measures reflecting the importance assigned to each in the current evaluation context. An example they give is a system that teaches girls to "talk to Barbie": here, speech recognition performance and task completion are of lesser importance than voice quality, responsiveness, and unpredictability. This system's value model is very different from one for the military, for example. Value models depend on who the stakeholders in a given evaluation are and vary from system to system and business to business. In a business context, it would depend on a company's business model and customer base and may even be proprietary. Möller and Ward's framework has yet to be developed and tested in full and on multiple systems.

SERVQUAL

Hartikainen *et al.* (2004) describe a purely subjective framework for evaluation (which involves *user like measures* only), adapting the SERVQUAL (SERVice QUALity) evaluation method developed by marketing academics. To a certain extent it allows definition of a value model; however, it is a less comprehensive evaluation framework than PARADISE and Möller and Ward's framework, in that it does not involve objective measures.

Hartikainen *et al.* analyze service quality in terms of "dimensions" (broadly, quality criteria) which are assessed in terms of (pre-test) expectations and (post-test) perceptions of quality. The dimensions of service quality criteria are tangibles, reliability, responsiveness, assurance, and empathy. Pre-test and post-test questionnaires are delivered to participants, with the pre-test questions covering the participants' expectations and the post-test questions their quality judgments. Hartikainen *et al.* tested their evaluation framework on an email system, showing how it facilitates discussion of different user group value priorities, such as the importance of system performance for males vs. females, and the zone of tolerance of system reliability. As yet, no one has looked at using SERVQUAL to examine the gap between expected and perceived NLG outputs.

Contender

Suendermann *et al.* (2010) describe a simple framework for evaluation of phone-based SDSs that involves *intrinsic automatic measures* and *extrinsic task success measures* only (no *user like measures*). Calls to an SDS number are routed to one of several versions of the SDS, and two measures are tracked: automation rate (number of calls that do not have to be transferred to an operator) and call duration. The different versions of the SDS are compared using these measures. Contender is designed for use by companies deploying SDSs on a large scale (hundreds of thousands of calls per day), so that small modifications to an SDS can be tested rapidly.

13.4.4   Shared tasks

In this section, two shared tasks for end-to-end SDS evaluations are discussed.

DARPA Communicator

The DARPA Communicator challenge was for flight-booking SDSs and comprised two stages of evaluation (performed in 2000 and 2001) involving nine and eight participating sites respectively (Walker *et al.*, 2001a,b, 2002a,b). This shared task was one of the first to attempt to use subjects with realistic goals from a target population of frequent travelers.

Task definition: In the 2000 challenge, task scenarios consisted of seven fixed and two open scenarios. The fixed scenarios consisted of planning three domestic one-way trips, two domestic round trips, and two international round trips. The open task scenarios were planning an intended business trip and planning a vacation. The fixed scenarios were designed to vary task complexity, defined as the number of constraints the user had (e.g., hotel, car hire). The open scenarios were defined by the user. As discussed by Walker *et al.* (2002b), having users define their own tasks did not give the expected results as users tended to change their task if faced with speech recognition errors.

In the 2001 experiment, the subjects also had to make travel arrangements with varying complexity (round vs. multi-leg, fixed airline, car-hire, hotel). This experiment ran for 6 months during which systems were continuously accessible via a toll-free number, requiring the systems to be stable. In addition to the fixed scenarios, participants had to make *real* travel arrangements that were actually booked through the system. The open scenario can be considered a form of evaluation "in the wild" (see also Section 13.4.2).

Evaluation: PARADISE was used for the Communicator evaluations. The measures that were assessed included dialogue efficiency measures such as time on task and system turn duration; dialogue quality measures such as response latency and word accuracy; task success (actual and perceived); and user satisfaction. Of these measures, system turn duration and system words per turn were both weighted positively in the regression models. The fact that utterance length is positively weighted in the regression model could be interpreted in one of two ways: (a) users prefer longer utterances, i.e., the longer the utterances the higher the average user satisfaction, or (b) users tend to get to a point in the dialogue where the system has understood well enough to give travel search results (e.g., flight options), which is typically a long utterance (most systems just gave results in a single long output).

Spoken dialogue challenge 2010 shared task

The Spoken Dialog Challenge (SDC) 2010 was a more recent exercise to investigate how different SDSs perform on the same task (Black *et al.*, 2011).

Task definition: The domain of the existing Let's Go! Pittsburgh bus information system was used. Four teams provided systems that were first tested in controlled conditions with speech technology researchers as subjects. For this initial evaluation, subjects were shown a simple web interface that presented eight different scenarios to callers; in each, the caller had to get a bus time. Through these controlled tests, three of the four systems were deemed robust and stable enough to be deployed to real callers in the second stage of the evaluation.

Evaluation: A set of PARADISE measures were assessed including different levels of task success (depending on whether there was actually a bus available at the desired time). Given the previously discussed difficulties with evaluations in the wild, no *intrinsic user like measures* were collected for the live results in the second stage so it is not known which system was preferred by users.

Black *et al.* (2011) compare the evaluation under controlled conditions with the evaluation involving real users; for example, dialogues with real users were shorter and had a higher word error rate (WER), a measure of speech recognition performance. With regard to inter-system differences, the system that achieved highest task success under controlled conditions did not perform the best with real users. Indeed, for some system pairs there was a big difference in WER with real users, but not under controlled conditions. As noted by Black *et al.* (2011), WER scores do not just reflect the language model used, but also the dialogue design and the NLG component, both of which influence the content and style of the system prompts, which will in turn influence user responses. More system-initiative, directed dialogue design seemed to contribute to a comparatively better speech recognition performance with real users.

13.4.5   Discussion

Möller and Ward (2008) argue that the PARADISE model does not reliably give valid predictions of individual user judgments, typically covering only around 40–50% of the variance in the data it is trained on. However, PARADISE has been shown to be able to predict certain quality criteria such as user satisfaction, which may indirectly reflect NLG module quality. In addition, frameworks such as PARADISE can be used to establish reward functions for data-driven NLG components. For example, working with dialogues collected in a Wizard of Oz setup as part of the GIVE Shared Task, Dethlefs and Cuayáhuitl (2011) use the PARADISE framework to identify the strongest predictors of user satisfaction in situated dialogue/NLG systems. They propose an approach that jointly optimizes content selection, utterance planning, and surface realization. For content selection and utterance planning, they use PARADISE to obtain a reward function that favors short interactions at maximal task success.

In summary, existing evaluation frameworks have their drawbacks and challenges. Nonetheless, they have been shown to go some way toward facilitating comparison of embedded NLG components and end-to-end interactive systems.

13.4.6   Concluding comments

In this section, we discussed realism vs. control in evaluation experiments for SDSs, and reviewed existing evaluation frameworks and shared tasks for SDSs. In the following section, we present a generic evaluation model structure that subsumes the evaluation frameworks presented in this section, and propose a specific evaluation model for comparative evaluation of NLG modules embedded in interactive systems.

Table 13.2. Steps involved in designing an **evaluation model:** a complete specification of the goals and implementation of a specific evaluation

	Objective/task	Example
1. System design	Achieve clear separation in terms of task specification and software architecture between components that will be evaluated separately and the remaining system.	Component to be evaluated: NLG module; embedding system: flight information system; NLG module performs all and only NLG tasks; module–system interaction specified by API.
2. Evaluation design	a. Specify the overall evaluative goal of the evaluation, identifying both the system and/or module to be evaluated, as well as which (subset of) its properties are to be evaluated.	Goal: to determine which of a set of alternative NLG modules performs better when embedded in a given flight information system; properties to be evaluated: module outputs and contribution to user satisfaction.
	b. Specify the quality criteria which the property to be evaluated breaks down into.	Module output quality: • Linguistic quality of NLG module outputs • Impact on task effectiveness and efficiency User satisfaction: • Impact on user satisfaction.
	c. For each quality criterion, select the evaluation measures that will be used to assess it.	Impact on user satisfaction: design questionnaire incorporating *user like measures* in the form of questions such as *Would you be willing to pay to use this system?*
	d. Create a composition model describing how the scores produced by the selected set of evaluation measures are to be combined.	Compute a weight function over the set of scores from the different evaluation measures, such as their weighted sum or weighted mean.

13.5   A methodology for comparative evaluation of NLG components in interactive systems

13.5.1   Evaluation model design

In Table 13.2, we provide a generic overview of the steps involved in designing an **evaluation model**, i.e., a complete specification of the goals and implementation of a specific evaluation. Note that the evaluation frameworks surveyed in Section 13.4 could all be described in these terms. Point 1 in the table simply expresses that in order to be able to evaluate multiple versions of a component, its task specification has to be clear enough for developers to work from, and it has to be architecturally separate enough to be plug-and-playable (so that different components can be plugged in and evaluated under the same conditions). This is independent of the granularity of the component, which could be a referring expression generation module, an NLG module generating all system turns, or a combined turn generator and dialogue manager.

Overall evaluative goals (2a) tend to be too abstract to be directly assessed by a set of measures, and it is often found helpful to break goals down into more fine-grained quality criteria (2b). This process could result in a hierarchical structure of more than two levels, with some of the quality criteria decomposed into further quality criteria. For example, in the PARADISE model the overall evaluative goal of user satisfaction is taken to be composed of quality criteria of low cost and task success, with low cost further composed of efficiency and quality. However this is done, at the leaves of the resulting (possibly flat) tree evaluation criteria have to be paired with individual evaluation measures that return a single score for each system (2c). If multiple measures are used, then a composition model can be created, defining how to combine them into a single score and/or to reflect their relative importance (2d), corresponding, for example, to the value model in SERVQUAL.

In the remainder of this section we look at each of the above points in turn, bearing in mind the specific case of evaluating NLG in the context of interactive systems (IS). We will consider three more specific cases: (a) an NLG module (generating system turns only) embedded within an IS; (b) a referring expression generation module embedded within (an NLG module within) an IS; and (c) a module within an IS which incorporates NLG as well as some other functionality of the IS. We will use the GIVE Task as an example of (a), a version of the GRUVE Task as an example of (c), and a new task (referring expression generation in dialogue context) as an example of (b). In our discussion, we will mostly focus on NLG modules (case a), but the intention is that what we say about these can be mapped, *mutatis mutandis*, to the other two cases. In Sections 13.5.2–13.5.7, we describe an evaluation model for the NLG module case, and in Sections 13.5.8 and 13.5.9 we describe evaluation models for the new referring expression generation task and a version of the GRUVE pedestrian navigation system task.

13.5.2 An evaluation model for comparative evaluation of NLG modules in interactive systems

The overall evaluative goal depends on who the stakeholders in the evaluation are, and on how the results will be used. In the present context, we are assuming a developers' perspective, and that we wish to evaluate (just) the NLG component, not the rest of the system. For this last reason, a small number of frozen versions of the embedding system suffice (although it cannot be ruled out that using other versions may produce different results).

To achieve encapsulation of NLG tasks, the embedding system should make available all the required input information, including some or all of the following types, depending on how tasks have been divided between NLG module and embedding system:

- Representations of the user's interaction with the system so far (including, e.g., user actions so far, previous dialogue turns, common ground, etc.)
- Representation of the virtual or real environment the user is in, if any
- Representation of current user state (including, e.g., most recent user dialogue act, current position and orientation of user, etc.)
- Representation of current and/or remaining options (plans) for how to achieve goal of interaction (where goal is, e.g., ticket booking, locating an item, etc.)
- User model.

The system should send requests to the module to produce a system turn, i.e., unlike in the GIVE-2 Task (see Section 13.3.6), all interaction management is performed outside of the NLG module.

A good general-purpose evaluative goal that embodies the developers' perspective is to assess NLG module performance. The following quality criteria can be seen as contributing to this property (also given are the evaluation measure *types* that can be used to assess the criteria; see Section 13.2):

1. **Context-independent intrinsic quality of NLG module outputs:** *intrinsic output-quality measures*, both automatic and human-assessed
2. **Context-dependent intrinsic quality of NLG module outputs:** *intrinsic output-quality measures*, both automatic and human-assessed
3. **Contribution to user satisfaction with system:** *intrinsic user like measures*
4. **Contribution to task effectiveness and efficiency:** *extrinsic user task success measures*
5. **Contribution to system purpose success:** *extrinsic system purpose-success measures*.

We discuss specific evaluation measures that can be used to assess each of the above quality criteria in Sections 13.5.3–13.5.7 below.

13.5.3   Context-independent intrinsic output quality

There are aspects to output quality that can be assessed in isolation (just considering the system turns generated by NLG components), including measures of grammaticality, fluency, and clarity. Such measures would be typically assessed by trained human evaluators, but can conceivably also be assessed automatically, e.g., by computing the likelihood of outputs according to a given language model (approximating fluency). The advantage of context-independent linguistic quality assessments are that their cost can be arbitrarily low; even for human assessment, arbitrarily small numbers of representative outputs can be selected for assessment.

13.5.4   Context-dependent intrinsic output quality

Important aspects of output quality, in particular in interactive systems, depend on the context in which outputs are generated. For example, requesting a piece of information from the user is appropriate only if it has not been requested before; referring to a landmark in an environment is not appropriate if the landmark is not currently within the user's field of vision; and if a landmark has so far been referred to by the user as *Big Ben* then a system reference to *the clock tower* causes unwanted implicatures. Such aspects of quality can be captured by measures of appropriateness, clarity, and efficiency. Some of these can be assessed directly by automatic methods (assessment of efficiency could be as simple as computing the average length of system turns, as in GIVE-2); others will require model data in order to be assessed automatically. The alternative is evaluation of outputs by trained human evaluators.

Model output data sets

As previously mentioned, having model outputs available is very useful in system development as they can be used (a) for automatically training and optimizing generators, and (b) in automatic measures of similarity between system and model outputs. Unlike other forms of evaluation, evaluation based on model outputs provides information about how to convert the results of evaluation into strategies for improving generators. Perhaps the most common way of producing model data in the context of interactive systems is to run systems in Wizard of Oz mode and record the interactions (see also Section 13.4.2). For the purposes of evaluating NLG components, the interaction record needs to be augmented with all the information that will be available to the generation module as potential input in ordinary running mode.

   While creating Wizard of Oz data is expensive, the above advantages potentially outweigh the cost, in particular because it provides a way of repeatedly evaluating systems during development at no additional expense.

Automatic measures

If model data is available then standard string comparison metrics such as BLEU and the NIST variant of BLEU can be applied. The great advantage of automatic metrics is, of course, their cheapness and replicability. Automatic metrics are particularly effective if multiple human model outputs are available for the same input, or at least if model outputs by multiple humans are available for different inputs.

Human assessment

The alternative to using model data is to train evaluators to assess quality criteria directly. System outputs are then evaluated by the evaluators in controlled experiments where each evaluator sees the same number of outputs from the same systems. The human evaluators, e.g., listen to the dialogue after the fact and assess each system turn. For example, Schmitt *et al.* (2011) annotated dialogues from the Let's Go! bus information system; raters were asked to annotate the quality of the interaction at each system–user exchange from 1 (very poor) to 5 (very good). A similar evaluation could be performed to compare several NLG components embedded within the same system, rating each pair of system and user turns in terms of a given set of measures. This method, however, is very labor intensive.

13.5.5   User satisfaction

Assessment of user satisfaction often involves asking users questions, and there are two alternatives for when to do this, either during the interaction with the system, or in an exit questionnaire at the end of the interaction. The former runs the risk of affecting, perhaps adversely, the interaction itself, and hence the evaluation results, whereas the latter, as previously mentioned, comes with the risk that the user exits early before completing all questions.

A good range of *user like measures* will include questions directly asking the user how much they liked different aspects of the system (*Was the system helpful?*, *Was the system clear and fluent?*, etc.), but also more indirect questions asking, for example, whether the user would like to use the system again, and whether they would be prepared to pay for it. User satisfaction measures can also be automatically computed from records of user–system interaction; for example, high early-exit rates can be viewed as a sign of user dissatisfaction.

13.5.6   Task effectiveness and efficiency

Task performance measures are essentially assessments of the success and speed with which users succeed in accomplishing given tasks while using the system. For example, in the DARPA Communicator Shared Tasks, one of the user tasks was to book a flight, and in GIVE-2, the overall task was to find a treasure. Perhaps the most straightforward approach to evaluating task performance is to compute the percentage of interactions that led to a successful task outcome (task effectiveness), and the average number of system and user turns in successful interactions (task efficiency).

With direction-giving shared tasks such as GIVE and GRUVE, a set of spatial task success metrics can be calculated, e.g., the distance travelled in the virtual world or the number of turns in the virtual world as a result of the directions given. Other measures such as time on task and number of restarts can also be indicative of task performance, although one has to factor out other variables such as internet speed for remotely run shared tasks.

13.5.7   System purpose success

How system purpose success is measured appropriately is clearly dependent on a specific system, but it will tend to involve both quantitative measures and asking users in some form whether they feel that the system accomplishes its purpose. For example, if a GIVE system's purpose is to entertain, one measure is asking users whether they find the system they are interacting with entertaining, and whether they would recommend the system to a friend. However, this should be complemented by measures that more objectively assess user entertainment, e.g., actual recommendations to friends, or whether users return to play the game again (and if so how many times). In tutoring systems, (part of) the system purpose is learning gain, so system purpose success measures will aim to assess learning gain in some form (Forbes-Riley and Litman, 2009).

13.5.8 A proposal for a shared task on referring expression generation in dialogue context

Referring expression generation (REG) is a particularly important subtask in interactive systems, because it is vital that the user is clear about what entities the system refers to, in particular in interactive systems that involve a virtual or real-world environment, such as games, pedestrian navigation, and "smart tourism." At the same time, several of the shared-task competitions that have been run in the NLG community over the past six years have addressed REG, both within discourse context and in isolation (the GREC and TUNA tasks, see Section 13.3.1), and these provide a solid basis for a new shared task addressing REG within dialogue.

Perhaps the quickest and simplest way to get such a shared task off the ground is to take existing corpora of human–human or human–system dialogues, annotate the referring expressions within them (and adapt the annotations in the corpus in other ways if necessary), and set participating systems the task of generating referring expression chains for the dialogues. In order to create test data, either all referring expressions, or referring expressions of a certain type (named entities, landmarks, etc.) could be deleted, the task for systems being to insert a suitable referring expression in each gap. For each gap, information such as a representation of the dialogue context, the entity ID, possible attributes, syntactic and semantic constraints, and perhaps a set of possible alternative referring expressions for the entity (as in the GREC Task) could be provided.

Among the existing annotated dialogue corpora that could be used for this purpose are the following:

- Tangrampuzzle-solving cooperative human–human dialogues collected for Japanese and English by Tokunaga *et al.* (2010), where an "operator" gives instructions to a "solver" about how to place puzzle pieces to form a shape.
- The COCONUT corpus (Jordan and Walker, 2005) of task-oriented dialogues in which two speakers collaboratively decide which furniture items from their two respective sets to use in furnishing a two-room house. The following is a brief excerpt (referring expressions highlighted in bold):

    G: *[36] That leaves us with 250 dollars. [37] I have **a yellow rug for 150 dollars**. [38] Do you have any other furniture left that matches for 100 dollars?*

    S: *[39] No, I have no furniture left that costs $100. [40] I guess you can buy **the yellow rug for $150**.*

    G: *[41] Okay. [42] I'll buy **the rug for 150 dollars**. [43] I have **a green chair** [44] that I can buy for 100 dollars [45] that should leave us with no money.*

- Troubleshooting dialogue data collected in a Wizard of Oz study (Janarthanam and Lemon, 2009). The following is an example dialogue in which NPs that refer to domain entities are highlighted:

  Sys 1:   *Is **your router** connected to **the computer**?*
  Usr 1:   *Uh. What's a router?*
  Sys 2:   *It's **the big black box**.*
  Usr 2:   *Ok. yes.*
  Sys 3:   *Do you see **a small white box** connected to **the router**?*
  Usr 3:   *Yes.*
  Sys 4:   *Ok. Is there **a flashing monitor symbol** at the bottom right of **the screen**?*
  Usr 4:   ***The network icon**?*
  Sys 5:   *Yes. Is **it** flashing?*
  Usr 5:   *Yes. **It** is flashing.*
  Sys 6:   *Ok. Please open **your browser**.*

Evaluation model

In a highly controlled shared-task setup such as the one described above, the evaluation model can be quite simple, because the dialogue context is fixed, and the effect of referring expression choice on user behavior and interactive patterns cannot be assessed. The overall evaluative goal could be to assess the quality of the referring expression module outputs (rather than their overall effectiveness, for example). This could be broken down into two top-level and four second-level quality criteria, assessed by evaluation measures as follows:

1. Context-dependent intrinsic quality of referring expressions:
   a. Humanlikeness: BLEU scores computed against model outputs from corpus;
   b. Language quality: appropriateness, clarity, and fluency assessments by trained human evaluators using rating scales.

2. Identifiability of referents of referring expressions: extrinsic identification experiment where participants "overhear" dialogues one turn at a time, and identify referents as they go along (perhaps from a visual representation of domain entities):
   a. Identification speed: average time taken to identify referent in identification experiment;
   b. Identification accuracy: percentage of correctly identified referents.

The task proposed here does not involve embedding referring expression modules in an actual system. Because based on recorded dialogues, there is no embedding system as such, and user and task are fixed. Consequently, any evaluation is restricted to assessing certain aspects of the referring expression modules only. The expectation would be that this sort of simple task would be followed by a more complex, embedded version. However, the proposed task has the great advantage that it can be set up in a short amount of time at relatively low cost, which is particularly important for research community-run tasks.

13.5.9   GRUVE: A shared task on instruction giving in pedestrian navigation

In this section, we describe the new GRUVE Shared Task (Janarthanam and Lemon, 2011), and propose an evaluation model for it. The GRUVE Task is based on generation of instructions for pedestrian users navigating in open-world virtual environments (here, a virtual version of the city of Edinburgh). GRUVE addresses three main research challenges. Firstly, how can the NLG module refer to landmarks to help with navigation? Secondly, how can the NLG module cope with varying levels of uncertainty about the user's location and about other aspects of the input? Thirdly, how can the NLG module use the user's "viewshed" (where they are looking) to improve instruction generation?

The envisaged task is a variant of direction generation similar to the GIVE challenge, but in GRUVE the system has varying amounts of information about the user. In the simplest case the system has total information about the user's location, heading/gaze direction, and history of interaction/behavior, and a clear and detailed set of feedback signals. At the other end of the spectrum, with a larger amount of uncertainty, the task is to generate directions for unknown users whose location is uncertain, where feedback signals are very noisy, and where directionality is not known for sure. The idea is that this latter scenario is analogous to real-world city navigation problems. The following is an example GRUVE dialogue:

User: *I'd like to go to the National Museum.*
System: *Right, the National Museum of Scotland. (a small pause) You first need to go to Charles Street. Do you know what direction that is?*
User: *No.*
System: *No problem. Can you see the Neuroscience building?*
User: *Ehh… yes!*
System: *Good. Please start walking in the direction of the Neuroscience building.*

In terms of the terminology from Section 13.2, the GRUVE Task would minimally require participants to develop an NLG module to be embedded in the GRUVE interactive system (provided by the organizers). The current intention is to have a fixed dialogue manager (DM) that determines the timing of the directions from an automatically generated route plan; the DM provides inputs to the NLG module including user location, viewshed, confidence scores, and button requests. An alternative version of the task would be to ask participants to develop both DM and NLG components and to evaluate these jointly. This would allow existing approaches that perform joint optimization of NLG with the DM (Lemon, 2008) to participate.

Joint evaluation of NLG and DM components implies that the components are treated as one module while all other aspects of the embedding system stay the same. The goal of such an evaluation would be to assess the combination of DM and NLG modules, and not attempt to separate them, which in turn implies joint development, and perhaps optimization, of the two components. For example, if the system proposes a long and complicated route then the optimal strategy might be for the DM to break it up into a number of turns with longer utterances from the NLG module. An alternative strategy might be to have one long turn with shorter utterances from the NLG module describing each step.

Evaluation model

The evaluation type is embedded evaluation of a system component (Section 13.2), where competing systems participating in the shared task are plugged into the same embedding system (the GRUVE system). The evaluative goal of the joint evaluation scenario could be to assess the performance of NLG/DM modules at the task of generating instructions when embedded in the GRUVE system. This evaluative goal could be broken down into the following quality criteria and corresponding evaluation measures:

1. Context-dependent intrinsic quality of system turns:
   a. Humanlikeness: BLEU scores computed against model outputs from corpus
   b. Language quality: appropriateness, clarity, and fluency assessments by trained human evaluators using rating scales
2. User satisfaction: intrinsic user like measures collected through post interaction questionnaires
3. Task effectiveness and efficiency: extrinsic user task success measures, e.g.:
   a. Time taken to reach destination
   b. Distance travelled
   c. Optimal paths
   d. Number of system turns
   e. Number of wrong turns (i.e., path errors)
   f. Overall task success (i.e., did the user get to their destination)
4. Robustness when faced with increasing uncertainty: e.g., compute how task effectiveness and efficiency measures change with increasing uncertainty.

Some of the measures above may reflect performance of one module (NLG vs. DM) more than the other; for example, the number of system turns may reflect the performance of the dialogue manager more than that of the NLG module. It is possible to run multiple evaluations, keeping one or the other module constant, and thereby gain insights into which module contributes more to different evaluation criteria. Varying uncertainty levels in different ways is also likely to affect the two modules differently. For example, varying the uncertainty of user information might affect the DM more, whereas varying the uncertainty of the viewshed might affect the NLG module more.

13.5.10 Concluding comments

In this section, we presented a generic evaluation model structure, proposed a specific evaluation model for comparative evaluation of NLG modules embedded in interactive systems, and outlined two future shared tasks each with a specific evaluation model. In the following, final section of this chapter, we provide some final concluding comments.

13.6 Conclusion

In this chapter, we started with an overview of different categories of evaluation measures, in order to provide us with a standard terminology for categorizing existing and new evaluation techniques. This was followed with some background on existing evaluation methodologies in the NLG and interactive systems fields separately, after which we moved on to presenting a methodology for evaluation of NLG components embedded within interactive systems, using specific tasks to provide illustrative examples.

Interactive systems have become an increasingly important type of application for deployment of NLG technology over recent years. At present, we do not yet have a commonly agreed terminology or methodology for evaluating NLG within interactive systems. We have attempted to take a step towards addressing this lack in this chapter.

Acknowledgments

Some of the research reported here has been funded by the Engineering and Physical Sciences Research Council (EPSRC), UK, under grant numbers EP/F059760/1, EP/G03995X/1, EP/H032886/1, and EP/I032320/1.

Other aspects of the research reported here have received funding from the EC's FP7 programmes: from FP7/2007-13 under grant agreement no. 216594 (CLASSiC); from FP7/2011-14 under grant agreement no. 248765 (Help4Mood); from FP7/2011-14 under grant agreement no. 287615 (PARLANCE); and from FP7/2011-14 under grant agreement no. 270019 (Spacebook).

The GRUVE Task was developed by Oliver Lemon and Srini Janarthanam, and we are grateful for their permission to use GRUVE as one of our example future shared tasks on NLG in interactive systems.

References

Ai, H., Raux, A., Bohus, D., Eskenzai, M., and Litman, D. (2007). Comparing spoken dialog corpora collected with recruited subjects versus real users. In *Proceedings of the SIGdial Workshop on Discourse and Dialogue (SIGDIAL)*, pages 124–131, Antwerp, Belgium. Association for Computational Linguistics.

Androutsopoulos, I., Kallonis, S., and Karkaletsis, V. (2005). Exploiting OWL ontologies in the multilingual generation of object description. In *Proceedings of the European Workshop on*

Natural Language Generation (ENLG), pages 150–155, Aberdeen, Scotland. Association for Computational Linguistics.

Angeli, G., Liang, P., and Klein, D. (2010). A simple domain-independent probabilistic approach to generation. In *Proceedings of the Conference on Empirical Methods in Natural Language Processing (EMNLP)*, pages 502–512, Boston, MA. Association for Computational Linguistics.

Bagga, A. and Baldwin, B. (1998). Algorithms for scoring coreference chains. In *Proceedings of the International Conference on Language Resources and Evaluation (LREC)*, pages 563–566, Granada, Spain. European Language Resources Association.

Basile, V. and Bos, J. (2011). Towards generating text from discourse representation structures. In *Proceedings of the European Workshop on Natural Language Generation (ENLG)*, pages 145–150, Nancy, France. Association for Computational Linguistics.

Belz, A. (2007). Probabilistic generation of weather forecast texts. In *Proceedings of Human Language Technologies: The Conference of the North American Chapter of the Association for Computational Linguistics (HLT-NAACL)*, pages 164–171, Rochester, NY. Association for Computational Linguistics.

Belz, A. (2009). Prodigy-METEO: Pre-alpha release notes. Technical Report NLTG-09-01, Natural Language Technology Group, CMIS, University of Brighton.

Belz, A. (2010). GREC named entity recognition and GREC named entity regeneration challenges 2010: Participants' pack. Technical Report NLTG-10-01, Natural Language Technology Group, University of Brighton.

Belz, A. and Kow, E. (2009). System building cost vs. output quality in data-to-text generation. In *Proceedings of the European Workshop on Natural Language Generation (ENLG)*, pages 16–24, Athens, Greece. Association for Computational Linguistics.

Belz, A., Kow, E., Viethen, J., and Gatt, A. (2009). The GREC main subject reference generation challenge 2009: Overview and evaluation results. In *Proceedings of the Workshop on Language Generation and Summarisation*, pages 79–87, Suntec, Singapore. Association for Computational Linguistics.

Belz, A. and Reiter, E. (2006). Comparing automatic and human evaluation of NLG systems. In *Proceedings of the Conference of the European Chapter of the Association for Computational Linguistics (EACL)*, pages 313–320, Trento, Italy. Association for Computational Linguistics.

Belz, A., White, M., Espinosa, D., Kow, E., Hogan, D., and Stent, A. (2011). The first surface realisation shared task: Overview and evaluation results. In *Proceedings of the Generation Challenges Session at the European Workshop on Natural Language Generation*, pages 217–226, Nancy, France. Association for Computational Linguistics.

Black, A. W., Burger, S., Conkie, A., Hastie, H., Keizer, S., Lemon, O., Merigaud, N., Parent, G., Schubiner, G., Thomson, B., Williams, J. D., Yu, K., Young, S., and Eskenazi, M. (2011). Spoken dialog challenge 2010: Comparison of live and control test results. In *Proceedings of the SIGdial Conference on Discourse and Dialogue (SIGDIAL)*, pages 2–7, Portland, OR. Association for Computational Linguistics.

Bohnet, B. and Dale, R. (2005). Viewing referring expression generation as search. In *Proceedings of the International Joint Conference on Artificial Intelligence (IJCAI)*, pages 1004–1009, Edinburgh, Scotland. International Joint Conference on Artificial Intelligence.

Bohnet, B., Wanner, L., Mille, S., and Burga, A. (2010). Broad coverage multilingual deep sentence generation with a stochastic multi-level realizer. In *Proceedings of the International Conference on Computational Linguistics (COLING)*, pages 98–106, Beijing, China. International Committee on Computational Linguistics.

Bonneau-Maynard, H., Devillers, L., and Rosset, S. (2000). Predictive performance of dialog systems. In *Proceedings of the International Conference on Language Resources and Evaluation (LREC)*, Athens, Greece. European Language Resources Association.

Boyd, A. and Meurers, D. (2011). Data-driven correction of function words in non-native English. In *Proceedings of the European Workshop on Natural Language Generation (ENLG)*, pages 267–269, Nancy, France. Association for Computational Linguistics.

Cahill, A. and van Genabith, J. (2006). Robust PCFG-based generation using automatically acquired LFG approximations. In *Proceedings of the International Conference on Computational Linguistics and the Annual Meeting of the Association for Computational Linguistics (COLING-ACL)*, pages 1033–1040, Sydney, Australia. Association for Computational Linguistics.

Callaway, C. B. (2003). Do we need deep generation of disfluent dialogue? In *Working Papers of the AAAI Spring Symposium on Natural Language Generation in Spoken and Written Dialogue*, pages 6–11, Stanford, CA. AAAI Press.

Cox, R., O'Donnell, M., and Oberlander, J. (1999). Dynamic versus static hypermedia in museum education: An evaluation of ILEX, the intelligent labelling explorer. In *Proceedings of the Conference on Artificial Intelligence in Education*, pages 181–188, Le Mans, France. International Artificial Intelligence in Education Society.

Dale, R. (1989). Cooking up referring expressions. In *Proceedings of the Annual Meeting of the Association for Computational Linguistics (ACL)*, pages 68–75, Vancouver, Canada. Association for Computational Linguistics.

Dale, R. (1990). Generating recipes: An overview of Epicure. In Dale, R., Mellish, C., and Zock, M., editors, *Current Research in Natural Language Generation*, pages 229–255. Academic Press, San Diego, CA, USA.

Dale, R. and Kilgarriff, A. (2011). Helping our own: The HOO 2011 pilot shared task. In *Proceedings of the European Workshop on Natural Language Generation (ENLG)*, pages 242–249, Nancy, France. Association for Computational Linguistics.

Dale, R. and Reiter, E. (1995). Computational interpretation of the Gricean maxims in the generation of referring expressions. *Cognitive Science*, **19**(2):233–263.

Dethlefs, N. and Cuayáhuitl, H. (2011). Combining hierarchical reinforcement learning and Bayesian networks for natural language generation in situated dialogue. In *Proceedings of the European Workshop on Natural Language Generation (ENLG)*, pages 110–120, Nancy, France. Association for Computational Linguistics.

Di Eugenio, B. (2007). Shared tasks and comparative evaluation for NLG: To go ahead, or not to go ahead? In *Proceedings of the NSF Workshop on Shared Tasks and Comparative Evaluation in Natural Language Generation*, Arlington, VA. National Science Foundation.

Di Eugenio, B., Glass, M., and Trolio, M. J. (2002). The DIAG experiments: Natural language generation for intelligent tutoring systems. In *Proceedings of the International Conference on Natural Language Generation (INLG)*, pages 120–127, Arden Conference Center, NY. Association for Computational Linguistics.

Doddington, G. (2002). Automatic evaluation of machine translation quality using n-gram co-occurrence statistics. In *Proceedings of the Human Language Technology Conference (HLT)*, pages 138–145, San Diego, CA. Morgan Kaufmann.

Eckert, W., Levin, E., and Pieraccini, R. (1997). User modeling for spoken dialogue system evaluation. In *Proceedings of the IEEE workshop on Automatic Speech Recognition and Understanding (ASRU)*, pages 80–87, Santa Barbara, CA. Institute of Electrical and Electronics Engineers.

Forbes-Riley, K. and Litman, D. (2009). Adapting to student uncertainty improves tutoring dialogues. In *Proceedings of the Artificial Intelligence in Education Conference (AIED)*, pages 33–40, Brighton, UK. IOS Press.

Foster, M. E. and White, M. (2005). Assessing the impact of adaptive generation in the COMIC multimodal dialogue system. In *Proceedings of the Workshop on Knowledge and Reasoning in Practical Dialogue Systems (KRPDS)*, pages 24–31, Edinburgh, Scotland. International Joint Conference on Artificial Intelligence.

Gatt, A. and Belz, A. (2010). Introducing shared tasks to NLG: The TUNA shared task evaluation challenges. In Krahmer, E. and Theune, M., editors, *Empirical Methods in Natural Language Generation*, pages 264–293. Springer, Berlin, Heidelberg.

Gatt, A., van der Sluis, I., and van Deemter, K. (2007). Evaluating algorithms for the generation of referring expressions using a balanced corpus. In *Proceedings of the European Workshop on Natural Language Generation (ENLG)*, pages 49–56, Saarbrücken, Germany. Association for Computational Linguistics.

Goldberg, E., Driedger, N., and Kittredge, R. I. (1994). Using natural-language processing to produce weather forecasts. *IEEE Expert: Intelligent Systems and Their Applications*, **9**(2): 45–53.

Gupta, S. and Stent, A. (2005). Automatic evaluation of referring expression generation using corpora. In *Proceedings of the Workshop on Using Corpora for Natural Language Generation (UCNLG)*, Birmingham, UK. ITRI, University of Brighton.

Hartikainen, M., Salonen, E.-P., and Turunen, M. (2004). Subjective evaluation of spoken dialogue systems using SERVQUAL method. In *Proceedings of the International Conference on Spoken Language Processing (INTERSPEECH)*, pages 2273–2276, Jeju Island, Korea. International Speech Communication Association.

Henderson, J., Lemon, O., and Georgila, K. (2008). Hybrid reinforcement/supervised learning of dialogue policies from fixed datasets. *Computational Linguistics*, **34**(4): 487–513.

Isard, A., Oberlander, J., Androutsopoulos, I., and Matheson, C. (2003). Speaking the users' languages. *IEEE Intelligent Systems Magazine: Special Issue "Advances in Natural Language Processing"*, **18**(1):40–45.

Janarthanam, S., Hastie, H., Lemon, O., and Liu, X. (2011). "The day after the day after tomorrow?": A machine learning approach to adaptive temporal expression generation: Training and evaluation with real users. In *Proceedings of the SIGdial Conference on Discourse and Dialogue (SIGDIAL)*, pages 142–151, Portland, OR. Association for Computational Linguistics.

Janarthanam, S. and Lemon, O. (2009). A Wizard of Oz environment to study referring expression generation in a situated spoken dialogue task. In *Proceedings of the European Workshop on Natural Language Generation (ENLG)*, pages 94–97, Athens, Greece. Association for Computational Linguistics.

Janarthanam, S. and Lemon, O. (2010). Learning to adapt to unknown users: Referring expression generation in spoken dialogue systems. In *Proceedings of the Annual Meeting of the Association for Computational Linguistics (ACL)*, pages 69–78, Uppsala, Sweden. Association for Computational Linguistics.

Janarthanam, S. and Lemon, O. (2011). The GRUVE challenge: Generating routes under uncertainty in virtual environments. In *Proceedings of the European Workshop on Natural Language Generation (ENLG)*, pages 208–211, Nancy, France. Association for Computational Linguistics.

Johansson, R. and Nugues, P. (2007). Extended constituent-to-dependency conversion for English. In *Proceedings of the 16th Nordic Conference on Computational Linguistics*, pages 105–112, Tartu, Estonia. Northern European Association for Language Technology.

Jordan, P. W. (2000). Can nominal expressions achieve multiple goals? An empirical study. In *Proceedings of the Annual Meeting of the Association for Computational Linguistics (ACL)*, pages 142–149, Hong Kong. Association for Computational Linguistics.

Jordan, P. W. and Walker, M. A. (2005). Learning content selection rules for generating object descriptions in dialogue. *Journal of Artificial Intelligence Research*, **24**(1): 157–194.

Jurčíček, F., Keizer, S., Gašić, M., Mairesse, F., Thomson, B., Yu, K., and Young, S. (2011). Real user evaluation of spoken dialogue systems using Amazon mechanical turk. In *Proceedings of the International Conference on Spoken Language Processing (INTERSPEECH)*, pages 3061–3064, Florence, Italy. International Speech Communication Association.

Karasimos, A. and Isard, A. (2004). Multi-lingual evaluation of a natural language generation systems. In *Proceedings of the International Conference on Language Resources and Evaluation (LREC)*, Lisbon, Portugal. European Language Resources Association.

Keeney, R. L. and Raiffa, H. (1976). *Decisions with Multiple Objectives: Preferences and Value Tradeoffs*. John Wiley & Sons, New York, NY.

Knight, K. and Langkilde, I. (2000). Preserving ambiguities in generation via automata intersection. In *Proceedings of the National Conference on Artificial Intelligence and the Conference on Innovative Applications of Artificial Intelligence (AAAI/IAAI)*, pages 697–702, Austin, TX. AAAI Press.

Koller, A., Striegnitz, K., Gargett, A., Byron, D., Cassell, J., Dale, R., Moore, J., and Oberlander, J. (2010). Report on the second NLG challenge on generating instructions in virtual environments (GIVE-2). In *Proceedings of the International Conference on Natural Language Generation (INLG)*, pages 243–250, Trim, Ireland. Association for Computational Linguistics.

Lamel, L., Rosset, S., Gauvain, J.-L., Bennacef, S., Garnier-Rizet, M., and Prouts, B. (2000). The LIMSI ARISE system. *Speech Communication*, **31**(4):339–354.

Langkilde-Geary, I. (2002). An empirical verification of coverage and correctness for a general-purpose sentence generator. In *Proceedings of the International Conference on Natural Language Generation (INLG)*, pages 17–24, Arden Conference Center, NY. Association for Computational Linguistics.

Langner, B. (2010). *Data-driven Natural Language Generation: Making Machines Talk Like Humans Using Natural Corpora*. PhD thesis, Language Technologies Institute, School of Computer Science, Carnegie Mellon University.

Lavie, A. and Agarwal, A. (2007). METEOR: An automatic metric for MT evaluation with high levels of correlation with human judgments. In *Proceedings of the ACL Workshop on Statistical Machine Translation*, pages 228–231, Prague, Czech Republic. Association for Computational Linguistics.

Lemon, O. (2008). Adaptive natural language generation in dialogue using reinforcement learning. In *Proceedings of the Workshop on the Semantics and Pragmatics of Dialogue (SEMDIAL)*, pages 141–148, London, UK. SemDial.

Liu, X., Rieser, V., and Lemon, O. (2009). A Wizard of Oz interface to study information presentation strategies for spoken dialogue systems. In *Proceedings of the Europe-Asia Spoken Dialogue Systems Technology Workshop*, Kloster Irsee, Germany.

Luo, X. (2005). On coreference resolution performance metrics. In *Proceedings of the Joint Human Language Technology Conference and Conference on Empirical Methods in Natural Language Processing (HLT-EMNLP)*, pages 25–32, Vancouver, Canada. Association for Computational Linguistics.

Marcus, M., Kim, G., Marcinkiewicz, M. A., MacIntyre, R., Bies, A., Ferguson, M., Katz, K., and Schasberger, B. (1994). The Penn Treebank: Annotating predicate argument structure. In *Proceedings of the Human Language Technology Conference (HLT)*, pages 114–119, Plainsboro, NJ. Association for Computational Linguistics.

Meyers, A., Reeves, R., Macleod, C., Szekely, R., Zielinska, V., Young, B., and Grishman, R. (2004). The NomBank project: An interim report. In *Proceedings of the NAACL/HLT Workshop on Frontiers in Corpus Annotation*, pages 24–31, Boston, MA. Association for Computational Linguistics.

Möller, S. and Ward, N. G. (2008). A framework for model-based evaluation of spoken dialog systems. In *Proceedings of the SIGdial Workshop on Discourse and Dialogue (SIGDIAL)*, pages 182–189, Columbus, OH. Association for Computational Linguistics.

Nakanishi, H., Miyao, Y., and Tsujii, J. (2005). Probabilistic models for disambiguation of an HPSG-based chart generator. In *Proceedings of the International Workshop on Parsing Technologies*, pages 93–102, Vancouver, Canada. Association for Computational Linguistics.

Palmer, M., Gildea, D., and Kingsbury, P. (2005). The Proposition Bank: An annotated corpus of semantic roles. *Computational Linguistics*, **31**(1):71–105.

Papineni, K., Roukos, S., Ward, T., and Zhu, W.-J. (2002). BLEU: A method for automatic evaluation of machine translation. In *Proceedings of the Annual Meeting of the Association for Computational Linguistics (ACL)*, pages 311–318, Philadelphia, PA. Association for Computational Linguistics.

Passonneau, R. (2006). Measuring agreement on set-valued items (MASI) for semantic and pragmatic annotation. In *Proceedings of the International Conference on Language Resources and Evaluation (LREC)*, Genoa, Italy. European Language Resources Association.

Rahim, M., Di Fabbrizio, G., Kamm, C., Walker, M. A., Pokrovsky, A., Ruscitti, P., Levin, E., Lee, S., Syrdal, A., and Schlosser, K. (2001). Voice-IF: A mixed-initiative spoken dialogue system for AT&T conference services. In *Proceedings of the European Conference on Speech Communication and Technology (EUROSPEECH)*, pages 1339–1342, Aalborg, Denmark. International Speech Communication Association.

Rambow, O., Rogati, M., and Walker, M. A. (2001). Evaluating a trainable sentence planner for a spoken dialogue system. In *Proceedings of the Annual Meeting of the Association for Computational Linguistics (ACL)*, pages 426–433, Toulouse, France. Association for Computational Linguistics.

Reiter, E. and Belz, A. (2006). GENEVAL: A proposal for shared-task evaluation in NLG. In *Proceedings of the International Workshop on Natural Language Generation (INLG)*, pages 136–138, Sydney, Australia. Association for Computational Linguistics.

Reiter, E. and Belz, A. (2009). An investigation into the validity of some metrics for automatically evaluating NLG systems. *Computational Linguistics*, **35**(4):529–558.

Reiter, E., Robertson, R., and Osman, L. M. (2003a). Lessons from a failure: Generating tailored smoking cessation letters. *Artificial Intelligence*, **144**(1–2):41–58.

Reiter, E., Sripada, S., Hunter, J., and Davy, I. (2005). Choosing words in computer-generated weather forecasts. *Artificial Intelligence*, **167**:137–169.

Reiter, E., Sripada, S., and Robertson, R. (2003b). Acquiring correct knowledge for natural language generation. *Journal of Artificial Intelligence Research*, **18**:491–516.

Rieser, V. and Lemon, O. (2010). Natural language generation as planning under uncertainty for spoken dialogue systems. In Krahmer, E. and Theune, M., editors, *Empirical Methods in Natural Language Generation*, pages 105–120. Springer, Berlin, Heidelberg.

Sambaraju, R., Reiter, E., Logie, R., McKinlay, A., McVittie, C., Gatt, A., and Sykes, C. (2011). What is in a text and what does it do: Qualitative evaluations of an NLG system – the BT-Nurse – using content analysis and discourse analysis. In *Proceedings of the European Workshop on Natural Language Generation (ENLG)*, pages 22–31, Nancy, France. Association for Computational Linguistics.

Schatzmann, J., Georgila, K., and Young, S. (2005). Quantitative evaluation of user simulation techniques for spoken dialogue systems. In *Proceedings of the SIGdial Workshop on Discourse and Dialogue (SIGDIAL)*, pages 45–54, Lisbon, Portugal. Association for Computational Linguistics.

Schmitt, A., Schatz, B., and Minker, W. (2011). Modeling and predicting quality in spoken human–computer interaction. In *Proceedings of the SIGdial Conference on Discourse and Dialogue (SIGDIAL)*, pages 173–184, Portland, OR. Association for Computational Linguistics.

Scott, D. and Moore, J. (2007). An NLG evaluation competition? Eight reasons to be cautious. In *Proceedings of the NSF Workshop on Shared Tasks and Comparative Evaluation in Natural Language Generation*, Arlington, VA. National Science Foundation.

Snover, M., Dorr, B., Schwartz, R., Micciulla, L., and Makhoul, J. (2006). A study of translation edit rate with targeted human annotation. In *Proceedings of the Association for Machine Translation in the Americas (AMTA)*, pages 223–231, Boston, MA. Association for Machine Translation in the Americas.

Sparck Jones, K. (1981). Retrieval system tests 1958–1978. In Sparck Jones, K., editor, *Information Retrieval Experiment*, pages 213–255. Butterworths, London.

Sparck Jones, K. (1994). Towards better NLP system evaluation. In *Proceedings of the Human Language Technology Conference (HLT)*, pages 102–107, Plainsboro, NJ. Association for Computational Linguistics.

Sripada, S. G., Reiter, E., Hunter, J., and Yu, J. (2002). SUMTIME-METEO: A parallel corpus of naturally occurring forecast texts and weather data. Technical Report AUCS/TR0201, Computing Science Department, University of Aberdeen.

Stent, A., Prasad, R., and Walker, M. A. (2004). Trainable sentence planning for complex information presentation in spoken dialog systems. In *Proceedings of the Annual Meeting of the Association for Computational Linguistics (ACL)*, pages 79–86, Barcelona, Spain. Association for Computational Linguistics.

Suendermann, D., Liscombe, J., and Pieraccini, R. (2010). Contender. In *Proceedings of the Spoken Language Technology Conference (SLT)*, pages 330–335, Berkeley, CA. Institute of Electrical and Electronics Engineers.

Surdeanu, M., Johansson, R., Meyers, A., Màrquez, L., and Nivre, J. (2008). The CoNLL-2008 shared task on joint parsing of syntactic and semantic dependencies. In *Proceedings of the Conference on Computational Natural Language Learning (CoNLL)*, pages 159–177, Manchester, UK. Association for Computational Linguistics.

Tokunaga, T., Iida, R., Yasuhara, M., Terai, A., Morris, D., and Belz, A. (2010). Construction of bilingual multimodal corpora of referring expressions in collaborative problem solving. In *Proceedings of the Workshop on Asian Language Resources*, pages 38–46, Beijing, China. Chinese Information Processing Society of China.

van Deemter, K., Gatt, A., van der Sluis, I., and Power, R. (2012). Generation of referring expressions: Assessing the Incremental Algorithm. *Cognitive Science*, **36**(5): 799–836.

van der Sluis, I., Gatt, A., and van Deemter, K. (2007). Evaluating algorithms for the generation of referring expressions: Going beyond toy domains. In *Proceedings of the International Conference on Recent Advances in Natural Language Processing (RANLP)*, Borovets, Bulgaria. Recent Advances in Natural Language Processing.

Viethen, J. and Dale, R. (2006). Algorithms for generating referring expressions: Do they do what people do? In *Proceedings of the International Conference on Natural Language Generation (INLG)*, pages 63–72, Sydney, Australia. Association for Computational Linguistics.

Vilain, M., Burger, J., Aberdeen, J., Connolly, D., and Hirschman, L. (1995). A model-theoretic coreference scoring scheme. In *Proceedings of the Message Understanding Conference*, pages 45–52, Columbia, MD. Defense Advanced Research Projects Agency.

Walker, M., Rudnicky, A. I., Aberdeen, J., Bratt, E. O., Garofolo, J., Hastie, H., Le, A., Pellom, B., Potamianos, A., Passonneau, R., Prasad, R., Roukos, S., Sanders, G., Seneff, S., and Stallard, D. (2002a). Darpa Communicator evaluation: Progress from 2000 to 2001. In *Proceedings of the International Conference on Spoken Language Processing (INTERSPEECH)*, pages 273–276, Denver, CO. International Speech Communication Association.

Walker, M. A., Aberdeen, J., Boland, J., Bratt, E. O., Garofolo, J., Hirschman, L., Le, A., Lee, S., Narayanan, S., Papineni, K., Pellom, B., Polifroni, J., Potamianos, A., Prabhu, P., Rudnicky, A. I., Sanders, G., Seneff, S., Stallard, D., and Whittaker, S. (2001a). Darpa Communicator dialog travel planning systems: The June 2000 data collection. In *Proceedings of the European Conference on Speech Communication and Technology (EUROSPEECH)*, pages 1371–1374, Aalborg, Denmark. International Speech Communication Association.

Walker, M. A., Kamm, C., and Litman, D. (2000). Towards developing general models of usability with PARADISE. *Natural Language Engineering*, **6**(3–4):363–377.

Walker, M. A., Passonneau, R., and Boland, J. (2001b). Quantitative and qualitative evaluation of Darpa Communicator spoken dialogue systems. In *Proceedings of the Annual Meeting of the Association for Computational Linguistics (ACL)*, pages 515–522, Toulouse, France. Association for Computational Linguistics.

Walker, M. A., Rudnicky, A. I., Prasad, R., Aberdeen, J., Bratt, E. O., Garofolo, J., Hastie, H., Le, A., Pellom, B., Potamianos, A., Passonneau, R., Roukos, S., Sanders, G., Seneff, S., and Stallard, D. (2002b). Darpa Communicator: Cross-system results for the 2001 evaluation. In *Proceedings of the International Conference on Spoken Language Processing (INTERSPEECH)*, pages 269–272, Denver, CO. International Speech Communication Association.

White, M. and Rajkumar, R. (2009). Perceptron reranking for CCG realization. In *Proceedings of the Conference on Empirical Methods in Natural Language Processing (EMNLP)*, pages 410–419, Singapore. Association for Computational Linguistics.

White, M., Rajkumar, R., and Martin, S. (2007). Towards broad coverage surface realization with CCG. In *Proceedings of the Workshop on Using Corpora for NLG: Language Generation and Machine Translation*. Association for Computational Linguistics.

Williams, J. D. and Young, S. (2007). Partially observable Markov decision processes for spoken dialog systems. *Computer Speech and Language*, **21**(2):393–422.

Williams, S. and Reiter, E. (2008). Generating basic skills reports for low-skilled readers. *Natural Language Engineering*, **14**(4):495–525.

Yang, Z., Li, B., Zhu, Y., King, I., Levow, G., and Meng, H. (2010). Collection of user judgments on spoken dialog system with crowdsourcing. In *Proceedings of the Spoken Language Technology Conference (SLT)*, pages 277–282, Berkeley, CA. Institute of Electrical and Electronics Engineers.

Zhong, H. and Stent, A. (2005). Building surface realizers automatically from corpora using general-purpose tools. In *Proceedings of the Workshop on Using Corpora for Natural Language Generation (UCNLG)*, Birmingham, UK. ITRI, University of Brighton.

Author index

Subject index